Combustion Science

Combustion Science

Principles and Practice

J.C.Jones

MILLENNIUM
BOOKS

First published in 1993 by
Millennium Books
an imprint of E.J. Dwyer (Australia) Pty Ltd
3/32–72 Alice Street
Newtown NSW 2042
Australia
Phone: (02) 550–2355
Fax: (02) 519–3218

National Library of Australia
Cataloguing-in-Publication data

Jones, J. C.
Combustion science: principles and practice.

Bibliography.
Includes index.
ISBN 0 85574 969 5.

1. Combustion. I. Title.

541.361

Copy-edited by Ken Tate
Cover design by Simon Leong
Text designed and typeset by Post
Typesetters, Brisbane in 10/12pt Times
Printed in Australia by Australian Print Group.
10 9 8 7 6 5 4 3 2 1

Contents

Foreword

In the last fifty or sixty years there have been considerable advances in both combustion science and engineering applications. This has arisen first because of the development of our understanding of the chain radical nature of the main combustion processes and secondly because of an increasing insight into the chemistry of complex hydrocarbons.

There have been various driving forces. Initially it was the scientific challenge, the magic of understanding the chemical processes that led to the chemical interpretation of flame and ignition processes. The work of Hinshelwood and Semenov set the scene for combustion kinetics and, almost simultaneously, the book, *Flame and Combustion in Gases* by Bone and Townend in 1927 established the technological background in respect of a wide range of applications, including engines and furnaces. This was further developed by Lewis and von Elbe in their classic book, *Combustion, Flames and Explosion of Gases* in 1951, particularly in respect of ignition, hazards and explosion. The later books typified by those by Glassman, Strehlow and many others, developed the chemical aspects of the combustion of gases, liquids and solids and built upon the mathematical techniques. The latter were further developed as exemplified in Forman Williams' book, *Combustion Theory* as well as by other authors, e.g. Kuo.

Additionally, there have been numerous books on related fuel technology commencing with Faraday's book *Chemical History of a Candle* through to more recent, and generally more extensive and comprehensive tomes of today. However, relatively few books have been published on combustion recently; the last major text was by Barnard and Bradley in 1985.

In recent years the applications of combustion, first to the gas turbine and the military rocket, and then to the automotive engine, became of increasing importance during oil price increases and as pollution control became a dominant factor. Now all types of combustors are subject to efficiency and environmental control, and indeed the two requirements have converged in at least one area—the need to reduce CO_2 emissions as an insurance against possible climate warming.

This book, which develops the chemistry of combustion, is welcomed. Combustion now plays an integral part in many courses in universities and similar teaching institutes. It is a complex subject that not only involves chemistry but also physics and fluid mechanics and is thus an ideal subject, not only because of its importance, but because of the intellectual challenge. It is a rapidly developing subject, partially because of the reasons set out above, but also as a result of developments in experimental and computational techniques.

The book commences with chapters on the basic combustion processes and in the first chapter it deals particularly with theories of chain branching reactions in relation to spontaneous ignition. The second chapter considers the framework and the chemical basis of the combustion of the traditional basic fuels, carbon monoxide, hydrogen and hydrocarbons. Cool flames and oscillating combustion are dealt with succinctly. In the light of the first two chapters, ignition by a heat source is then covered.

Laminar flames are next considered with an analytical approach to premixed flame structure, burning velocities and turbulence. Chapters 5, 6 and 7 deal with gaseous fuels, coal and solid coal products, and less common fuels respectively.

The next block of chapters deals with pollutants: sulphur, nitrogen, and other pollutants, principally carbon and PAH. This sets the scene for the chapter on the combustion of liquid fuels and solid materials.

'Fire science' is also tackled, first in terms of fire modelling and then explosion hazards. This is a rapidly expanding field and the examples taken cover a number of topical situations and events. Finally the last chapter deals with propellants, explosives and pyrotechnics and a good survey of current problems is given.

This book is suitable for undergraduate and postgraduate students in Departments of Chemistry, Energy, and Environmental Science as well as in part for students in Mechanical Engineering. In each chapter, interesting examples are given from contemporary applications of combustion. It is naturally also suitable for many people engaged in research in industrial and government organizations who are entering the fields of combustion, environment and safety.

This book gives an up-to-date coverage of the chemistry of this wide-ranging topic and as such I greatly welcome it.

Professor Alan Williams
Livesey Professor
Department of Fuel and Energy
The University of Leeds

Preface

My aim in writing this book has been to review a wide variety of combustion topics by drawing in each case on research literature. It has been my approach to develop in detail the formulations and discussions from the selected literature. It is hoped that this will enable the reader to obtain not only a grasp of the material but also a sense of its appeal.

In undertaking a task of this sort one is very dependent on the co-operation of colleagues. Professor B.F. Gray of the University of Sydney read each chapter draft as it became available and made many helpful suggestions. Useful comments on three of the chapters were also obtained from Dr M.F.R. Mulcahy. The diagrams are mainly the work of L. Kortvelyesy to whom thanks are due. Thanks are also expressed to Professor A. Williams of the University of Leeds for his Foreword. Dr J.F. Griffiths, also of Leeds, gave valuable advice as to content during the early planning of the book. Finally I wish to thank the staff of Millennium Books, in particular Anthony Dwyer, Catherine Hammond and Katrina Rendell, from whom I received the utmost support at every stage of the book's path to publication.

"We would be astonished if this book were free of errors." So wrote R.M. Fristrom and A.A. Westenberg in the preface to their 1965 book *Flame Structure*. In releasing this book I wish to make those words my own and any reader identifying an error is asked to bring it to my notice.

J.C. Jones
School of Chemical Engineering and Industrial Chemistry
University of NSW
PO Box 1, Kensington
NSW 2033, Australia

Preface

CHAPTER I

Thermal, Branched Chain and Chain-thermal Ignition

Abstract

The Semenov model of thermal ignition is presented and conditions for its application, with a number of examples, are discussed. The Frank–Kamenetskii model is also presented, giving examples of its application. For both the Semenov and Frank–Kamenetskii treatments, possible effects of fuel consumption are dealt with quantitatively. The branched chain route to ignition, with a suitable example, is discussed and the principles of the Unified Theory, which accounts for both thermal effects and chain branching, are given in detail.

1.1 Thermal ignition

1.1.1 Introduction

Heat release by a reacting mass, and heat transfer to the surroundings, are both important in determining the course of a combustion process. As recognized by Taffanel and Le Floch as far back as 1913 [1], this precludes the assignment of an *ignition temperature* to a particular exothermically reacting mixture: whether or not ignition occurs does not depend solely on the reactants but also on heat exchange with the surroundings. This is the principle of thermal ignition: ignition owing its existence to thermal imbalance. If not predominantly thermal, an ignition may be due to chain branching, leading to rapid multiplication of reactive intermediates. Many ignitions have both thermal and chain-branching contributions. Beginning with thermal theory, we will discuss each of them. All ignition phenomena discussed in this chapter are spontaneous, that is, they occur without introduced heat (e.g. a spark). Spark ignition will be discussed in Chapter 3.

1.1.2 The Semenov model

- **Basic formulation**

The first quantitative description of thermal ignition was by Semenov [2] in the late 1920s. His model has the following features:

(a) The heat release rate has an Arrhenius temperature dependence.

(b) The reactant has a uniform temperature T distinct from the ambient temperature T_o. There is therefore a temperature step $(T - T_o)$ at the boundary between the reactant and the surroundings.

(c) The heat transfer to the surroundings is convective and linked with the temperature difference solely by a constant of proportionality.

It is instructive to consider practical circumstances such that (b) and (c) hold. The uniform temperature requirement within the reactant mass is entirely analogous to *lumped capacity analysis* in heat transfer, which applies to bodies with low thermal resistance. This might be due to a high thermal conductivity or simply a small physical size. With fluid reactants, the uniform temperature condition may be achieved by stirring. The criterion for approximating the internal temperature of a body to a single value, in heat transfer or in ignition theory, is that there shall be a low value of the Biot number, defined as:

$$\mathrm{Bi} = \frac{h(V/S)}{k}, \tag{1.1}$$

where h = convection coefficient (W m^{-2} K^{-1})
V = reactant volume (m^3)
S = reactant surface area (m^2)
k = reactant thermal conductivity (W m^{-1} K^{-1}).

Hence a low Biot number is favored by a high k (as discussed) and a low value for h, as might apply to natural rather than forced convection. For a particle suspended in air, for example, such a coefficient will itself depend in principle on the temperature step as well as on the particle size, and the heat loss rate is given by:

$$q_- = hS(T - T_o). \tag{1.2}$$

We can write down the heat balance equation for a reacting system fulfilling the Semenov conditions as:

$$\text{Rate of heat gain by reacting mass} = \text{rate of heat release} - \text{rate of heat loss to surroundings}$$

$$= q_+ - q_- \tag{1.3}$$

$$c\sigma V \frac{\mathrm{d}T}{\mathrm{d}t} = V\sigma QA \exp(-E/RT) - hS(T - T_o), \tag{1.4}$$

where c = heat capacity (J kg^{-1}K^{-1})
σ = density (kg m^{-3})
A = pre-exponential factor (s^{-1})
E = activation energy (J mol^{-1})
Q = exothermicity (J kg^{-1}).

We now introduce the dimensionless temperature θ, given by:

$$\theta = \frac{E}{RT_o^2}(T - T_o), \tag{1.5}$$

where R is the Universal Gas Constant = 8.314 J K^{-1} mol^{-1}.

Since the density appears to the power 1, the reaction rate as written is in the formal sense for a first-order process, hence A has units s^{-1}. Some formulations of this problem omit the density from the heat release term, in which case A has units appropriate to a zero-th order reaction.

It is straightforwardly shown that:

$$\exp(-E/RT) = \exp\frac{\theta}{1+\epsilon\theta}\exp(-E/RT_o), \tag{1.6}$$

$$\text{where } \epsilon = RT_o/E, \text{ usually} \ll 1.$$

Making use of the approximation that $1 + \epsilon\theta \approx 1$ gives:

$$\exp(-E/RT) = \exp\theta \exp(-E/RT_o). \tag{1.7}$$

By this substitution, the heat balance equation becomes:

$$\frac{RT_o^2 c \exp(E/RT_o)}{EQA} \times \frac{d\theta}{dt} = \exp\theta - \frac{RT_o^2 hS \exp(E/RT_o)}{EV\sigma QA}\theta. \tag{1.8}$$

Defining a dimensionless time as:

$$\tau = \frac{EQA}{RT_o^2\, c \exp(E/RT_o)} \times t, \tag{1.9}$$

and a dimensionless heat transfer coefficient as:

$$a = \frac{RT_o^2\, hS \exp(E/RT_o)}{EV\sigma QA}, \tag{1.10}$$

the heat balance equation becomes:

$$\frac{d\theta}{d\tau} = \exp\theta - a\theta. \tag{1.11}$$

This equation encapsulates the principles of the Semenov model very conveniently, although it cannot in fact be integrated analytically. It can of course be easily integrated by numerical means. The exponential term represents heat release and the linear term heat loss and it is convenient to plot these on a thermal diagram as shown in Figure 1.1. On the vertical axis are plotted $\exp\theta$ and also $a\theta$ for various values of a. Equality of the heat release and heat loss terms marks a thermal steady state such that:

$$q_+ = q_-. \tag{1.12}$$

Physically, such a steady state will be characterized by a temperature excess of the reacting mass of a few tens of degrees above ambient. Intersection of the

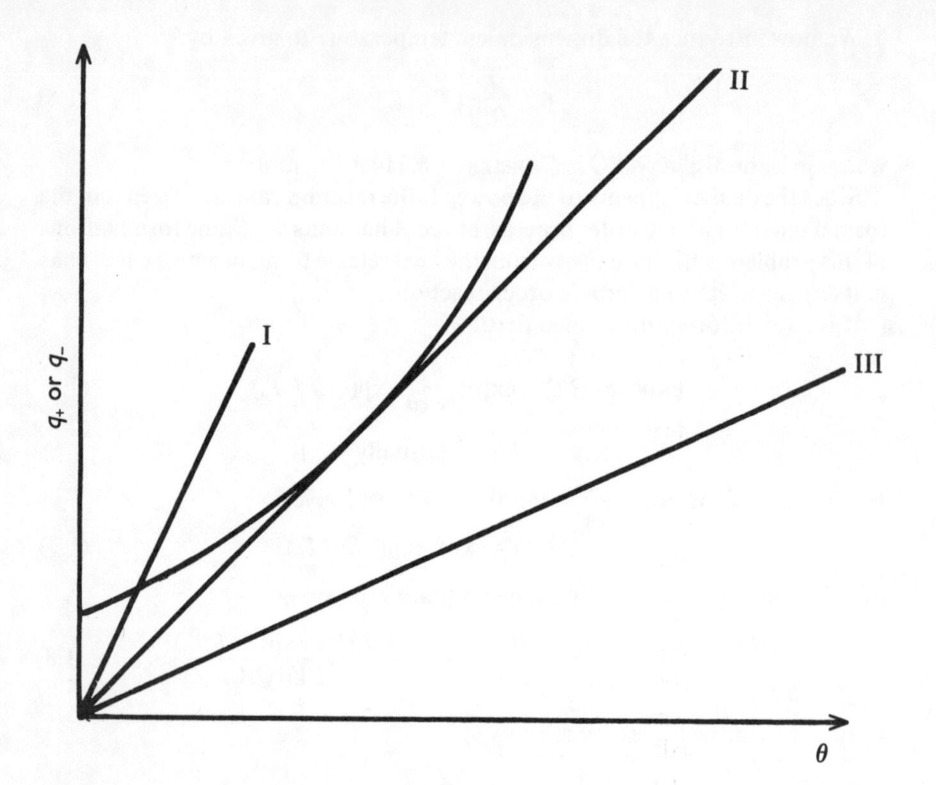

Figure 1.1 Thermal diagram showing expθ and $a\theta$ against θ for various a. Curve I, subcritical. Curve II, critical. Curve III supercritical.

curves for q_+ and q_-, as shown for Case I in the figure, denotes such a steady state. Case III on the other hand has no intersection. There is therefore no solution of the steady state equation 1.12 above, and ignition results. The threshold between one type of behavior and the other is Case II, in which the q_+ and q_- terms intersect tangentially. At this point we have the following conditions:

$$q_+ = q_-, \text{ and} \tag{1.13}$$

$$\frac{dq_+}{d\theta} = \frac{dq_-}{d\theta}. \tag{1.14}$$

These conditions are called the critical conditions and lead directly to the results that:

$$a_{\text{critical}} = e \tag{1.15}$$

$$\theta_{\text{critical}} = 1. \tag{1.16}$$

So for $a < e$ there is ignition and for $a > e$ there is a steady state. Returning to the dimensioned equations, equation 1.16 gives a value of the maximum temperature excess above ambient in the absence of an ignition, that is:

$$(T - T_o) = RT_o^2/E. \tag{1.17}$$

Arbitrarily putting $T_o = 700\text{K}$ (see example below) and $E = 100 \text{ kJ mol}^{-1}$ gives $(T - T_o)$ as 41K. Ignitions are of course characterized by much higher temperature excesses than this. Two important points about the Semenov treatment now need to be addressed:

(a) Is there any need to modify the critical conditions to allow for reactant consumption, not accounted for in the treatment above?

(b) What practical systems conform to the model and what information can be gained by applying the model to them?

- **Reactant consumption**

In addressing question (a) we must first recognize that, owing to there being no allowance for reactant consumption, equation 1.11 when integrated always gives temperature–time trajectories in disagreement with experiment at long times. When supercritical, the trajectories go to larger and larger values well beyond physical significance and the computer must be instructed to stop integrating. When subcritical, the trajectories plateau at a temperature corresponding to $\theta \leq 1$ and this persists indefinitely. In both cases in practice, reactant consumption brings the temperature–time trajectory back to ambient.

Nevertheless, at small extents of fuel consumption as θ approaches its steady value, or just exceeds unity, corrections to allow for reactant consumption are often small. Clearly the effect of reactant consumption will be to stabilize the system, causing a_{critical} to drop below its value, in the absence of reactant consumption, of 2.718.

When discussing reactant consumption it is convenient to return to the heat balance equation in dimensioned quantities and to combine with it an equation for reactant consumption. For a reaction of order m:

$$\sigma c \frac{dT}{dt} = QAw^m \exp(-E/RT) - \frac{hS}{V}(T - T_o), \tag{1.18}$$

where w = reactant concentration
$w = w_o$ at $t = 0$.

Recalling the dimensionless temperature θ (equation 1.5) and dividing through by w_o^m we can write:

$$\frac{\sigma c RT_o^2 \exp(E/RT_o)}{EQAw_o^m}\frac{d\theta}{dt} = \lambda^m \exp\frac{\theta}{1 + \epsilon\theta} - a\theta, \tag{1.19}$$

where λ = dimensionless concentration w/w_o.

$$a = \frac{hSRT_o^2 \exp(E/RT_o)}{w_o^m EVQA}, \tag{1.20}$$

(as defined previously in equation 1.10, but now containing w_o^m).

If we put $t_R = \dfrac{\exp(E/RT_o)}{Aw_o^{(m-1)}}$ and $B = \dfrac{w_oQE}{\sigma cRT_0^2}$, we can write equation 1.19 as:

$$\frac{t_R}{B}\frac{d\theta}{dt} = \lambda^m \exp\frac{\theta}{1+\epsilon\theta} - a\theta, \tag{1.21}$$

and the equation for reactant consumption is easily shown to be:

$$t_R\frac{d\lambda}{dt} = -\lambda^m \exp\frac{\theta}{1+\epsilon\theta}. \tag{1.22}$$

Now $\lambda = 1$ at $t = 0$. The quantity t_R is the time taken for the reaction to go to completion ($\lambda = 0$) isothermally at T_o. More importantly, the dimensionless quantity B, containing the heat of reaction divided by the heat capacity, has the nature of an adiabatic (peak) flame temperature, such as is discussed more fully in Chapter 5 on gaseous fuels. In thermal ignition theory literature (e.g. Boddington, Gray and Wake [3]) corrections to the critical condition for reactant consumption are given in terms of B or its reciprocal, and in their derivation the approximation made in obtaining equation 1.8 above, that $\epsilon\theta$ is negligible compared to unity, is frequently used. A number of such corrections have been made by various approaches, for example the one due to Frank–Kamenetskii:

$$a_{\text{critical}} = e[1 - 2.703(m/B)^{2/3}]. \tag{1.23}$$

For zero-th order, or very high B, this will reduce to equation 1.15. Note that this value of a_{critical} gives a margin of safety. A system set up in such a way that it is stable even if reactant consumption effects are ignored will be safer, not less safe, if reactant consumption does in fact influence the critical behavior. Among the practical materials believed to lend themselves to analysis by means of the Semenov model are certain liquid explosives (see below), which, if well stirred, fulfill the requirement of a uniform reactant temperature. They release heat by decomposition with Q values of 1 to 1.5 MJ kg^{-1}, an order of magnitude or more lower than Q values for oxidation combustion processes. Such a liquid explosive will typically have E = 150 kJ mol^{-1}, c = 1200 J kg^{-1} K^{-1}, and at an arbitrary ambient temperature of 315K this gives a value for B of 225. The corrected value of a_{critical} for a first-order reaction is therefore obtainable from equation 1.23 as 2.521, some 7 percent lower than the value in the absence of reactant consumption.

- **Practical systems**

Experimental determination of critical conditions will usually involve changing the ambient temperature T_o while leaving the other quantities unchanged. A combustion test at a particular ambient temperature, if it fails to result in ignition, can be repeated with an identical sample at a higher temperature. Families of supercritical results and families of subcritical results will be obtained from multiple experiments and it will be possible to estimate the value of $(T_o)_{crit}$, the ambient temperature corresponding to $a_{critical}$. Such determinations can be made for different S/V ratios.

Experiments of this sort were carried out [4] for the liquid explosive dinitroxydiethylnitramine (DINA) when well stirred, affording temperature uniformity. Figure 1.2 shows critical behavior for the explosive in a cylindrical vessel of 3.65 cm diameter and 7.3 cm length. The value of $(T_o)_{crit}$ is clearly $(158 \pm 0.5)°C$. The investigators had values for Q, E and A from independent work and had also measured the heat transfer coefficient. Treating the reaction as zero-th order ($a_{critical} = e$), they were able to calculate $(T_o)_{crit}$ for each sample vessel size by solving the expression for $a_{critical}$ (equation 1.10) as a transcendental equation in $(T_o)_{crit}$. The experimental and calculated values were then compared. In five such experiments with different size vessels the experimental and calculated values of $(T_o)_{crit}$ varied by not more than 2° C.

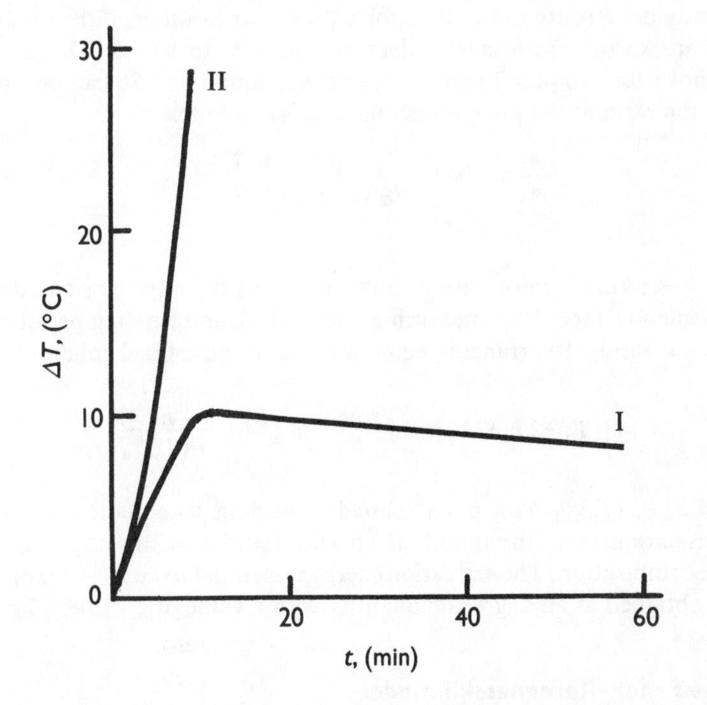

Figure 1.2 Sub- and supercritical temperature–time curves for the explosive DINA. Curve I: T_o = 157.5° C. Curve II: T_o = 158.5° C. The vertical axis represents the temperature excess of the reactant with respect to ambient (Merzhanov et al. [4])

- **Kinetic information**

As well as stirred systems, those with very low thermal resistance by reason of their small size may have nearly uniform internal temperatures and can therefore be treated satisfactorily by the Semenov approach, yielding kinetic information. We might consider very thin discs of wood in thermobalance experiments at temperatures of about 700K. Suppose that one such wood (sample 1) ignites at a furnace (ambient) temperature of 700K and the other (sample 2) at 715K. Applying equation 1.10 to each:

$$\frac{R(T_o)_2^2}{E_2}\frac{S}{V}\frac{h}{\sigma QA_2}\exp[E_2/R(T_o)_2] = e, \tag{1.24}$$

$$\text{and } \frac{R(T_o)_1^2}{E_1}\frac{S}{V}\frac{h}{\sigma QA_1}\exp[E_1/R(T_o)_1] = e, \tag{1.25}$$

and h, σ, S, V and Q are assumed not to vary between the samples. Combining the above equations:

$$\frac{E_2A_2}{E_1A_1}\frac{\exp[E_1/R(T_o)_1]}{\exp[E_2/R(T_o)_2]} = \frac{(T_o)_2^2}{(T_o)_1^2} = 1.043. \tag{1.26}$$

This provides a route to kinetic information. For example, if the activation energies are known from independent experiments to be 120 kJ mol^{-1} and 100 kJ mol^{-1} for samples 2 and 1 respectively, equation 1.26 can be used to estimate the ratio of the pre-exponential factors as follows:

$$\frac{A_2}{A_1} = 1.043 \times \frac{E_1}{E_2}\frac{\exp[E_2/R(T_o)_2]}{\exp[E_1/R(T_o)_1]} \tag{1.27}$$
$$= 17.5.$$

More direct kinetic information can be obtained from systems conforming to the Semenov model by measuring the critical ambient temperature for various S/V ratios. Rearranging equation 1.10 at the critical condition $a = e$ gives:

$$\ln[(T_o)_{crit}^2 h(S/V)] = \ln\frac{E\sigma e}{R} + \ln QA - \frac{E}{R(T_o)_{crit}}. \tag{1.28}$$

Hence S/V, $(T_o)_{crit}$ data pairs plotted according to equation 1.28 yield Arrhenius parameters. An example of a reaction studied in this way is cellulose nitrate decomposition. The activation energy was found to be 204 kJ mol^{-1} and QA was obtained as an aggregate quantity with a value of 2×10^{25} J kg^{-1} s^{-1}.

1.1.3 The Frank–Kamenetskii model

• Background

In situations where there is considerable resistance to heat transfer in an exothermically reacting mass, there will be a corresponding temperature gradient, making the Semenov model inapplicable. The Frank–Kamenetskii model considers systems with a high Biot number, heat transfer being conductive within the reacting mass. The heat balance equation for a steady state is therefore:

$$k\nabla^2T + QA\sigma\exp[-E/RT] = 0, \tag{1.29}$$

where k = thermal conductivity W m^{-1} K^{-1}
∇ denotes the Laplacian operator
$T = T_o$ at the surface.

The dimensionless temperature θ (equation 1.5) is used, together with the exponential approximation ($\epsilon\theta \ll 1$), to give the heat-balance equation in its dimensionless form:

$$\nabla^2\theta + \delta \exp\theta = 0, \qquad (1.30)$$

$$\text{where } \delta = \frac{r_o^2 \, Q\sigma EA \exp(-E/RT_o)}{kRT_o^2}, \qquad (1.31)$$

where r_o = reactant dimension (see below).

It can be seen from equation 1.31 that δ is the dimensionless form of the heat release rate at ambient temperature. Solutions to the steady state equation (1.30) exist only for certain values of δ. A system set up in such a way that δ is outside the range for which steady state solutions exist will therefore ignite. The critical value of δ above which ignitions occur is denoted δ_{crit} and is dependent on the reactant shape. The simplest systems to analyze by this approach initially are those with a Laplacian dependent on one space coordinate only. Exact values for δ_{crit} for these shapes, termed class A geometries, have been deduced and are given in Table 1.1. The table also gives the maximum temperature excesses for stability in terms of θ. It will be recalled (equation 1.16) that this condition under Semenov conditions is $\theta = 1$. In steady states with conductive heat transfer the body is always hottest at the center, and the temperature profile, along a diameter or width as the case may be, is parabolic.

Table 1.1 Critical conditions for Class A geometries*

Shape	*Physical meaning*	δ_{crit}	θ_{max}
Sphere	Sphere	3.32	1.61
Infinite cylinder	Cylinder lagged at its flat ends so that all conduction is radial	2.00	1.39
Infinite slab	Slab lagged across two pairs of parallel faces so that all conduction is between the remaining pair	0.88	1.19

*For derivations of these values see Gray and Lee [5].

The Frank–Kamenetskii approach, with purely conductive heat transfer within the reacting mass, and high Biot number, has been applied fairly extensively to two types of system: gas-phase reactions where convection is excluded, and assemblies of solid fuels or explosives. Various geometries have been used. Selected examples from the literature will be discussed.

- **Gas-phase processes in spherical vessels**

The exclusion of natural convection requires a low value of the Rayleigh number, defined as:

$$\mathrm{Ra} = \frac{g\beta(T - T_o)r_o^3 \, c\sigma^2}{k\mu}, \tag{1.32}$$

where g = acceleration due to gravity (9.81 m s^{-2})
β = coefficient of expansion (K^{-1})
μ = dynamic viscosity (kg m^{-1} s^{-1})
r_o = radius (m) for a spherical container of gas.

The threshold value of Ra above which heat transfer ceases to be purely conductive was examined for the oxidation of nitric oxide [6] in spherical vessels and found to be about 600. At this value of the Rayleigh number the steady state temperature profile along a diameter is parabolic and in conformity with the calculated profile. At higher values of the Rayleigh number the profile becomes unsymmetrical, indicating convective as well as conductive heat transfer. This threshold value is believed to be appropriate to other gas-phase processes.

The decomposition of diethyl peroxide ($C_2H_5 - O - O - C_2H_5$) has been studied as an example of thermal ignition [7] and tested for conformity with the Frank–Kamenetskii model. As with the above work on liquid explosives, the procedure was for identical samples of the gas to be examined at different ambient temperatures, so that $(T_o)_{crit}$, corresponding to δ_{crit}, could be determined. The experiments were performed in a spherical vessel ($\delta_{crit} = 3.32$) and $(T_o)_{crit}$ was obtained for several pressures of reactant.

When the expression defining δ (equation 1.31) is applied to gases, the density is given its ideal gas law value and the expression becomes:

$$\delta = \frac{r_o^2 \, Q(P/RT_o)AE \exp(-E/RT_o)}{RT_o^2 \, k}, \tag{1.33}$$

where P = pressure (N m^{-2}).

It can be deduced from equation 1.33 that for a particular gas-phase reaction performed at different pressures a graph of $\ln(P/(T_o)^3{}_{crit})$ against $1/(T_o)_{crit}$ yields a straight line of slope $-E/R$, and results from this kind of work are often expressed in this way.

Figure 1.3 shows the explosion limit for diethyl peroxide, which has the form characteristic of thermal ignition. For each of the several points on the limit, δ_{crit} was calculated from the experimental pressure/temperature pairs, together with values for the other quantities of which δ is composed. For example, a sample at 10 torr (1329 N m^{-2}) pressure gave a critical ambient temperature of ignition of 451.4K, the vessel radius in this and all the other experiments being 6.63×10^{-2} m. Other relevant quantities are:

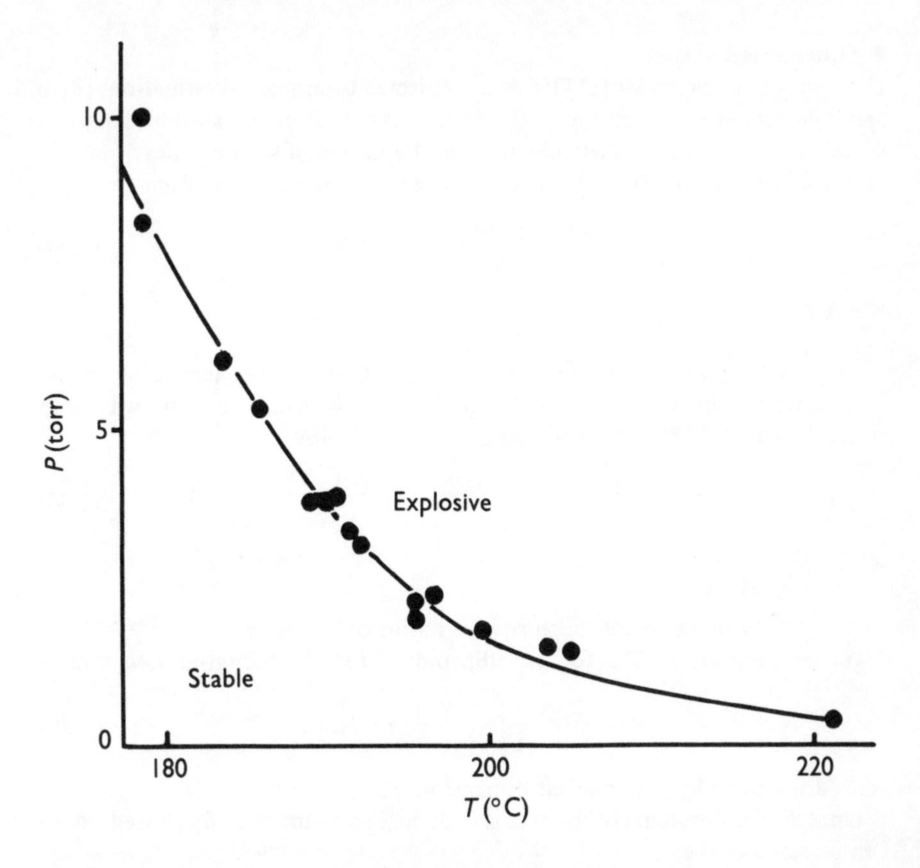

Figure 1.3 Explosion limit for diethyl peroxide (Fine, Gray and MacKinven [7])
(Reproduced with the permission of the Royal Society and Professor P. Gray.)

$$Q = 197 \text{ kJ mol}^{-1}$$
$$E = 143 \text{ kJ mol}^{-1}$$
$$k = 0.027 \text{ W m}^{-1}\text{ K}^{-1}$$
$$A = 1.6 \times 10^{14} \text{ s}^{-1}.$$

Substitution into equation 1.33 gives δ_{crit} = 4.3. This is higher than the theoretical value and the fifteen values of δ_{crit} calculated for diethyl peroxide gave a mean of 4.0. The fact that the measured values are higher means that ignition is occurring less easily than would be expected on the basis of the theory and the discrepancy is attributed to reactant consumption. Corrections for this are possible and depend on the dimensionless adiabatic flame temperature B (see equation 1.19). Such corrections for diethyl peroxide decomposition give a value of δ_{crit} of 3.95, in excellent agreement with the mean of the measured values. Agreement with theory was also excellent in terms of steady state temperature excesses. The maximum excess measured was 20° C, indistinguishable from the predicted maximum of 1.61 RT_0^2/E (Table 1.1).

- **Other vessel shapes**

Ditertiary butyl peroxide (DTBP) was subjected to similar investigations [8] in spherical and other vessel shapes. This requires definition of a suitable length in δ and calculation of the critical condition by means of scaling rules.

For Class A geometries the heat balance equation can be written:

$$\frac{d^2\theta}{dr'^2} + \frac{j}{r'}\frac{d\theta}{dr'} + \delta \exp\theta = 0, \tag{1.34}$$

where $r' = r/r_o$

j = shape factor

= 0, 1 and 2 for infinite slab, infinite cylinder and sphere respectively.

Geometries other than Class A can be dealt with in the same way with suitable values for j. These can be calculated as follows:

$$j = \frac{3R_o^2}{R_s^2} - 1, \tag{1.35}$$

where R_s = Semenov radius

$= 3V/S$

R_o = harmonic root mean square radius of the body.

As an example of R_o, for an ellipsoid of axis half-lengths a, b and c,

$$\frac{1}{R_o^2} = \frac{1}{3}\left[\frac{1}{a^2} + \frac{1}{b^2} + \frac{1}{c^2}\right]. \tag{1.36}$$

R_o is dominated by the smallest dimension.

Once R_o for the shape of interest is deduced, an estimate of δ_{crit} based on R_o can be calculated as:

$$\delta_{crit}(R_o) = \frac{6(j+3)}{(j+7)}. \tag{1.37}$$

These scaling rules apply to stellate bodies (all of the surface visible from the origin).

Glass vessels of unusual shape were used in addition to spheres in the DTBP work, for example an oblate spheroid for which $R_s = 5.19$ cm and $R_o = 5.11$ cm. Substitution of these values into equation 1.35 gives $j = 1.91$ and from equation 1.37, $\delta_{crit}(R_o) = 3.31$. For the six vessels used, the results were plotted according to equation 1.33 with $r_o = R_o$ and $\delta = \delta(R_o)$ and the points from the different vessels were found to lie on a good single straight line, validating the use of $\delta(R_o)$. However, agreement between the values of $\delta(R_o)$ and values of δ_{crit} calculated from the pressure–temperature results plus thermal conductivity values, etc. was only semi-quantitative. This is attributed not to inadequacies of the scaling rules but rather to difficulties inherent in obtaining δ_{crit} from its constituent quantities.

- **Solid explosives**

Cylindrical charges of benzoyl peroxide at temperatures below its melting point (105° C) were performed [9]. Similar experiments were carried out on a paste comprising 65 percent of the peroxide and 35 percent plasticizer. Whereas the applications of Frank–Kamenetskii previously discussed have used values of δ_{crit} appropriate to an effectively infinite Biot number, in the work on benzoyl peroxide, δ-values themselves a function of the Biot number were estimated and used. This is defined in this work as:

$$\mathrm{Bi} = \frac{hr}{k}, \tag{1.38}$$

where r = cylinder radius.

The method used to calculate δ_{crit} makes use of an effective heat transfer coefficient in place of the Laplacian operator in the heat-balance equation and gives results of the form $\delta_{crit}(\mathrm{Bi}, L/D)$ where L is the cylinder height and D its diameter. Bi was calculated from a knowledge of the convection heat transfer coefficient and of the explosives' thermal conductivities. Criticality tests to determine $(T_o)_{crit}$ were performed and the results checked for conformity with thermal theory in the following ways.

First, the measured $(T_o)_{crit}$ was in each case compared with a value calculated from the equation for δ_{crit} (equation 1.31). Values of A and E were taken from previous isothermal work. Agreement was very reasonable, as the following example illustrates. A 0.08 m diameter charge of the 65 percent paste with L/D = 1.6 and Bi = (4.3 ± 1.4) gave an experimental $(T_o)_{crit}$ of 59.9° C and a calculated value of 59.8° C. Not all the values agreed so closely but the separation of each measured and calculated pair was less than 3° C.

Secondly, activation energies of decomposition were calculated from the thermal measurements and compared with values obtained isothermally. This requires plotting of $\ln[\delta_{crit}(T_o)^2_{crit}/r^2]$ against $1/(T_o)_{crit}$. It will be seen by inspection of equation 1.31 that such a graph will have a slope of $-E/R$. Here again the agreement was very satisfactory (values ≈ 200 kJ mol^{-1}), although the thermally derived values tended to be a few percent lower than the isothermal values.

- **Basket heating of solid samples**

Equation 1.31 rearranges to:

$$\ln\left[\frac{\delta_{crit}(T_o)^2_{crit}}{r_o^2\sigma}\right] = \ln\frac{EQA}{Rk} - \frac{E}{R(T_o)_{crit}}. \tag{1.39}$$

Hence if samples of a combustible solid such as wood or carbon are examined in different-size samples in a shape for which δ_{crit} is known (e.g. a cube, δ_{crit} = 2.57, with r_o = half-width) and $(T_o)_{crit}$, r_o pairs determined, a graph of the logarithmic term on the left against $1/(T_o)_{crit}$ gives a straight line which can be extrapolated to values of $1/(T_o)_{crit}$ characteristic of storage temperatures. Putting, for example, $(T_o)_{crit}$ = 313K, the value of r_o corresponding to this gives the maximum half-width of a cubic assembly of the particular material which it would be safe to store out-of-doors on a warm day. Alternatively, the temperature at which a particular size of assembly would be critical can be deduced graphically. Experimental implementation of this approach will involve the heating of samples of the material of interest in gauze baskets up to about 10 cm side in laboratory ovens and determining $(T_o)_{crit}$ by trial and error with multiple subsamples for each basket size, to give the $(T_o)_{crit}$, r_o pair required.

This was the method used by Cameron and MacDowall [10] in their assessment of the shipping safety of certain carbon products. As an example of their results, a particular carbon in a cube of side 51 mm and bulk density 354 kg m^{-3} failed to ignite at an oven temperature of 122°C but, with a fresh subsample, ignited at an oven temperature of 132°C, therefore $(T_o)_{crit} = 127 \pm 5$°C for $r_o = 25.5$ mm. Results of this type were plotted according to equation 1.39 for 10 carbon products and the temperature at which a 3 m cubic shipping stow ($\approx$ 10 tonnes) of each material would be critical was deduced graphically. Two were found to be critical below 40°C and therefore classified unsafe for shipping. A value of $(T_o)_{crit}$ of 55°C was arbitrarily fixed above which shipping in a 3 m hold was permissible, and 7 of the 10 carbon materials were discovered to fall in this category. Five carbons had $(T_o)_{crit}$ (r_o = 1.5 m) actually in excess of 100°C.

The basket heating method has in fact been quite widely used in the assessment of storage safety of combustible solids [11] and, if supplemented with exothermicities and thermal conductivities, can be used to obtain kinetic information in the shape of A and E values.

1.2 Branched-chain ignition

1.2.1 Principles

Proliferation of active centers occurs via branching steps, for example:

$$H + O_2 \rightarrow OH + O$$

and

$$O + H_2 \rightarrow OH + H.$$

Removal of active centers is caused by termination steps, for example:

$$H + O_2 + M \rightarrow HO_2 + M \text{, where } M \text{ is a third body,}$$

followed by:

$$HO_2 \rightarrow \text{wall.}$$

These steps are discussed more fully in Chapter 2 when describing the hydrogen/oxygen reaction. In general terms, imagine an active center x branching according to first order kinetics in x and terminating also according to a first order scheme. We can then write:

$$x + \ldots \xrightarrow{k_b} x + \ldots, \quad \text{branching,}$$

and

$$x + \ldots \xrightarrow{k_t} \text{inert products,} \quad \text{termination,}$$

so the kinetic equation is:

$$\frac{dx}{dt} = (k_b - k_t)x = \phi x, \tag{1.40}$$

where $\phi = k_b - k_t$ = branching factor.

Integrating, and putting $x = x_0$ at $t = 0$,

$$x = x_0 \exp(\phi t). \tag{1.41}$$

Therefore if ignition is attributed to proliferation of x, the ignition condition is:

$$\phi > 0, \text{ or } k_b > k_t,$$

and this is the simple criterion for an ignition due solely to chain branching. However, it has long been known [12] that even where an ignition is predominantly a chain-branching ignition, thermal effects are not in general absent and cannot be eliminated from detailed analysis. The main current significance of branched-chain theory is therefore that it is an introduction to the much more far-reaching Unified Theory in which both energy balance and reactive intermediate kinetic equations are examined to deduce critical conditions when thermal and branching effects operate. Ignitions with demonstrable participation by both thermal imbalance and chain branching are termed *chain thermal.*

1.2.2 Chlorine dioxide decomposition

Notwithstanding the need to view many ignition processes from both chain and thermal points of view, it is true that a few do appear to be predominantly due to chain branching and therefore describable as *chain-isothermal.* An example [13] is the explosive decomposition of chlorine dioxide. When, for example, this substance is admitted to a spherical vessel at 46° C at a pressure of 5.5 torr, there is an induction period of over an hour during which a fast-response thermocouple at the center of the vessel registers precisely the same temperature as the outside of the vessel wall, indicating zero self-heating. The temperature trace is almost vertical at the ignition itself. This is consistent with an isothermal branched-chain ignition and branching intermediates are believed to be ClO and Cl, with steps including:

$$\begin{array}{lll}
 & & Cl + O_2 \\
 & & \uparrow \\
\text{Initiation} \rightarrow ClO & & \\
ClO + ClO_2 \rightarrow Cl_2O_3 & & ClOO + Cl \\
\quad\quad\quad \downarrow ClO & \uparrow & \\
2ClO \xleftarrow{Cl} ClO_2 + ClOOCl & &
\end{array}$$

in which Cl and ClO are involved in branching.

1.3 The Unified Theory [14]

1.3.1 Principles and formulation

The Unified Theory in its most general form considers a reactant mixture at ambient (vessel outside wall) temperature T_o with a reactant temperature T and active center concentration x, both spatially uniform. The theory retains the kinetic equation for active centers:

$$\frac{\mathrm{d}x}{\mathrm{d}t} = (k_b - k_t)x, \tag{1.40}$$

and adds to it the heat-balance equation:

$$\frac{\mathrm{d}T}{\mathrm{d}t} = (k_b h_b + k_t h_t)x - L, \tag{1.42}$$

where h_b and h_t are exothermicity (J mol^{-1}) divided by reactant heat capacity c (J m^{-3} K^{-1}) for the branching and termination steps respectively, and:

$$L = \frac{HS}{cV}(T - T_o) \qquad \mathrm{K\ s^{-1}}, \tag{1.43}$$

where H = heat transfer coefficient (W m^{-2} K^{-1})
S = vessel surface area (m^2)
V = vessel volume (m^3),
that is, $L = f(T)$.

Equations 1.40 and 1.42 are autonomous (the time variable does not appear outside the differentials) so they can be combined to give:

$$\frac{\mathrm{d}T}{\mathrm{d}x} = \frac{(k_b h_b + k_t h_t)x - L}{(k_b - k_t)x}, \tag{1.44}$$

and the singularities of this system are given by the solutions of:

$$(k_b h_b + k_t h_t)x - L = 0 \tag{1.45}$$

and

$$(k_b - k_t)x = 0. \tag{1.46}$$

There are two singular points, (x_{s1}, T_{s1}) and (x_{s2}, T_{s2}), generalized to (x_{si}, T_{si}) and the following transformation of variables can be effected:

$$x' = x - x_{si}, \tag{1.47}$$

and

$$T' = T - T_{si}. \tag{1.48}$$

Putting $k_b = A_b \exp[-E_b/RT]$,

and $k_t = A_t$

(that is, no temperature dependence of this step) gives as the revised form of equation 1.44:

$$\frac{\mathrm{d}T'}{\mathrm{d}x'} = \frac{x'RT_{si}^2(k_t h_t + k_b h_b) + T'(x_{si}k_b E_b h_b - \alpha)}{x'RT_{si}^2(k_b - k_t) + T'(x_{si}k_b E_b)}, \tag{1.49}$$

where $\alpha = RT_{si}^2\left[\frac{\mathrm{d}L}{\mathrm{d}T}\right]$ at $T = T_{si}$.

Singular point (x_{s2}, T_{s2}) is simply $(0, T_o)$, that is, zero concentration of active centers at ambient temperature. For ignition behavior it is necessary that (x_{s1}, T_{s1}) be a saddle point, and the separatrices through it distinguish conditions such that either:

(a) the system tends to x_{s2}, T_o and there is no ignition, or

(b) x and T both go to higher and higher values and there is ignition.

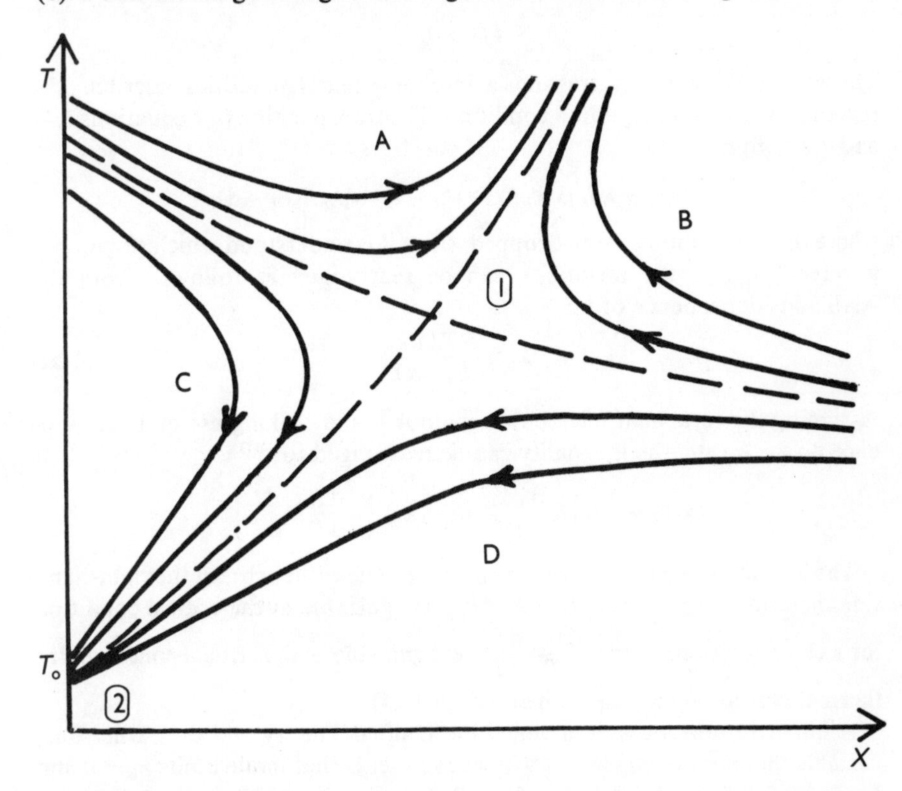

Figure 1.4 Phase-plane diagram for chain-thermal ignition based on a scheme comprising a branching and a termination step (Gray and Yang [14])

(Copyright 1965, American Chemical Society. Reproduced with permission.)

Figure 1.4 shows the ignition diagram in the (x, T) plane, with (x_{s2}, T_{s2}) shown as the point to which an experiment set up in quadrants C or D will go. An experiment set up in quadrants A or B will ignite, and, clearly, if an analytical expression of the conditions for the formation of a saddle point as shown is obtainable, it will represent the ignition condition for the system analyzed: a branching step, a termination step and heat balance and branching intermediate kinetic equations.

We can rewrite equation 1.49 as:

$$\frac{dT'}{dx'} = \frac{Ax' + BT'}{Cx' + DT'}, \tag{1.50}$$

and the nature of the singularity (x_{s1}, T_{s1}) depends on the roots of the characteristic equation:

$$\beta^2 - (B + C)\,\beta + (BC - AD) = 0. \qquad (1.51)$$

The condition for a saddle point is that the roots of this equation be both real (not complex) and differing in sign and this will be true provided that:

$$AD > BC.$$

Therefore this inequality, transposed into the original quantities describing the reacting system, is the ignition condition. The transposition (see equations 1.49 and 1.50) gives:

$$(k_b h_b + k_t h_t) x_s k_b E_b > (k_b - k_t)(x_s k_b E_b h_b - \alpha),$$

where the subscript has been dropped, since it is understood which singularity it refers to, and the inequality can be rearranged as follows. From the Arrhenius dependence of k_b:

$$k_b = \frac{RT_s^2}{E_b}\frac{dk_b}{dT}. \qquad (1.52)$$

Assigning the total heat release rate symbol $\mathfrak{R}$ and making use of its dependence on k_b, h_t, etc. the inequality can be rearranged to:

$$(k_b h_b + k_t h_t) x_s \frac{\mathrm{d}k_b}{\mathrm{d}T} > (k_t - k_b)\left\{-\frac{\mathrm{d}\mathfrak{R}}{\mathrm{d}T} + \frac{\mathrm{d}L}{\mathrm{d}T}\right\}.$$

The left-hand side must always be positive. The value zero for the right-hand side therefore means either that $k_t = k_b$, recognizable as the critical condition for a chain ignition, or that $\frac{\mathrm{d}\mathfrak{R}}{\mathrm{d}T} = \frac{\mathrm{d}L}{\mathrm{d}T}$, recognizable as the critical condition for thermal ignition (compare with equation 1.14).

Another important way in which the Unified Theory and the earlier one-variable theories are consistent with each other is that in the limits $h_b \to 0$ and $h_t \to 0$, $\mathrm{d}T/dx \to 0$ and therefore the separatrix in Figure 1.4 becomes horizontal and can be identified with the classical chain critical condition $k_b = k_t$. The other limiting case is where there is no branching mechanism for the intermediate x, in which case the counterparts of equations 1.40 and 1.42 are:

$$\frac{\mathrm{d}x}{\mathrm{d}t} = k_1 - k_2 x \qquad (1.53)$$

and

$$\frac{\mathrm{d}T}{\mathrm{d}t} = k_1 h_1 + k_2 h_2 x - L, \qquad (1.54)$$

where the subscripts 1 and 2 denote formation and removal of x.

Transforming variables as before:

$$\frac{\mathrm{d}T'}{\mathrm{d}x'} = \frac{RT_s^2 k_2 h_2 x' + (k_1 h_1 E_1 - \alpha)T'}{k_1 E_1 T' - RT_s^2 k_2 x'}, \qquad (1.55)$$

and the condition for a saddle point is:

$$k_1E_1(h_1 + h_2) > \alpha.$$

Recalling the definition of α, this means that for $L = 0$, $\alpha = 0$ and therefore ignition results, a rediscovery of the fact that an exothermic process inevitably ignites under adiabatic conditions. However, if the net process is not exothermic and therefore $(h_1 + h_2)$ is negative, then since α cannot be negative there can be no ignition and therefore the Unified Theory has, in this limiting case, predicted another fact thoroughly consistent with basic principles: that thermal ignition cannot occur unless the reaction is exothermic.

1.3.2 Application to hydrogen sulfide oxidation

In Section 1.3.1 we have given a fairly full description of the principles of the Unified Theory and in following chapters we will see how it describes and predicts combustion phenomenology that a single-variable model cannot account for.

We conclude this discussion of the theory with a description of one of its early triumphs: its prediction that a temperature–time trajectory can be displaying a negative slope immediately before ignition. This is easily seen by inspection of quadrant A in Figure 1.4, where at first T drops before rising sharply owing to the ignition, although x increases along any of the curves in quadrant A. This indicates that while the reactant temperature actually drops, the branching intermediate builds up to a degree where ignition occurs and therefore the temperature ceases to drop, rising sharply in response to the ignition. Such behavior was unknown when the Unified Theory was first presented, but was subsequently found in hydrogen sulfide oxidation [15]. A stoichiometric H_2S/O_2 mixture at ambient temperature 625K and total pressure 4518 Pa in a 0.29 litre spherical vessel displayed the temperature–time behavior shown in Figure 1.5. The temperature is actually dropping prior to ignition and this was the first experimental discovery of an ignition of the type in quadrant A of Figure 1.4. The branching intermediate is believed to be S_2O.

1.4 Conclusion

We have discussed thermal, chain-isothermal and chain-thermal ignitions with suitable examples. In subsequent chapters we see how the principles discussed, especially the Unified Theory, throw light on the features of gas-phase oxidation processes including hydrocarbon oxidation. We will also consider ignitions which, unlike those discussed in this chapter, require introduced heat (for example a spark).

References

[1] Taffenel and Le Floch, 'Sur la combustion des melanges gazeux et les temperatures d'inflammation', *Compt. Rend.* **157** 469 (1913)
[2] Semenov N. N. (1928), cited by Boddington et al. [3] *inter alia.*
[3] Boddington T., Gray P., Wake G. C., 'Criteria for thermal explosions with and without reactant consumption', *Proceedings of the Royal Society* **A357** 403–422 (1977)

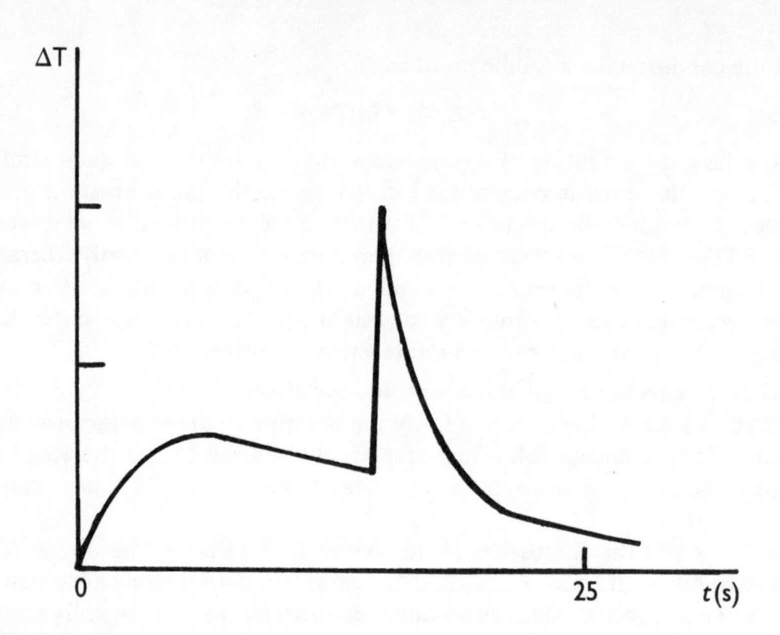

Figure 1.5 Temperature excess against time for a stoichiometric H_2S/O_2 in a 0.29 litre spherical vessel mixture at vessel temperature 625K, a total pressure 4518 Pa (Gray and Sherrington [15]) *(Reproduced with the permission of the Royal Society of Chemistry.)*

[4] Merzhanov A. G., Barzykin V. V., Abramov V. G., Dubovitskii F. I., 'Thermal explosion in the liquid phase with heat transfer by convection only', *Russian Journal of Physical Chemistry* **35** 1024–1027 (1961)

[5] Gray P., Lee P. R., 'Thermal explosion theory', *Oxidation and Combustion Reviews* **2** 1–183 (1967)

[6] Tyler B. J., 'An experimental investigation of conductive and convective heat transfer during exothermic gas-phase reactions', *Combustion and Flame* **10** 90–91 (1966)

[7] Fine D. H., Gray P., MacKinven R., 'Thermal effects accompanying spontaneous ignition. I, The explosive decomposition of diethyl peroxide', *Proc. Roy. Soc.* **A316** 255–268 (1970)

[8] Egeiban O. M., Griffiths J. F., Mullins J. R., Scott S. K., 'Explosion hazards in exothermic materials: critical conditions and scaling rules for masses of different geometry', *Nineteenth Symposium (International) on Combustion*, 825–833, Pittsburgh: The Combustion Institute (1983)

[9] Bowes P. C., 'Thermal explosion of benzoyl peroxide', *Combustion and Flame* **12** 289–301 (1968)

[10] Cameron A., MacDowall J. D., 'The self-heating of commercial active carbons', *Journal of Applied Chemistry and Biotechnology* **22** 1007 (1972)

[11] Beever P., Thorne P. F., 'Isothermal methods for assessing combustible powders—theory and experimental procedures', *Symposium Series* No. 68, *Runaway Reactions*, Rugby: Institution of Chemical Engineers.

[12] Ben-Aim R., Lucquin M., 'Application of the theory of branched chain reactions in low-temperature combustion', *Oxidation and Combustion Reviews* **1** 1 (1965)

[13] Gray P., Ip J. K. K., 'Spontaneous ignition supported by chlorine dioxide. I, Chlorine dioxide alone and with diluents', *Combustion and Flame* **18** 361 (1972)

[14] Gray B. F., Yang C. H., 'On the unification of thermal and chain theories of explosion', *Journal of Physical Chemistry* **69** 2747 (1965)

[15] Gray P., Sherrington M. E., 'Explosive oxidation of hydrogen sulphide: self-heating, chain-branching and chain-thermal contributions to spontaneous ignition', *Journal of the Chemical Society, Faraday Transactions* I **70** 2336 (1974)

CHAPTER 2

Gas-phase Oxidation Processes

Abstract

Examples of laboratory investigations of gas-phase oxidation processes are discussed in detail. These include hydrocarbon oxidations, for example C_3H_8, which display oscillating cool flames. Interpretations of oscillating cool flames in terms of the older theories invoking decomposition of a reactive intermediate and the more rigorous treatment on the basis of the Unified Theory are presented. In addition to hydrocarbon oxidation, CO/O_2 *and* H_2/O_2 *are covered. In the latter, both classical and recent work are reviewed, bringing out their significance. The Unified Theory is also applied to the* H_2/O_2 *reaction.*

2.1 Introduction

In the purely thermal treatment of ignition, the kinetics component of the modelling is quite simple, consisting of a single Arrhenius expression. Such an approach is often insufficient to describe gas-phase oxidations. One reason for this is that as well as thermal imbalance, the proliferation of highly reactive intermediates may be a decisive factor in the global behavior of such reactions, hence the value of the Unified Theory in understanding gaseous oxidations. Moreover, combustion phenomenology other than simply slow reaction or ignition can be observed in many gas-phase reactions, notably multistage ignition and oscillating cool flames. None of these can be understood by reference to a one-variable (temperature or branching intermediate concentration) model. We are concerned in this chapter with experimental and theoretical studies of hydrocarbon, hydrogen and carbon monoxide oxidations. Exothermicities of these oxidations are high, for example 27 MJ kg^{-1} fuel for the combustion of acetaldehyde to carbon dioxide and water, compared with 1.5 MJ kg^{-1} for the explosive decomposition of ClO_2, discussed in Chapter 1.

For oxidation as opposed to decomposition, ignition diagrams analogous to Figure 1.3 may show a region corresponding to thermal ignition at certain temperatures and pressures, and also a region of ignition well outside the predicted thermal limit, as well as regions representing neither ignition nor slow combustion but one of the other behavior types. In this chapter the important principles and features of gas-phase oxidations are discussed against a background of selected examples from the research literature.

2.2 Purely thermal behavior

An example of a gas-phase oxidation which does appear to conform to simple thermal theory is the ethylamine–oxygen reaction [1]. In Figure 2.1 are shown the ignition limits for 25% and 16.7% ethylamine in oxygen and each curve has the same basic shape as that for the decomposition of diethyl peroxide (Figure 1.3). No features other than ignitions and slow reaction were detected in the pressure–temperature region employed. Other amines appear to behave predominantly according to thermal theory when reacted with oxygen, although some also display cool flames outside the ignition region. The nature of cool flames is discussed below.

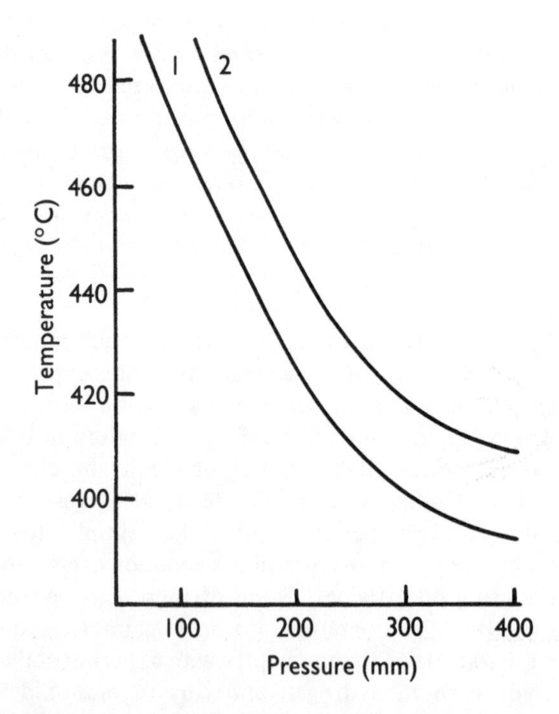

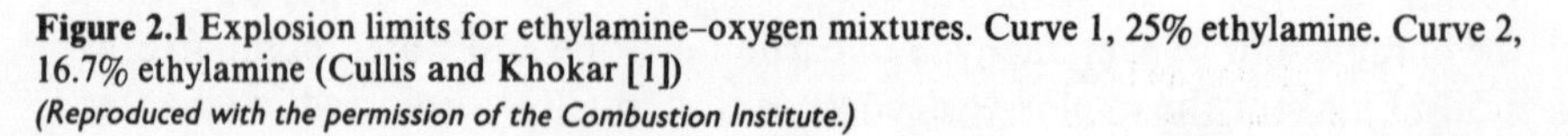
Figure 2.1 Explosion limits for ethylamine–oxygen mixtures. Curve 1, 25% ethylamine. Curve 2, 16.7% ethylamine (Cullis and Khokar [1])
(Reproduced with the permission of the Combustion Institute.)

2.3 Classical work on hydrocarbon oxidation

2.3.1 Butane oxidation

As an example of classical work on gas-phase oxidations displaying complex behavior we may examine the work on butane oxidation by Bardwell [2]. Figure 2.2 shows the experimental ignition limits obtained in a cylindrical vessel mounted in a furnace. In each experiment the mixture was 50% butane. The temperature in the graph is that of the furnace. The following features are evident in Figure 2.2:

(a) *Slow combustion.* This is characterized by absence of light emission or of pressure pulsations.

(b) *Single cool flame.* This involves a flicker of blue light and a moderate pressure pulsation.

(c) *Multiple cool flames.* Up to four were observed (up to five with a 33% butane/oxygen mixture) separated in time by periods from one to one hundred seconds depending on pressure and temperature.

(d) *Simple ignition.* This is characterized by a single intense flame, a large pressure rise and a small acoustic effect. The limit for this type of behavior does not have the simple shape expected for purely thermal behavior (e.g. Figure 2.1) but has a lobe at lower temperatures.

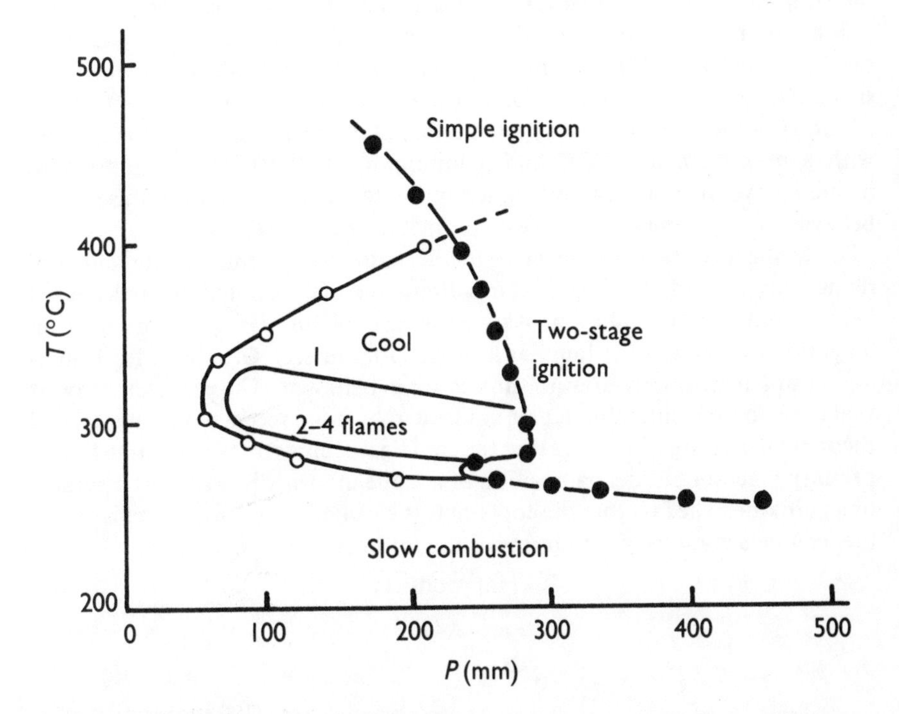

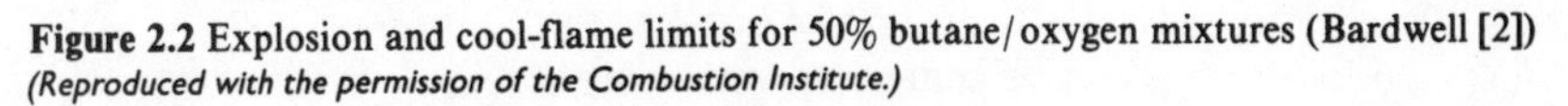

Figure 2.2 Explosion and cool-flame limits for 50% butane/oxygen mixtures (Bardwell [2]) *(Reproduced with the permission of the Combustion Institute.)*

(e) *Two-stage ignition.* A hot flame as described under (d) preceded by a cool one as described under (b).

Many other substrates including hexane, butanone, propane and acetaldehyde have long been known to display cool flame behavior and the following facts were known to the experimenters responsible for the classical work.

Only a small proportion of the available heat is released during a cool flame (otherwise multiple cool flames would not be possible without reactant replenishment) and the temperature maximum of a cool flame is only 100–200° C above the initial temperature. The reaction products are largely partial oxidation products of the fuel, for example, peroxides and aldehydes. No solid carbon is formed. The light emission is due to formaldehyde fluorescence. The ratio of quanta emitted to fuel molecules reacting is extremely small: 10^{-6} to 10^{-16}.

2.3.2 Early interpretations of cool flame behavior

Also known to relatively early investigators of hydrocarbon oxidation was the existence under some conditions of a negative temperature coefficient (n.t.c.) of rate, that is, behavior such that the oxidation rate decreases with temperature. This was reported for propane oxidation in 1949 [3] in regions in pressure–temperature space where there are no cool flames or ignitions. The investigators used pressure rise as an indication of reaction extent and regarded the steepest slope of the pressure–time graph as the maximum rate.

Maximum rates were deduced for a number of vessel temperatures and the results plotted as log [maximum rate] against $1/T$. Clearly a reaction obeying a single Arrhenius rate law will give a straight line for such a graph, with slope $-E/R$. However, a propane oxygen mixture (12% basis propane) gave a graph with a maximum at 315° C and a minimum at 360° C. Other substrates, including cyclopentane [4], examined in the same way, displayed this sort of behavior. Cool flames and n.t.c. are in fact intimately related.

The connection between negative temperature coefficients of rate and cool flames was pointed out in 1951 [5] in a discussion of cool flames in methyl ethyl ketone (butanone) oxidation. As well as a cool flame region, the ignition diagram also showed the familiar lobe (see Figure 2.2). Graphs of log [maximum rate] against temperature showed n.t.c. behavior. The prevalent view at that time, to a considerable degree vindicated by more recent experimental and theoretical investigations, was that the cool flame represents not combustion of primary reactant but decomposition of a labile intermediate such as a peracid or a peroxide. The fact that the cool flame is extinguished before exhaustion of fuel can be accounted for in terms of the following type of scheme:

$$\begin{array}{c} \text{Final products} \\ \uparrow k_1(T) \\ A \overset{v_0}{\rightarrow} X \\ \downarrow k_2(T) \\ \text{Active centers regenerating } X, \end{array}$$

where v_o = rate of initiation (mol m^{-3} s^{-1})

k_1, k_2 = respective rate constants (s^{-1}) and are Arrhenius functions of temperature.

The labile intermediate X (most likely a peroxide) can react according to two paths of markedly different activation energies. If the regeneration reaction leads to a number v of X molecules, we can write the following equation for the rate of formation of X:

$$\frac{d[X]}{dt} = v_o + \left\{k_2(T)(v-1) - k_1(T)\right\}[X]. \tag{2.1}$$

If the above reaction 1, leading to stable final products, has the higher activation energy, it will become more prevalent at higher temperatures and k_1 will accelerate more rapidly than k_2. As a result, during the temperature rise accompanying a cool flame a stage will be reached where:

$$k_2(T)(v-1) - k_1(T) = 0. \tag{2.2}$$

Proliferation of X will therefore cease and the reaction rate and hence the heat release rate will drop. This leads to extinction of the cool flame. The resulting drop in temperature allows reaction 2 to increase in rate, leading to a fresh buildup of X. The same principle applies to slow combustion where n.t.c. is observed, the drop in reaction rate at higher temperatures being due to the interplay of $k_1(T)$ and $k_2(T)$ with their different temperature dependences.

In the butanone [5] work peroxide as a group of chemicals (that is, not a specific peroxide) was examined by removal of reacting mixture samples by an evacuated pipette and analysis by classical chemical methods. It was found that the peroxide concentration built up exponentially before a cool flame and decreased rapidly during the flame itself. This is entirely consistent with the scheme proposed above for the mechanism of a cool flame and the fact that it is extinguished before there is depletion of reactants, with X as peroxide.

In concrete terms we might envisage a peroxy intermediate X decomposing along reaction route 2 to give an oxygen atom or a hydroxyl radical, and alkyl (for example, methyl CH_3) or alkoxy (for example methoxy CH_3O) radicals. One or more of them will be involved in the generation of more peroxide, giving v molecules of X per molecule reacting along reaction route 2.

2.4 Modern interpretive and experimental work

2.4.1 Introduction

For a system behaving thermally, we can calculate the explosion limit in pressure–temperature terms from a knowledge of the quantities comprising δ_{crit} or a_{crit} as the case may be. This treatment is based on one differential equation—heat balance—as described in Chapter 1. Calculation of more complex limits involving a region of cool flames and an ignition peninsula, or of n.t.c. behavior, requires a model in which a differential equation describing formation and reaction of a branching intermediate such as peroxide is coupled with the one for heat balance, that is, the Unified Theory previously outlined. We will see how the Unified Theory can treat cool flames and n.t.c., which are not amenable to a one-variable model.

2.4.2 Use of coupled energy and branching intermediate equations

Yang and Gray [6] proposed the following generalized scheme for hydrocarbon oxidation:

$$A \xrightarrow[h_i]{k_i} x \qquad \text{initiation}$$

$$x \xrightarrow[h_b]{k_b} 2x \qquad \text{branching}$$

$$\left.\begin{array}{l} x \xrightarrow[h_{t1}]{k_{t1}} \text{stable products} \\ x \xrightarrow[h_{t2}]{k_{t2}} \text{stable products} \end{array}\right\} \text{termination,}$$

where k_1, etc. denote rate constants with an Arrhenius temperature dependence and h_1, etc. denote heats of reaction (J mol^{-1}). Clearly x is the branching intermediate in the above scheme. The Law of Mass Action and the First Law of Thermodynamics give us respectively equations 2.3 and 2.4 below in which a reactant temperature T, uniform throughout the reaction vessel, is used:

$$\frac{\mathrm{d}[x]}{\mathrm{d}t} = k_i - (k_{t1} + k_{t2} - k_b)[x] = y(x,T) \tag{2.3}$$

$$\frac{\mathrm{d}T}{\mathrm{d}t} = \frac{1}{C_v}\{k_i h_i + (k_{t1}h_{t1} + k_{t2}h_{t2} + k_b h_b)[x] - L\} = z(x,T), \tag{2.4}$$

where C_v = specific heat of the reacting mixture at constant volume (J m^{-3} K^{-1})
L = rate of conductive heat loss to the surroundings (W m^{-3})
$L \propto (T - T_o)$ where T_o = ambient temperature (K).

In bringing out the significance of the above equations we first consider slow combustion. If following the precedents of Semenov and Frank–Kamenetskii we neglect fuel consumption, slow combustion corresponds to a steady state. We should also note at this stage that, in slow combustion, more recent investigators have used $(T - T_o)$ as an indicator of reaction rate rather than the maximum rate of pressure rise, and this is preferable. At a steady state $y(x,T) = 0$ this gives:

$$[x]_s = \frac{k_i}{k_{t1} + k_{t2} - k_b}, \tag{2.5}$$

where $[x]_s$ denotes the steady value. This can be used to evaluate the heat release rate. Neglecting the heat effect of the initiation step compared to those of branching and termination, and assigning temperature independence to termination step 1, gives for the steady-state heat release R_s:

$$R_s = \frac{k_i(k_{t1}h_{t1} + k_{t2}h_{t2} + k_b h_b)}{k_{t1} + k_{t2} - k_b}. \tag{2.6}$$

Recalling that k_{t2} and k_b are both Arrhenius functions of temperature, then for different steady states with different values of the steady temperature T_s (obtainable in principle by setting $z(x,T)$ to zero) the numerator of equation 2.6 will rise consistently with temperature. However, there is the clear potential for the denominator to pass through a minimum with respect to temperature, reflected as a maximum in R_s. Experimentally, of course, this corresponds to n.t.c. behavior.

More advanced mathematical techniques were also applied by Yang and Gray to investigate cool flames. In the absence of fuel consumption, these are oscillations. Equations 2.3 and 2.4 do not contain t on the right-hand side (that is, they are autonomous) and so it is both possible and convenient to work in (x,T) space, denoted the phase plane. The conditions such that:

$$y(x,T) = 0 \tag{2.7}$$

and

$$z(x,T) = 0, \tag{2.8}$$

need to be investigated, these being singularities. We have already discussed how such a singularity characterized by (x_s, T_s) can represent a steady state in the sense of slow combustion, but this is not the only possibility. Another is that the phase plane graph of $[x]$ against T will be a closed curve called a limit cycle, the physical meaning of which is the occurrence of temperature and concentration oscillations recognizable as cool flames. In the phase plane a limit cycle will surround a singular point. The conditions for oscillations are developed by Gray and Yang in terms of the characteristic equation:

$$\beta^2 - (B + C)\beta + (BC - AD) = 0, \tag{2.9}$$

$$\text{where } A = \frac{\partial z}{\partial x} \qquad C = \frac{\partial y}{\partial x}$$

$$B = \frac{\partial z}{\partial T} \qquad D = \frac{\partial y}{\partial T}.$$

A condition for oscillatory behavior in the (x,T) plane is that equation 2.9 has complex roots. From elementary treatment of quadratic equations, this condition is readily shown to be:

$$(B - C)^2 + 4AD < 0.$$

Hence the equality:

$$(B - C)^2 + 4AD = 0, \tag{2.10}$$

represents the cool flame boundary in a way entirely analogous to that in which δ_{crit} represents the ignition boundary for a thermal process with conductive heat transfer. *A*, *B*, *C* and *D* can readily be obtained from equations 2.3 and 2.4. The total pressure *P* will appear via the rates, and the heat capacity and the ambient temperature T_0 via the heat conduction term. If literature values for the rate constants are used it is possible to solve equations 2.7, 2.8 and 2.10 simultaneously for selected *P* to give the corresponding T_0 and this is one point on the predicted cool flame limit. This can be done for various *P* values to obtain the full limit, corresponding to the experimental limit shown for butane in Figure 2.2.

Yang and Gray obtain such a limit with the aid of selected rate constants and also compute the temperature–time behavior. Fulfillment of equations 2.7 and 2.8 in the oscillatory case is strictly for the singular point that the limit cycle surrounds, and the temperature–time behavior once a P, T_0 point on the limit is deduced can be obtained by insertion into equations 2.3 and 2.4 and integrating

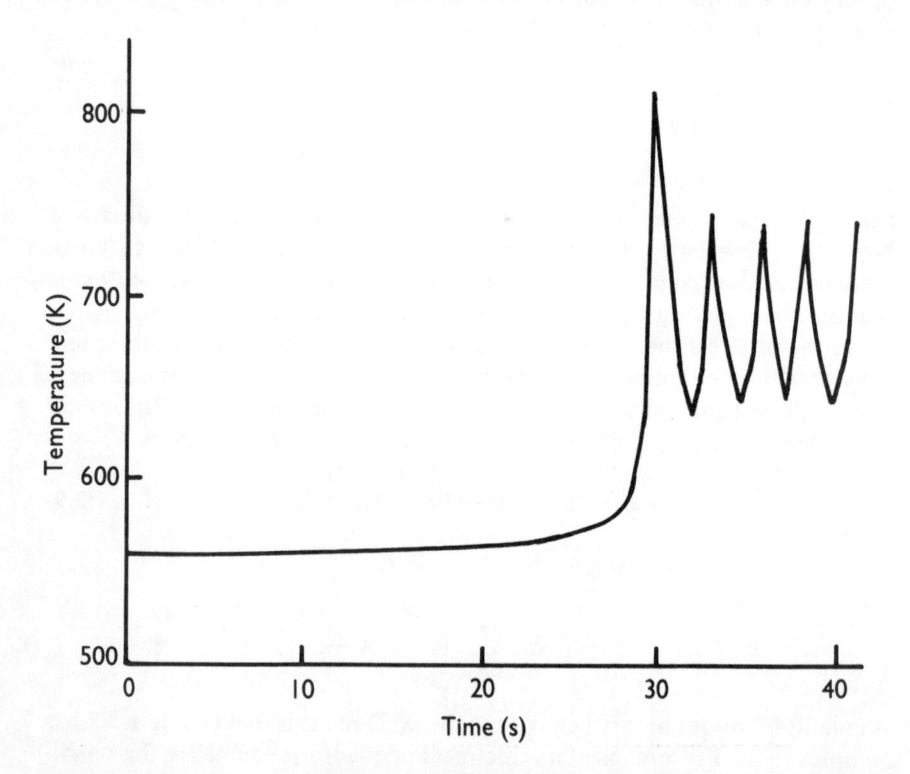

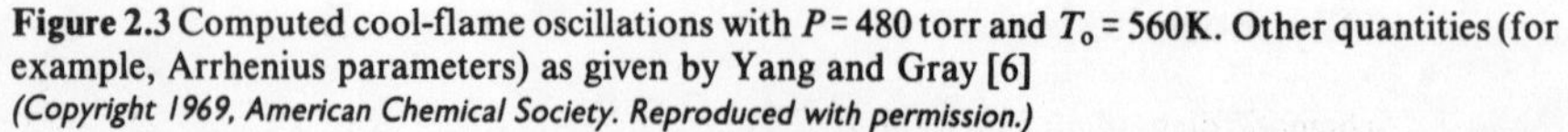

Figure 2.3 Computed cool-flame oscillations with $P = 480$ torr and $T_0 = 560$K. Other quantities (for example, Arrhenius parameters) as given by Yang and Gray [6]
(Copyright 1969, American Chemical Society. Reproduced with permission.)

numerically. The result is shown in Figure 2.3 where the computed oscillations can be seen to have periods of 2–3 s and amplitudes of around 100K.

Figure 2.4 is a schematic diagram of a limit cycle surrounding a singular point $\mathfrak{P}$ and if we trace the variables around one cycle in the direction indicated by the arrows, we can understand how this description is quite consistent with the earlier qualitative ones discussed in section 2.3.2. Imagine a hydrocarbon/oxygen mixture at point $\mathfrak{Q}$ where it is at the low end of the temperature scale with a high x concentration. The x concentration drops because of reaction, which is heat-releasing, therefore the drop in x is followed by a rise in temperature as shown and this leads to a cool flame. At the high temperature that the system has now reached, removal of x by reaction is rapid and its depletion causes a sharp temperature drop and extinction of the cool flame. Because of the lower temperature, x is again able to build up, taking the system to the place where we began our discussion and on to another cycle.

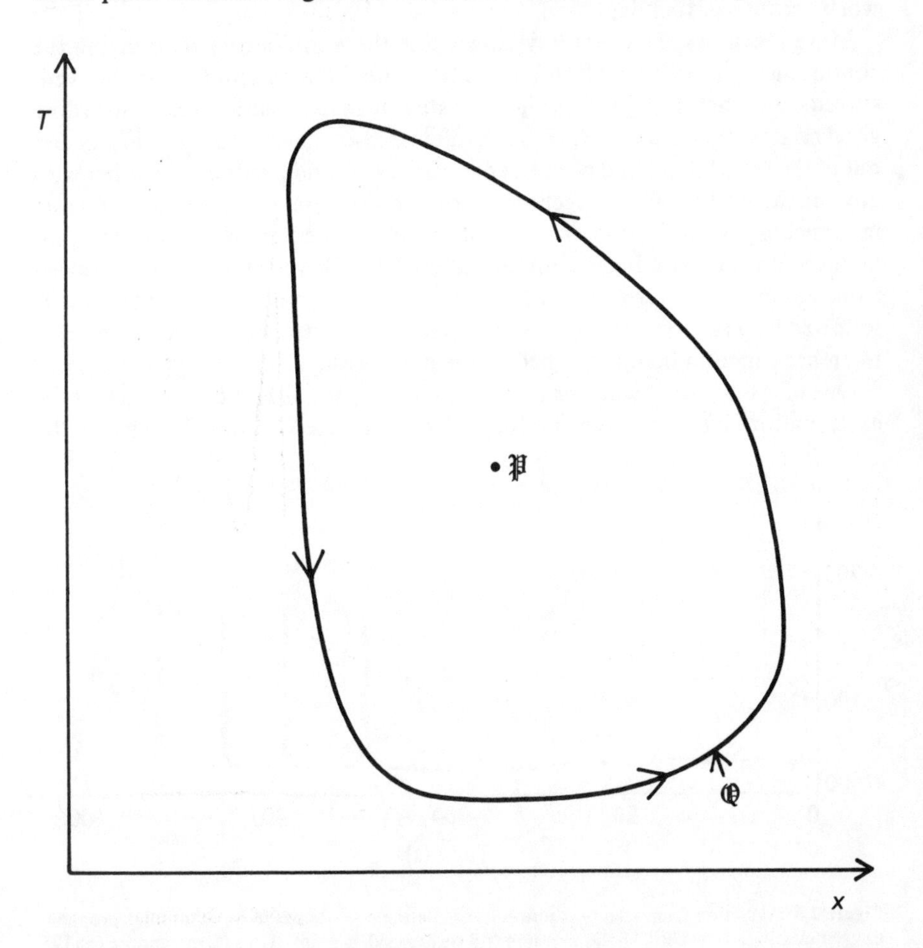

Figure 2.4 Schematic diagram of a limit cycle in the (x, T) phase plane for hydrocarbon oxidation

The condition for fully-fledged ignition is:

$$AD = BC, \tag{2.11}$$

and so from selected values of the rate constants etc., the ignition limit can be calculated.

Returning to cool flames, with closed-vessel experiments such as those of Bardwell [2], sustained oscillations are not possible because of fuel depletion. A strictly limited number of cool flames—up to about five—can be observed in closed systems and this represents a serious shortcoming of the classical experiments. Moreover, as we have already seen, slow combustion in the terms of the Yang and Gray treatment corresponds to a steady state. The temperature excess in a steady state is a direct measure of the heat release rate, linked to it by a constant of proportionality which is a weak function of the total pressure. However, potential calorimetric information of this kind is lost in closed vessel work because of fuel depletion.

More recent experiments have overcome these difficulties by utilizing the continuous stirred-tank reactor (CSTR), sometimes referred to as the well-stirred flow reactor. As well as vigorous stirring in the reaction vessel affording good mixing of fuel and oxygen, such a device also incorporates gas flow in and out of the vessel. Fuel and oxygen enter the vessel, and products, together with any unreacted fuel and oxygen, flow out. In this type of experiment, factors influencing thermal behavior are pressure, ambient (vessel outside wall) temperature and residence time of material in the vessel, commonly varied from about 3 s upward in this type of work. Temperature–time behavior is followed by a fast-response (<100 ms) rare metal thermocouple in the vessel, a reference junction being attached to the outer wall.

One of the first applications of the CSTR to hydrocarbon oxidation was the examination by Gray and Felton [7] of propane. Figure 2.5 shows the

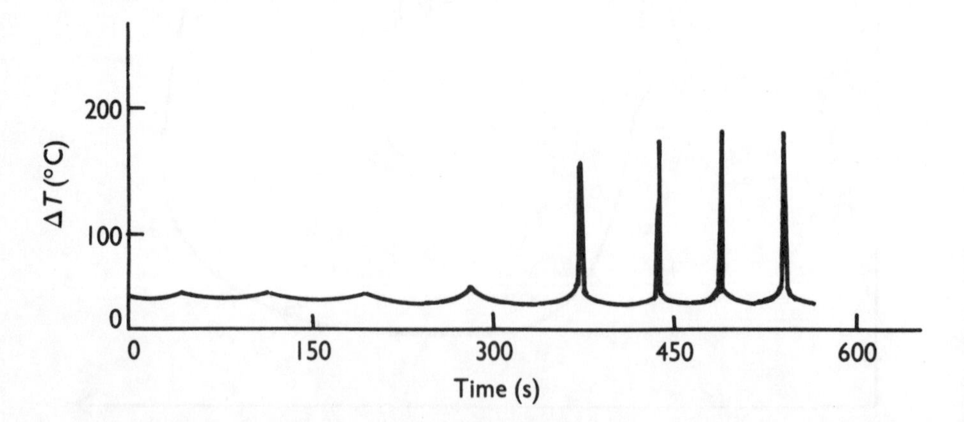

Figure 2.5 Transition from steady reaction to oscillating cool flames in an equimolar propane/oxygen mixture. T_0 = 572K. Total pressure 619 torr. Residence time 125 s (Gray and Felton [7]) *(Reproduced with the permission of the Combustion Institute.)*

transition from steady reaction to cool flames in response to an adjustment of ambient temperature. Oscillations of the type shown can be sustained indefinitely in a CSTR. Figure 2.6 shows the heat release rate as a function of reacting temperature, calculated from the temperature excess in steady state experiments as previously described. Those steady states on the low-temperature end of the upper branch are stable foci (S.F.) in contrast to those on the lower branch which are stable nodes (S.N.). Stable nodes occur at the higher temperature end of the upper branch under these conditions. Whether a particular steady state will be one or the other can in principle be deduced from the roots of the characteristic equation (2.9). However, Figure 2.6 represents an experimental distinction between the two sorts of singularity. If a stable node is perturbed, for example by pumping out some reactant to cause cooling, it will return to the steady state along a monotonic path. A stable focus similarly perturbed will return along an oscillatory path (though a very fast-response temperature measuring and recording system will be needed to detect the oscillatory behavior). Between two steady-state branches there occur limit cycles surrounding a singularity, that is, oscillating cool flames, and steady states are experimentally unobtainable.

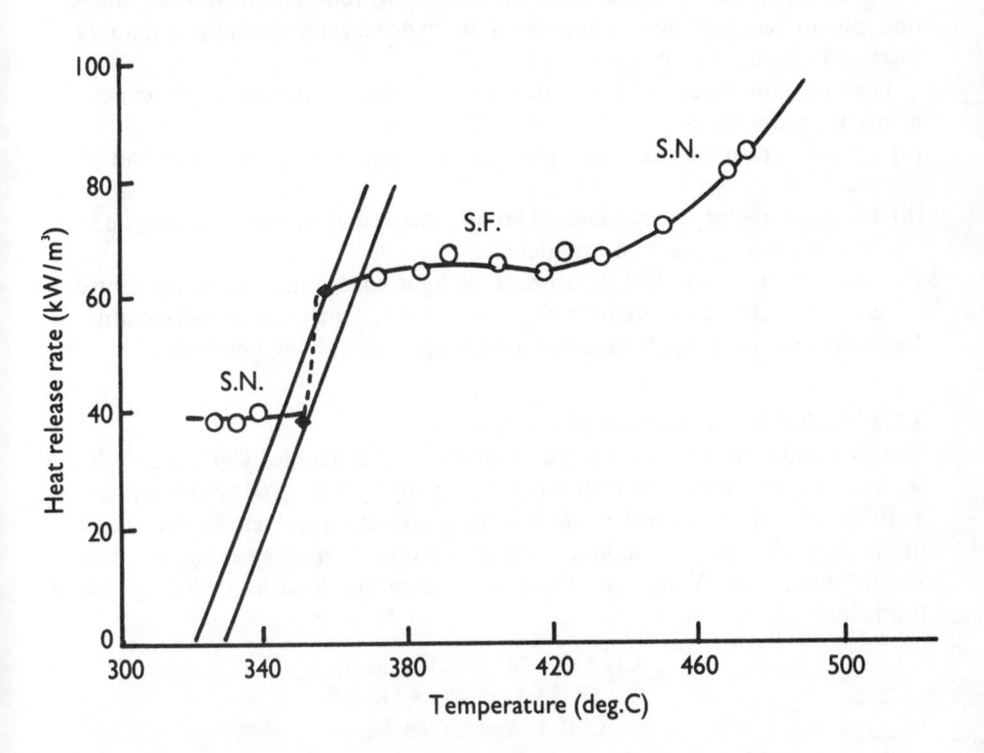

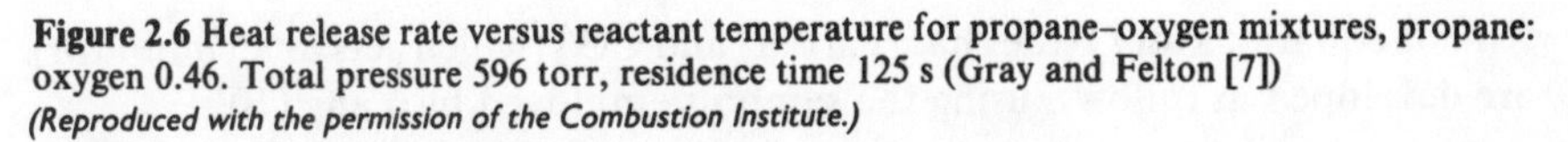
Figure 2.6 Heat release rate versus reactant temperature for propane–oxygen mixtures, propane: oxygen 0.46. Total pressure 596 torr, residence time 125 s (Gray and Felton [7])
(Reproduced with the permission of the Combustion Institute.)

Hysteresis occurs at the boundaries between stable steady states and oscillatory behavior. The S.F. point shown as a filled-in diamond shape at the low-temperature extreme of the upper branch is accessible only from a higher temperature and the S.N. point shown as a filled-in diamond at the high-temperature extreme of the lower branch is accessible only from a lower temperature.

Many other substrates have been examined by means of a CSTR, including acetaldehyde [8,9] and concurrently with the temperature measurements, light emission was measured (by photomultiplier tube) and reaction products analyzed (by mass spectrometry). In the application of chain-thermal theory as outlined above to the behavior of acetaldehyde in a CSTR, the all-important branching intermediate was identified as peracetic acid.

2.5 Carbon monoxide oxidation

2.5.1 Phenomenology

Knowledge of the combustion chemistry of carbon monoxide is important for a number of reasons. It is a primary combustion product of many practical fuels, going on to form carbon dioxide with further oxygen. It is a devolatilization product of coals, especially low-rank coals. Moreover, it displays combustion phenomenology not encountered in hydrocarbon oxidation, notably sustained and oscillatory glow.

The behavior types in carbon monoxide oxidation in laboratory experiments are as follows:

(a) Ignition, characterized by light emission and very rapid depletion of reactants.
(b) Glow, characterized by a faint blue emission. It may be steady, lasting up to 20 s in a closed vessel, or oscillatory.
(c) Slow reaction, marked by absence of light emission and a temperature excess of the reactants over the vessel outside wall due to self-heating.

The behavior is extremely sensitive to hydrogen-containing impurities.

2.5.2 Mechanism and phase-plane analysis

The glow is due to excited carbon dioxide molecules, denoted CO_2^*, and limits for its occurrence have been deduced theoretically by Yang [10] by phase-plane analysis. This analysis involves not one energy equation and one kinetic one as in the discussion of ignition limits and cool flames of hydrocarbons, but two kinetic equations. Yang constructs a ten-step mechanism including the following:

$$CO + O + M \rightarrow CO_2^* + M$$
$$CO_2^* + O \rightarrow CO + O_2$$
$$CO_2^* + M \rightarrow CO_2 + M.$$

Kinetic equations for O (the chain carrier) and CO_2^* (which acts as an inhibitor) are developed as follows, using the symbols employed by Yang [10]:

$$\frac{d[O]}{dt} = K_1[O] - k_2[O][CO_2^*] \tag{2.12}$$

$$= P([O], [CO_2^*])$$

$$\frac{d[CO_2^*]}{dt} = K_3[O] - k_2[O][CO_2^*] - K_4[CO_2^*]$$

$$= Q([O], [CO_2^*]). \tag{2.13}$$

The quantities k_2 and k_3 refer respectively to the second and third reaction steps listed above, and K_1, K_3 and K_4 contain constants for the other steps not shown, and also pressures. Examination of the singularities of equations 2.12 and 2.13 provides a theoretical basis of the ignition and glow phenomena. The singularity types depend on A, B, C and D, where these are defined as:

$$\frac{\partial P}{\partial [O]} = C \qquad \frac{\partial P}{\partial [CO_2^*]} = D$$

$$\frac{\partial Q}{\partial [O]} = A \qquad \frac{\partial Q}{\partial [CO_2^*]} = B,$$

and are easily obtainable in terms of the rate constants and [O] and $[CO_2^*]$. Conditions for the various types of behavior are obtainable from the roots of the characteristic equation (compare with equation 2.9) in terms of A, B, C and D. The condition for steady glow is that the singularities of equations 2.12 and 2.13 be a stable node, for which the conditions are that:

$$AD < \frac{(B + C)^2}{4}.$$

Hence the equality:

$$AD - \frac{(B + C)^2}{4} = 0, \tag{2.14}$$

defines the glow limit in pressure–temperature space. For the *ignition* limit we consider the solution for [O] of equations 2.12 and 2.13 with the derivatives set to zero, easily shown to be:

$$[O] = \frac{K_4 K_1}{k_2(K_3 - K_1)}. \tag{2.15}$$

Recalling that K_3 and K_1 are composed of rate constants and pressures, the condition that $(K_3 - K_1) \rightarrow 0$ will, from equation 2.15, clearly signal explosive reactivity, so in pressure–temperature space the explosion limit can be calculated from:

$$K_3 - K_1 = 0. \tag{2.16}$$

To calculate conditions for a limit cycle (sustained oscillatory glow), Yang modified his kinetic model to include the following processes:

$$O + \text{wall} \xrightarrow{k_a} \text{complex}$$

$$\text{complex} + CO \xrightarrow{k_p} CO_2.$$

The modified differential equation in [O] (compare with equation 2.12) becomes (continuing to use the notation of Yang [10]):

$$\frac{d[O]}{dt} = \left\{K_1' - \frac{k_8}{1 + r_{ap}[O]}\right\}[O] - k_2[O][CO_2^*]$$

$$= P'([O], [CO_2^*]), \tag{2.17}$$

where $r_{ap} = k_a/k_p$ [CO], and k_8 refers to the step in the original ten-step mechanism:

$$O \rightarrow \text{wall}.$$

The quantity K_1' depends on rate constants and pressures. The equation for $[CO_2^*]$ is unchanged. Partial differentials A, B, C and D must be calculated as before for this modified scheme. For sustained oscillations, in addition to the condition for a singularity:

$$P'([O], [CO_2^*]) = 0 \tag{2.18}$$

$$Q[O], [CO_2^*]) = 0, \tag{2.19}$$

the following further conditions are requisite:

$$AD < 0$$

$$B + C = 0, \tag{2.20}$$

where A, etc. are now obtained from the modified mechanism. The main experimental features of this system are therefore predicted by a relatively simple binary model considering the phase plane behavior of two intermediate species.

2.5.3 Experimental observations

Experimental results for this system include the relatively recent ones of Bond et al. [11]. Conditions for glow in a stoichiometric ($2CO + O_2$) mixture in a cylindrical vessel of 330 cm^3 volume are shown in Figure 2.7. The stoichiometric mixtures used were *dry*, that is, with no hydrogen-containing impurities. The influence of such impurities on both glow and ignition limits is very great. For example, in the cylindrical vessel used by Bond et al. [11] an equimolar mixture of carbon monoxide and oxygen at a pressure of 25 torr required a vessel temperature of 770K for ignition with 0.09% hydrogen present: with hydrogen present at 0.12% the vessel temperature required for ignition was 753K. Effects of hydrogen on the other forms of thermal behavior are equally

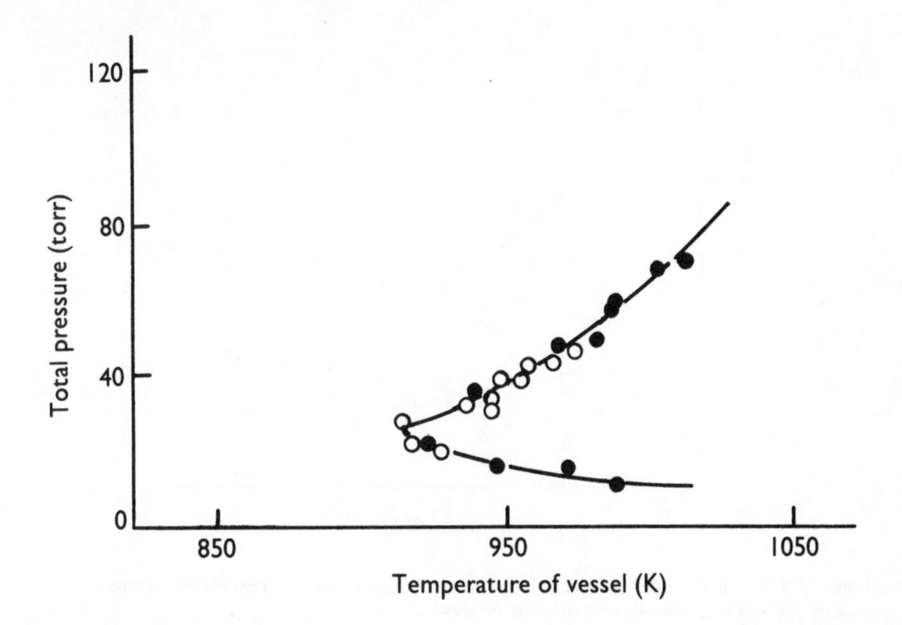

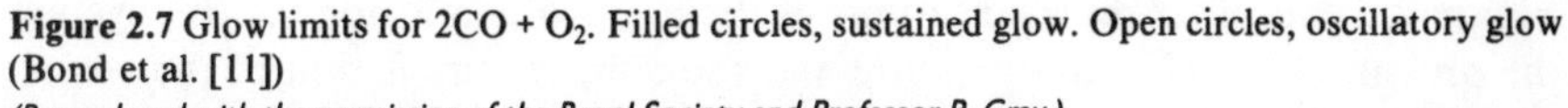

Figure 2.7 Glow limits for $2CO + O_2$. Filled circles, sustained glow. Open circles, oscillatory glow (Bond et al. [11])
(Reproduced with the permission of the Royal Society and Professor P. Gray.)

marked and the current view, in conflict with some earlier findings, is that ignition of a completely dry CO/O_2 mixture is not possible. Classical work on dry mixtures is believed to have mistaken glow for ignition.

In the binary model outlined in section 2.5.2 the concentration of H atoms is taken to be in its steady state. Simulation of the effect of hydrogen on the behavior requires abandonment of this approximation and coupling of $d[H]/dt$ to the kinetic equations for the other species involved (OH, O, HO_2, etc.), the use of literature values for the rate constants, and numerical integration [12].

2.6 The hydrogen–oxygen reaction

2.6.1 Explosion limits in classical experiments

The H_2/O_2 reaction has been very extensively studied [13] and in pressure–temperature space the explosion limit has the form shown in Figure 2.8. The exact positions of the limits in a pressure–temperature diagram depend on vessel size and coating, fuel/oxygen ratio and amount of diluent if any. An experiment set up so as to fall within the shaded areas will explode. The limits are commonly referred to as the first, second and third limits, denoted by P_1, P_2 and P_3 on Figure 2.8. An experiment set up so as to fall outside the shaded areas will undergo slow, non-explosive reaction. It should be borne in mind that in

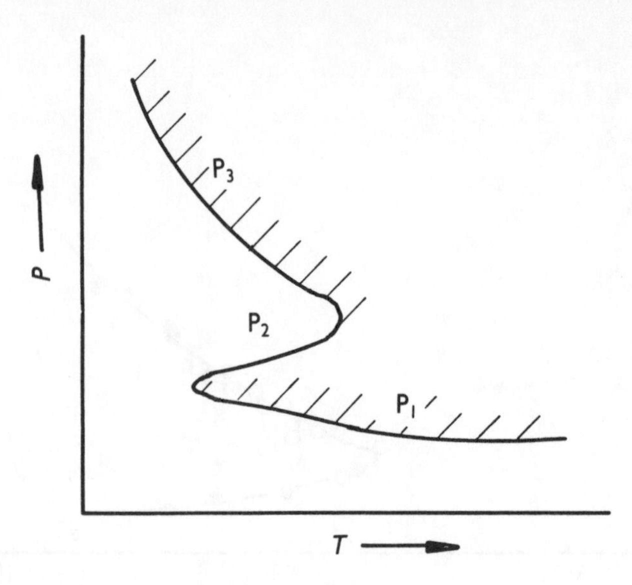

Figure 2.8 Explosion limits (generalized form) for H_2/O_2 reaction (Minkoff and Tipper [13])
(Reproduced with the permission of Butterworth Heinemann.)

Figure 2.8 and other representations of the limits of explosion for this system the pressure is total initial pressure and the temperature is that of the vessel. The nature of the vessel wall is also of importance. Walls may be either chain breaking (for example a potassium chloride-coated wall) or chain reflective (for example acid-washed Pyrex).

For the first limit the pressure–temperature variation can be expressed as follows:

$$P_1 = K \exp[E/RT], \tag{2.21}$$

with $E \approx$ 40–60 kJ mol^{-1}. Proportions of hydrogen and oxygen can be varied without the need to deduce different limits, provided that the following hyperbolic equation is satisfied:

$$P_{\text{oxygen}} \times P_{\text{hydrogen}} = \text{a constant}, \tag{2.22}$$

where P denotes partial pressure. Explosion is favored by an increase in vessel diameter (that is, such an increase reduces P_1 for given T), indicating that a decrease in surface/volume ratio favors explosion. Similarly, other things being equal, the presence of an inert diluent favors explosion. Both observations are consistent with the view that wall termination of active centers (for example, H, O, OH) plays a part in the first limit. The inert gas molecules will retard transport of active centers to the wall.

The second limit has a more marked temperature dependence than the first. The counterpart of equation 2.21 for the second limit is:

$$P_2 = K'\exp[-E'/RT], \tag{2.23}$$

with $E' \approx 70$–100 kJ mol^{-1}. The partial pressures at the limit are linked by a linear relation:

$$P_{\text{hydrogen}} + (a \times P_{\text{oxygen}}) = \text{a constant.} \tag{2.24}$$

Vessel diameter and surface coating are less influential than for the first limit (the former having hardly any effect at all) and a diluent inhibits explosion. The second limit represents the pressure maximum of the lower shaded (explosion) area for a given vessel temperature and this, together with the fact that a diluent inhibits explosion, points to a gas-phase origin of this limit. It is in fact due to competition between two possible fates of a chain-branching intermediate. Suppose that the hydrogen atom, an important chain carrier in this process, can react as follows with an oxygen molecule:

$$H + O_2 \rightarrow OH + O.$$

Alternatively, in the presence of a third body M, the following process may occur:

$$H + O_2 + M \rightarrow HO_2 + M.$$

HO_2 is not as reactive as OH or H and is much more likely to travel to the vessel wall and be terminated there. Hence increased pressure favors HO_2 formation and so inhibits explosion. The effect of diluent can be explained in the same way.

The third limit certainly has some thermal character as would be expected from its general shape. Some investigators have pointed out that it is possible to explain, in the following way, the existence of a third limit in this system without invoking self-heating. We saw how the second limit involves formation of HO_2 radicals and their loss to the wall. However, at higher pressures their loss is prevented sufficiently for the following process to take place to a significant degree:

$$HO_2 + H_2 \rightarrow H + H_2O_2,$$

and the chain carrier is regenerated.

2.6.2 Mechanistic discussion of the limits

That the first limit, occurring at pressures of a few torr, and the second, occurring at typically 150 torr, are due to wall and gas-phase termination respectively was known to the early investigators of the reaction, notably Hinshelwood [14], whose treatment of the first explosion limit is essentially as follows:

We have already discussed the step:

$$H + O_2 \rightarrow OH + O,$$

to which we add another branching step:

$$O + H_2 \rightarrow OH + H.$$

These represent *alternate* formation of O and H; formation of one involves removal of the other. They are the chain carriers, whose migration to the wall causes there to be no explosion below the first limit.

Let n_1 = number of collisions between O and H_2
n_2 = number of collisions between O and O_2
n_3 = number of collisions between H and H_2
n_4 = number of collisions between H and O_2
$n = n_1 + n_2 + n_3 + n_4$.

In view of the alternate formation of H and O, we can see that:

$$n_1 = n_4, \tag{2.25}$$

and the following is easily shown to be approximately true:

$$n_1 = n_4 = \frac{P_{\text{oxygen}}\, P_{\text{hydrogen}}}{(P_{\text{oxygen}} + P_{\text{hydrogen}})^2}\, n. \tag{2.26}$$

Consider an active center H or O travelling to the wall. The collisions it makes en route will be proportional to the square of the distance ℓ travelled and inversely proportional to the square of the mean free path λ.

$$n \propto \frac{\ell^2}{\lambda^2}$$

$$n = C \times \frac{\ell^2}{\lambda^2}, \tag{2.27}$$

where C is a constant of proportionality.

Now
$$\lambda = \frac{\lambda_o}{P_{\text{hydrogen}} + P_{\text{oxygen}}}, \tag{2.28}$$

where λ_o = mean free path at unit pressure. Putting together equations 2.26, 2.27 and 2.28 gives:

$$n_1 = n_4 = P_{\text{hydrogen}}\, P_{\text{oxygen}} \times \frac{C\ell^2}{\lambda_o^2}. \tag{2.29}$$

This expression relates to collisions experienced by the chain carriers with potential for branching en route to the wall. There is a fractional probability μ that a collision will lead to branching, and the number of collisions actually resulting in branching will therefore be:

$$n' = \mu n_1 = \mu \times P_{\text{hydrogen}}\, P_{\text{oxygen}} \times \frac{C\ell^2}{\lambda_o^2}. \tag{2.30}$$

This is the number of times the chains branch before termination. There will be a minimum value n'_{crit} necessary for explosion. So the explosion limit is given by:

$$P_{\text{hydrogen}}\, P_{\text{oxygen}} = \frac{n'_{\text{crit}}\, \lambda_o^2}{\mu C \ell^2} = \text{a constant}. \tag{2.31}$$

This is equivalent to equation 2.22, and is clearly amenable to experimental test. For particular values of the temperature, total pressure and vessel size on the first explosion limit, a point ought to be reproducible with different relative pressures of hydrogen and oxygen, these in each case fulfilling equation 2.31. The pressure pairs, when plotted against each other, should give a rectangular hyperbola. Such experimental results by Hinshelwood and Moelwyn-Hughes [15] using a single vessel for a point on the limit at 550° C with total pressure (3.8 ± 0.5) torr and various relative pressures of the two reactants, are plotted in Figure 2.9. The graph is seen to be of the expected form. If we consider ℓ to be the vessel dimension, the observation that a bigger vessel favors explosion is consistent with equation 2.31.

The second limit can be treated in the following way. This is due to gas-phase loss of the chain carrier H as a result of ternary collisions:

$$H + O_2 + H_2 \rightarrow HO_2 + H_2$$

$$H + O_2 + O_2 \rightarrow HO_2 + O_2,$$

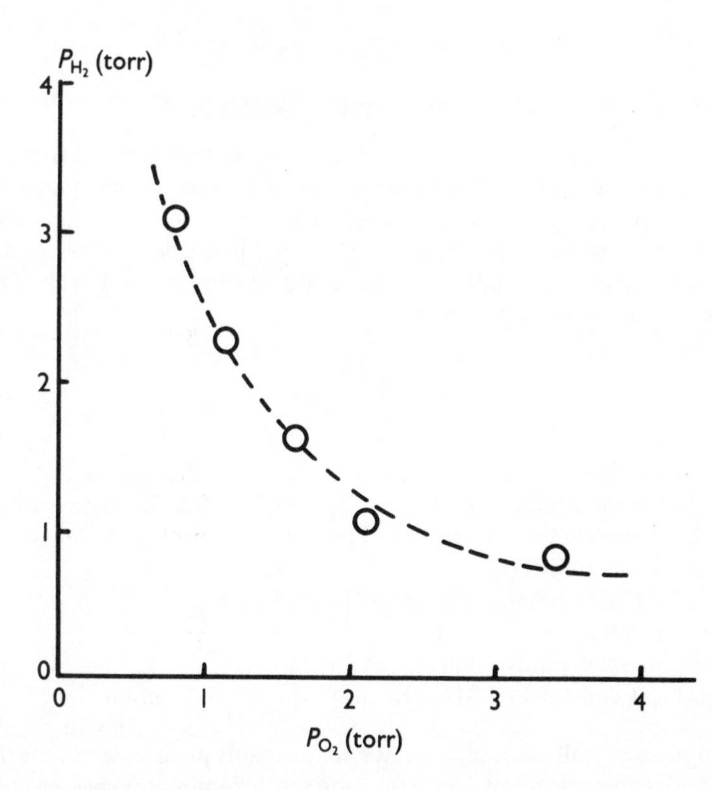

Figure 2.9 P_{hydrogen} against P_{oxygen} for a point on the first explosion limit of the hydrogen–oxygen reaction. Temperature 550° C, total pressure (3.8 ± 0.5) torr. Plotted from results tabulated by Hinshelwood and Moelwyn-Hughes [15]

that is, the only difference in the above two processes is the identity of the third body. Actually on the limit the combined rates of the above will be equal to that of the branching step:

$$H + O_2 \xrightarrow{k} OH + O,$$

so the following equality will apply:

$$k\,[H][O_2] = Z_1[H][O_2][H_2] + Z_2[H][O_2]^2, \qquad (2.32)$$

where k is the rate constant pertaining to the branching step shown. The quantities Z_1 and Z_2 are proportional to the respective collision frequencies.

Rearranging:

$$k = Z_1[H_2] + Z_2[O_2] \qquad (2.33)$$

$$\frac{k}{Z_1} = [H_2] + \{\text{constant} \times [O_2]\}, \qquad (2.34)$$

where the constant has the value Z_2/Z_1. This is clearly consistent with equation 2.24:

$$P_{\text{hydrogen}} + (a \times P_{\text{oxygen}}) = \text{a constant},$$

previously given as applying to the second explosion limit.

The third limit, occurring at 200–600 torr, needs to be examined from both kinetic and thermal viewpoints and experimental work by Barnard and Platts [16] was interpreted by a model in which self-heating was combined with a kinetic scheme. The experiments were performed in a spherical vessel under conductive heat transfer conditions, so a modified form of the Frank–Kamenetskii (see Chapter 1) model applies:

$$\delta_{\text{crit}} = \frac{QEr^2w}{kR(T_o)^2}, \qquad (2.35)$$

where δ_{crit} = Frank–Kamenetskii parameter
Q = exothermicity (J mol^{-1})
E = *overall* activation energy (J mol^{-1})
r = vessel radius (m)
T_o = ambient (vessel outside wall) temperature
w = reaction rate (mol m^{-3} s^{-1})
k = thermal conductivity (W m^{-1} K^{-1})
R = gas constant = 8.314 J K^{-1} mol^{-1}.

Although an overall value of E is inserted (probably justifiable in view of the temperature independence of E) the reaction rate w cannot be expected to have a simple Arrhenius temperature dependence for this reaction which involves chain branching. Barnard and Platts develop a rate expression similar to that previously given by Hinshelwood in a purely kinetic treatment of the limit. Using their notation, this rate expression is:

$$w = \frac{3k_o k_5[H_2][O_2] \times k_4[M]}{k_4[M](k_5 - 2k_{11}[H_2]) - 2k_2(k_5 + k_{11}[H_2])}, \tag{2.36}$$

where k_o pertains to: $H_2 + O_2 \rightarrow 2OH$
k_1 pertains to: $OH + H_2 \rightarrow H_2O + H$
k_2 ($\equiv$ k in Eq. 2.32) pertains to: $H + O_2 \rightarrow OH + O$
k_4 pertains to: $H + O_2 + M \rightarrow HO_2 + M$
k_5 pertains to: $HO_2 \rightarrow$ wall
k_{11} pertains to: $HO_2 + H_2 \rightarrow H_2O_2 + H$.

These rate constants and their temperature dependences are known from independent work, and equations 2.35 and 2.36 can be combined to give an expression for w in terms of the rate constants, gas concentrations and Q, k, etc. Since $[O_2]$, $[H_2]$ and $[M]$ all appear in the numerator and are all multiples of the total pressure P, the combined equation becomes a cubic in P, whose two positive roots are taken to be the second and third explosion limits. This provides a basis for calculation of the third explosion limit and comparison with experiment. Using a T_o of 567° C and various fuel:oxygen ratios, the limit was calculated in terms of total pressure required for explosion and the results are shown in Figure 2.10, where they are seen to be in fair agreement with experiment.

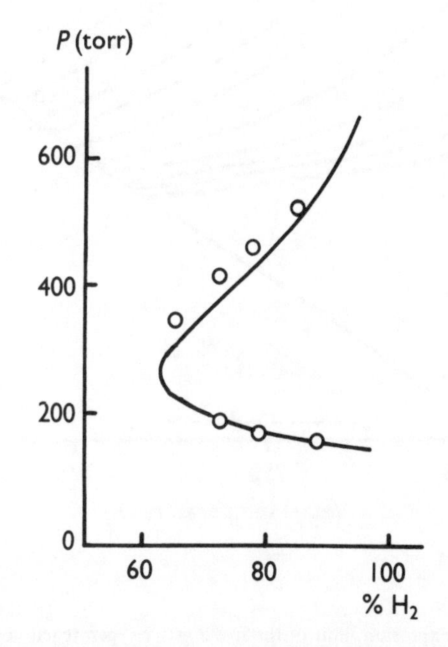

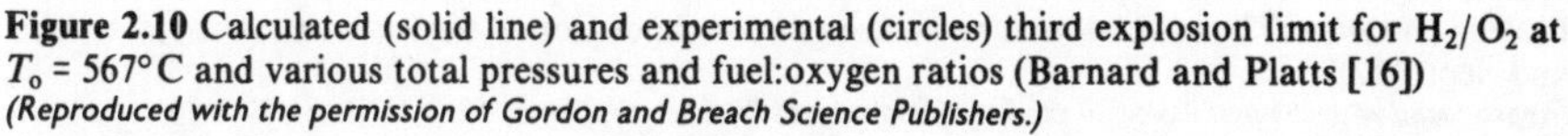
Figure 2.10 Calculated (solid line) and experimental (circles) third explosion limit for H_2/O_2 at T_o = 567° C and various total pressures and fuel:oxygen ratios (Barnard and Platts [16]) *(Reproduced with the permission of Gordon and Breach Science Publishers.)*

2.6.3 The slow reaction

The rate of slow reaction between the second and third limits has been estimated by pressure change and is known to be very sensitive to water present. This region was once believed to be rather featureless. Accurate thermocouple measurements between the second and third limits for hydrogen/oxygen mixtures both in closed vessels and in a CSTR [17] indicate that a *small amount* of self-heating can occur.

Figure 2.11 shows self-heating in the slow reaction for equimolar H_2/O_2 in a CSTR. The solid line is the experimental second limit and in a CSTR this is marked by apparently oscillatory ignition because of a depletion/replenishment cycle of reactants. The numbers on the contour lines are degrees Kelvin temperature excess, and the larger values are observed at pressures closer to the third limit, known (see above) to involve some self-heating. Temperature excesses at the lower pressures are very small or zero.

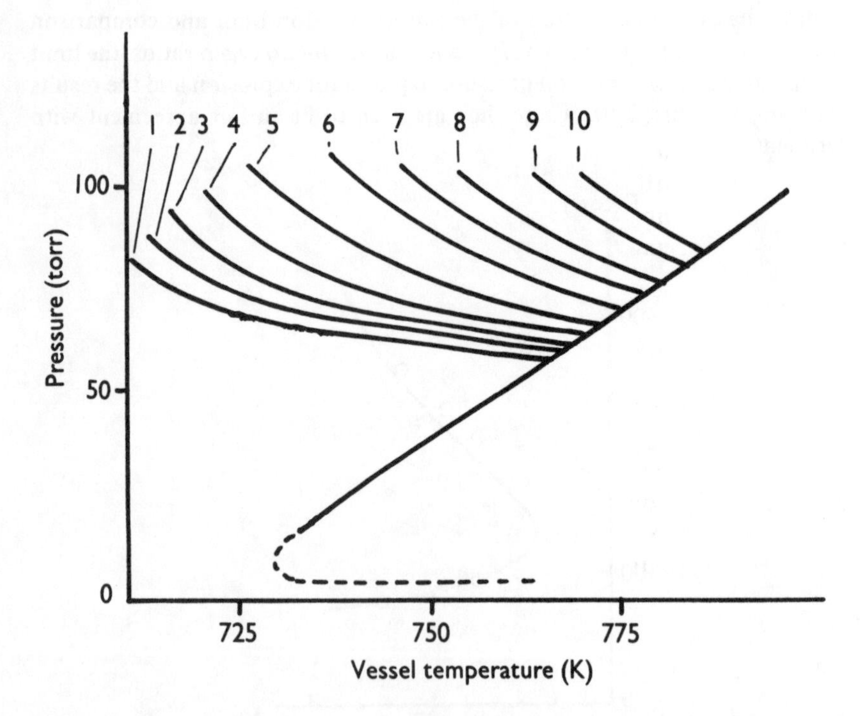

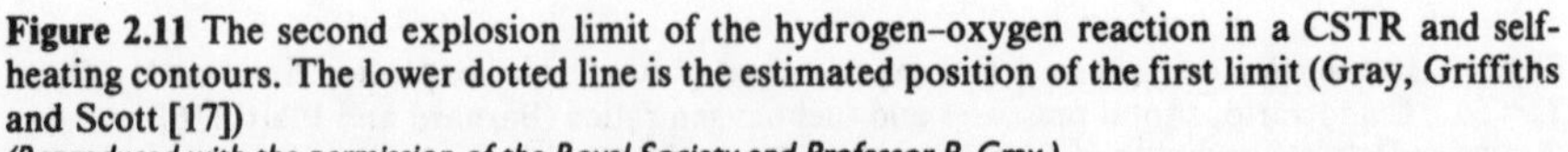
Figure 2.11 The second explosion limit of the hydrogen–oxygen reaction in a CSTR and self-heating contours. The lower dotted line is the estimated position of the first limit (Gray, Griffiths and Scott [17])
(Reproduced with the permission of the Royal Society and Professor P. Gray.)

2.6.4 Treatment of the hydrogen/oxygen reaction by the Unified Theory [18]

The involvement of thermal effects at the second limit, indicated by the self-heating contours in Figure 2.11, was predicted in application of the Unified Theory to the hydrogen/oxygen system [18]. As discussed in Chapter 1, the procedure for application of the Unified Theory was the determination of $y(x,T)$ and $z(x,T)$ from branching-intermediate and energy equations respectively and their setting to zero to investigate singular points. Between the second and third limits, with H as the branching intermediate, these equations can be summarized using the notation of Yang and Gray:

$$[\mathrm{H}]_s = \frac{k_{10}P^2}{k_5 + k_6P^2 - k_2'P}, \tag{2.37}$$

where k_5 pertains to: H + wall → $0.5H_2$

$k_6 \equiv k_4$ in the notation of Barnard and Platts

$k_2' = 2k_2$ in the notation of Barnard and Platts

P is the pressure

k_{10} = an amalgamation of rate constants for steps not shown.

Also,

$$(k_bh_b + k_th_t + k_ih_i)[\mathrm{H}]_s - L = 0, \tag{2.38}$$

where the notation is as in equations 2.4ff. and subscripts i, b and t denote, respectively, initiation, branching and termination.

Consider an experiment set up between the second and third limits. We can approach the third limit by increasing P (reactant addition) or the second limit by reducing P (reactant removal). Rewriting equation 2.37 in the form:

$$[\mathrm{H}]_s = \frac{k_{10}}{k_6 - (k_2'/P) + (k_5/P^2)}. \tag{2.39}$$

then, as $P \to \infty$, $[\mathrm{H}]_s \to k_{10}/k_6$. Thus, if an ignition of branched-chain origin is identified with larger and larger values of $[\mathrm{H}]_s$, this cannot possibly occur, as the third limit is reached on this model and the ignition is therefore thermal, and is due to accelerating reaction and hence heat release rates stemming from the Arrhenius dependence of the rate constants in the heat-balance equation. However, if for an experimental set-up initially between the second and third limits the pressure is dropped, then from equation 2.40, which is equation 2.37 rearranged:

$$[\mathrm{H}]_s = \frac{k_{10}P^2}{(k_6P^2 - k_2'P) + k_5}. \tag{2.40}$$

$[\mathrm{H}]_s \to \infty$ as $k_6P^2 \to k_2'P - k_5$ and there can be no value of $[\mathrm{H}]_s$, therefore there is ignition. However, because of the dependence of the heat release rate on $[\mathrm{H}]_s$, the lack of a finite value means that there can be no solution to the steady heat balance equation 2.38 either, and the correct conclusion is that according to the Unified Theory the second limit is both thermal and chain branching in origin.

The observation of self-heating between the two limits [17] is consistent with this conclusion.

2.7. Conclusion

We have examined details of several thoroughly researched gas-phase oxidations and, apart from their intrinsic interest, the principles constitute a sound basis for further chapters on flames, spark ignition and practical gaseous fuels.

References

[1] Cullis C. F., Khokar B. A., 'The spontaneous ignition of aliphatic amines', *Combustion and Flame* **4** 265 (1960)
[2] Bardwell J., 'Cool flames in butane oxidation', *Fifth Symposium (International) on Combustion*, 529, New York: Reinhold (1955)
[3] Mulcahy M. F. R., 'The oxidation of hydrocarbons. Some observations on the induction period', *Transactions of the Faraday Society* **45** 575 (1949)
[4] McGowan I. R., Tipper C. F. H., 'The slow combustion of cyclopentane. I, Kinetics in coated and uncoated vessels', *Proceedings of the Royal Society of London* **A246** 52 (1958)
[5] Hinshelwood Sir C. N., Bardwell J., 'The cool flame of methyl ethyl ketone', *Proceedings of the Royal Society of London* **A205** 375 (1951)
[6] Yang C. H., Gray B. F., 'On the slow oxidation of hydrocarbon and cool flames', *Journal of Physical Chemistry* **73** 3395 (1969)
[7] Gray B. F., Felton P. G., 'Low-temperature oxidation in a stirred-flow reactor. I, Propane', *Combustion and Flame* **23** 295 (1974)
[8] Gray P., Griffiths J. F., Hasko S. M., Lignola P. G., 'Oscillatory ignitions and cool flames accompanying the non-isothermal oxidation of acetaldehyde in a well-stirred flow reactor', *Proceedings of the Royal Society of London* **A374** 313 (1981)
[9] Felton P. G., Gray B. F., Shank N., 'Low-temperature oxidation in a stirred-flow reactor. II, Acetaldehyde (theory)', *Combustion and Flame* **27** 363 (1976)
[10] Yang C. H., 'On the explosion, glow and oscillation phenomena in the oxidation of carbon monoxide', *Combustion and Flame* **23** 97–108 (1974)
[11] Bond J. R., Gray P., Griffiths J. F., 'Oscillations, glow and ignition in carbon monoxide oxidation. I, Glow and ignition in a closed reaction vessel and the effect of added hydrogen', *Proceedings of the Royal Society of London* **A375** 43–64 (1981)
[12] Yang C. H., Berlad A. L., 'Kinetics and kinetic oscillation in carbon monoxide oxidation', *Journal of the Chemical Society, Faraday Transactions* I **70** 1661 (1974)
[13] Minkoff G. J., Tipper C. F. H., *Chemistry of Combustion Reactions*, London: Butterworths (1962)
[14] Hinshelwood Sir C. N., Williamson A. T., *The Reaction between Hydrogen and Oxygen*, Oxford: The University Press (1934), and references therein.
[15] Hinshelwood Sir C. N., Moelwyn-Hughes E. A., 'The Lower pressure limit in the chain reaction between hydrogen and oxygen', *Proceedings of the Royal Society of London* **A138** 311 (1932)
[16] Barnard J. A., Platts A. G., 'The hydrogen–oxygen reaction in a potassium chloride-coated vessel in the region of the third explosion limit', *Combustion Science and Technology* **6** 177–186 (1972)
[17] Gray P., Griffiths J. F., Scott S. K., 'Branched-chain reactions in open systems: theory of the oscillatory ignition limit for the hydrogen–oxygen reaction in a CSTR', *Proceedings of the Royal Society of London* **A394** 243–258 (1984)
[18] Yang C. H., Gray B. F., 'The determination of explosion limit from a unified thermal and chain theory', *Eleventh Symposium (International) on Combustion*, 1099–1106, Pittsburgh: The Combustion Institute (1967)

CHAPTER 3

Ignition by a Heat Source

Abstract

Ignitions requiring input of energy, for example as a spark, are discussed and various treatments presented including a fundamental one based on energy and material balance. The significance of the minimum energy for ignition is brought out in the discussion and values for a number of fuel/oxidant mixtures are given. Other topics covered include the excess energy *approach, a critique of which is given, and hot-spot theory, which is essentially the theory for thermal ignition, presented in Chapter 1, extended to ignition by a heat source.*

3.1 Introduction

In our examples of thermal, chain and chain-thermal ignition we have encountered ignition processes which are spontaneous in the sense that they occur without a pilot flame or other introduced heat (although there may be a long induction time). Many practical combustion applications are piloted with a spark or other heat source and the energy input has to be sufficient for ignition. A related feature of gas-phase ignitions is the existence of lower and upper (lean and rich) fuel/air mixture limits of ignition, limits outside which propagation will not occur. For example [1], the respective limits for benzene vapor in air at atmospheric pressure are often given as 1.4 and 7.4% molar.

3.2 Minimum ignition energy and lean limits

3.2.1 Treatment by ignition theory

This problem can be treated according to ignition theory in the following way which has been adapted from the theory by Fenn [2]. Heat balance at the front of a flame at the lean limit, where heat release and heat transfer rates just equal each other, gives:

$$Qx_f x_o \exp[-E/RT_f] = \beta[T_f - T_o], \tag{3.1}$$

where Q = heat of reaction (J mol^{-1})

x_f, x_o = mole fraction of fuel and oxygen respectively, and dependence of the heat release rate on the first power of each of these has been assumed

β = a constant incorporating the pre-exponential factor and the coefficient of heat transfer (J K^{-1} mol^{-1})

T_f, T_o = flame and influx gas temperatures respectively.

In this formalism the heat released in the flame is taken up entirely as sensible heat in the combustion products on inflammation, and some of this heat is subsequently transferred to the surroundings at the flame front (see Section 4.2). T_f is therefore the adiabatic flame temperature. The heat balance equation 3.1 can be used to establish a link between the flame temperature T_f at the lean limit and the activation energy of the reaction.

Rearranging equation 3.1:

$$\exp[-E/RT_f] = \frac{\beta(T_f - T_o)}{Qx_f x_o} = \hbar. \tag{3.2}$$

Now since T_f is the adiabatic temperature, $(T_f - T_o)$ and $Qx_o x_f$ are proportional to each other (with a constant of proportionality depending on the heat capacity):

$$\hbar \approx \text{a constant for a particular mixture at a particular } T_o.$$

Rearranging further and applying to the lean limit:

$$E = -[R \ln\hbar](T_f)_{\text{lean}} = \alpha(T_f)_{\text{lean}}. \tag{3.3}$$

The significance of equation 3.3 is that at a lean limit the flame temperature and activation energy are proportional to each other. Hence if the lean limit is determined for a variety of fuels in the same apparatus (because of the apparatus dependence on the heat transfer coefficient and hence of $\hbar$) and flame temperatures either measured or calculated from exothermicities and heat capacities, $(T_f)_{\text{lean}}$ can be used in place of E, which is more difficult to deduce from first principles.

The concept of minimum ignition energy is treated on the principle that for ignition to occur, a critical spherical volume of the gas must be supplied with enough energy to take it from T_o to T_f. Heat balance for a three-dimensional volume element is:

$$\frac{4}{3}\pi r^3 A Q x_o x_f \sigma^2 \exp[-E/RT_f] = 4\pi r^2 k \frac{dT}{dr} \approx 4\pi r k(T_f - T_o), \tag{3.4}$$

where A = pre-exponential factor (mol m^{-3})$^{-1}$ s^{-1}
r = volume radius (m)
k = thermal conductivity (W m^{-1} s^{-1})
σ = density (mol m^{-3}).

Rearranging:

$$r = \left[\frac{3k(T_f - T_o)\exp(E/RT_f)}{AQx_ox_f\sigma^2}\right]^{0.5}$$

$$= [(\gamma/x_o\sigma^2)\exp(E/RT_f)]^{0.5}, \qquad (3.5)$$

where $\gamma = \dfrac{3k(T_f - T_o)}{AQx_f}$.

If the critical radius of the volume requiring heating to T_f for the mixture to ignite is denoted r_{crit}, the minimum ignition energy H_{min} is:

$$H_{min} = \frac{4}{3}\pi r_{crit}^3 C_p\sigma(T_f - T_o) \quad \text{J}, \qquad (3.6)$$

where C_p = specific heat at constant pressure (J mol^{-1} K^{-1}).
Substituting for r_{crit} from equation 3.5 gives:

$$H_{min} = Zx_o^{-1.5}\sigma^{-2}(T_f - T_o)\exp[3E/2RT_f], \qquad (3.7)$$

where $Z = \frac{4}{3}\pi C_p\gamma^{1.5}$.

We have seen in equation 3.3 that:

$$E = \alpha(T_f)_{lean},$$

and by making use of the Ideal Gas Laws we can substitute for σ the quantity P_o/RT_o, where P_o = initial pressure (N m^{-2}).

Putting together equations 3.3 and 3.7 and taking logs (to base 10) we obtain:

$$\log\frac{H_{min}x_o^{1.5}}{T_f - T_o} = Z' + 2\log T_o - 2\log P_o + \frac{0.65\alpha(T_f)_{lean}}{RT_f}, \qquad (3.8)$$

where Z' is an amalgamation of the constants.

The quantity H_{min} can be measured for a particular mixture from knowledge of the minimum electrical energy supply to a spark gap needed to cause ignition. $(T_f)_{lean}$ and T_f can be calculated from thermodynamic information, so the assertion that the basis of H_{min} is identifiable with the supply of enough energy to heat a spherical element of gas of radius r_{crit} to T_f is verifiable by experiment.

3.2.2 Experimental examination of the theory

The variables in equation 3.8 are T_o, $(T_f)_{lean}$, $(T_f)_{lean}/T_f$, H_{min} and P_o. The quantity H_{min} is for a mixture strength corresponding to T_f. Two data sets which are potential input to equation 3.8 are given in Table 3.1. The equivalence ratio is the actual proportion of fuel divided by the stoichiometric proportion, and in both entries T_o = 298K.

Table 3.1 Ignition energy and lean limit data for fuel/air mixtures [2]

Fuel	*Equivalence ratio*		$(T_f)_{lean}$ (K)	T_f (K)	$\frac{(T_f)_{lean}}{T_f}$	$10^4 H_{min}/J$
	Lean limit	*Ignition mixture*				
CH_3CHO	0.557	1.00	1675	2300	0.729	3.8
C_2H_2	0.307	1.00	1275	2580	0.495	0.2

Hence at 298K influx temperature the acetaldehyde/air system has a lean limit of 0.557 and a corresponding flame temperature of 1675K. At stoichiometric conditions, T_f is 2300K and a minimum spark energy of 0.38 mJ is required for ignition of the stoichiometric mixture.

Conformity to equation 3.8 was checked [2] by performing experiments to determine H_{min} and the lean equivalence ratio for a number of fuels at different mixture strengths and T_o values but constant P_o. Plotting:

$$\log \frac{H_{min} x_o^{1.5}}{T_f - T_o} - C_1 \frac{(T_f)_{lean}}{T_f} + C_2 \text{ against } \log T_o,$$

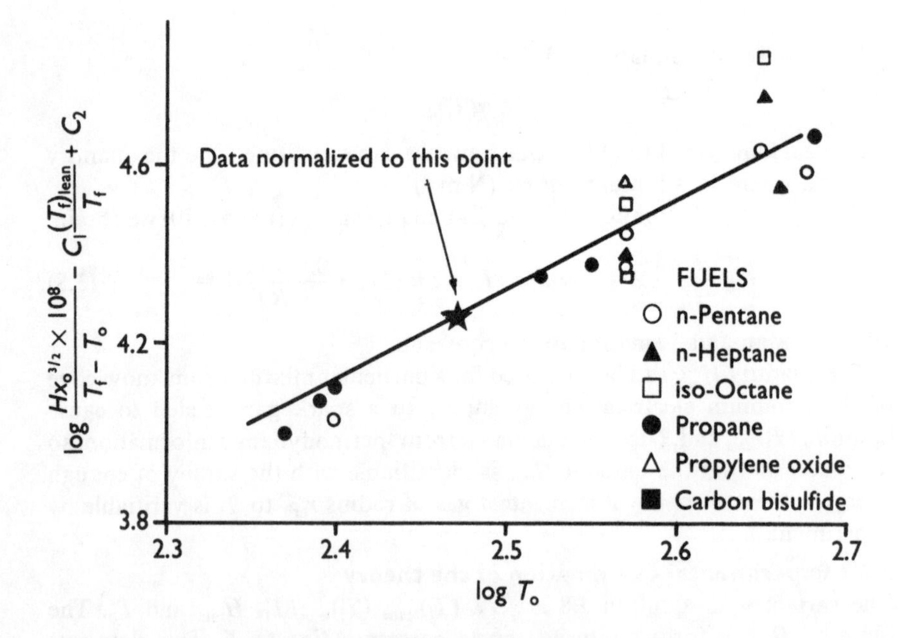

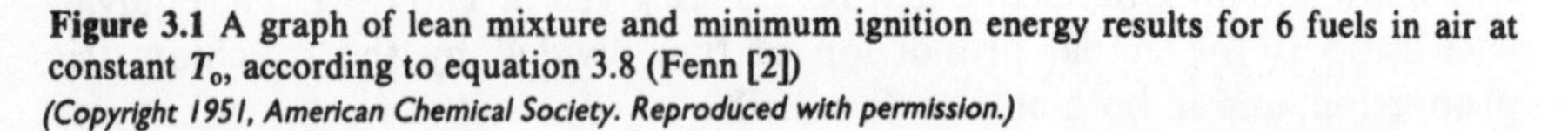

Figure 3.1 A graph of lean mixture and minimum ignition energy results for 6 fuels in air at constant T_o, according to equation 3.8 (Fenn [2])

(Copyright 1951, American Chemical Society. Reproduced with permission.)

where C_1 and C_2 are constants, is expected to give a straight line. The constants have different values for different fuels and these were determined by fitting results from all fuels to the same point at T_o = 298K. Figure 3.1 shows results treated in this way for six fuels examined with T_o in the range –40°C to 200°C and the line drawn has a slope of 2, the expected value if equation 3.8 holds. Agreement is certainly reasonable, holding promise for the view that the raising of the temperature of a critical volume of mixture to T_f is the origin of the minimum ignition energy.

For a different set of fuels, the experiments were performed at constant T_o and different pressures, constants in the equation being obtained by fitting all the data to one point at P_o = 1 atmosphere. Plotting:

$$\log P_o \quad \text{against} \quad \log \frac{H_{min} x_o^{1.5}}{T_f - T_o} - C_1 \frac{(T_f)_{lean}}{T_f} + C_3,$$

where C_3 is a further constant, is expected to yield a straight line of slope –0.5, and in Figure 3.2 the graph is shown to conform remarkably well.

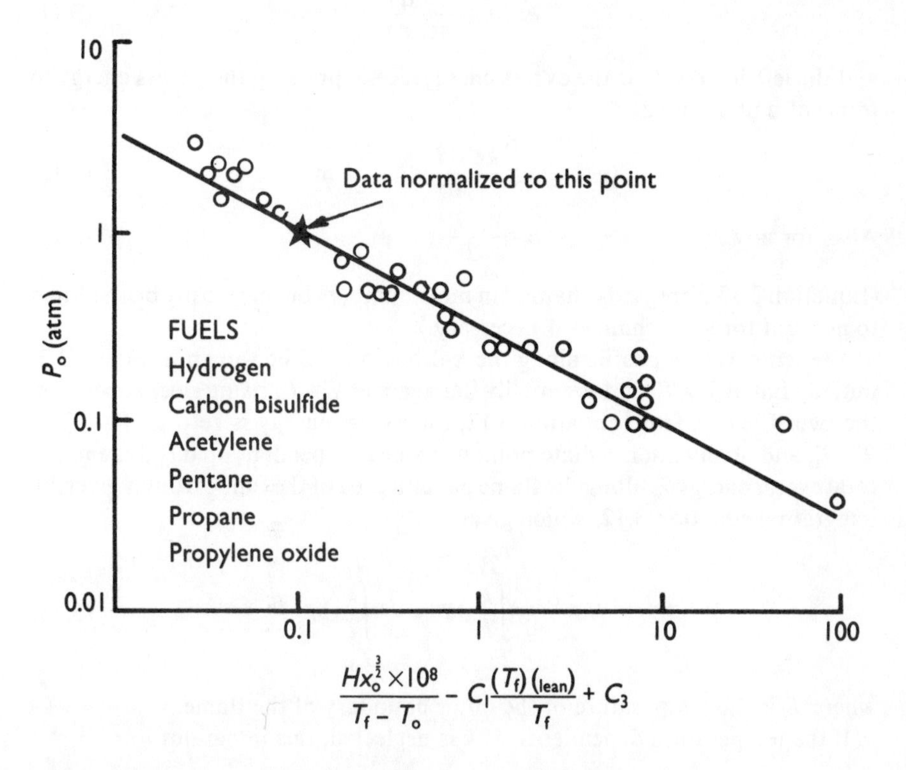

Figure 3.2 A graph of lean mixture and minimum ignition energy results for 6 fuels in air at constant P_o according to equation 3.8 (Fenn [2])
(Copyright 1951, American Chemical Society. Reproduced with permission.)

3.3 The excess energy approach [3]

3.3.1 Basic equations

Consider a flame front with propagation coordinate x. We have at any value of x the continuity condition:

$$\sigma v = \sigma_o v_o = m', \tag{3.9}$$

where σ = density (kg m^{-3})
v = velocity (m s^{-1})
m' = mass flow rate (kg m^{-2} s^{-1}), and
subscript o denotes $x = 0$.

Energy balance gives:

$$\sigma_o v_o e_o = \sigma v e - k\frac{dT}{dx}, \tag{3.10}$$

where e = energy content of the gas (J kg^{-1}).

Substituting from the continuity condition and rearranging:

$$e - e_o = \frac{k}{m'}\frac{dT}{dx}, \tag{3.11}$$

and the left-hand side is the excess energy. Re-expressing the excess energy in terms of unit volume:

$$(e - e_o)\sigma = \frac{k\sigma}{m'}\frac{dT}{dx} \qquad \text{J m}^{-3}. \tag{3.12}$$

Also, for any x:

$$\sigma = \frac{\sigma_o T_o}{T}. \tag{3.13}$$

(Equation 3.13 disregards changes in mole numbers but can easily be modified to account for such changes if necessary.)

The temperature profile along the x-direction will be shown in Figure 3.3: initially flat at $T = T_o$ and eventually flat again at $T = T_f$, rising steeply between the two. Clearly, from equation 3.12, the excess energy is zero at $T = T_o$ or $T = T_f$, and at any intermediate point its value is dependent on dT/dx, and the total excess energy H along the flame per unit area of the flame front is given by integrating equation 3.12, which gives:

$$H = \frac{\sigma_o T_o}{m'}\int_{T_o}^{T_f}\frac{k}{T}\,dT = \frac{T_o}{v_o}\int_{T_o}^{T_f}\frac{k}{T}\,dT, \tag{3.14}$$

where T_f is the temperature of the outer boundary of the flame.

If the temperature dependence of k is neglected, this integrates to:

$$H = \frac{kT_o}{v_o}\ln\frac{T_f}{T_o} \qquad \text{J m}^{-2}, \tag{3.15}$$

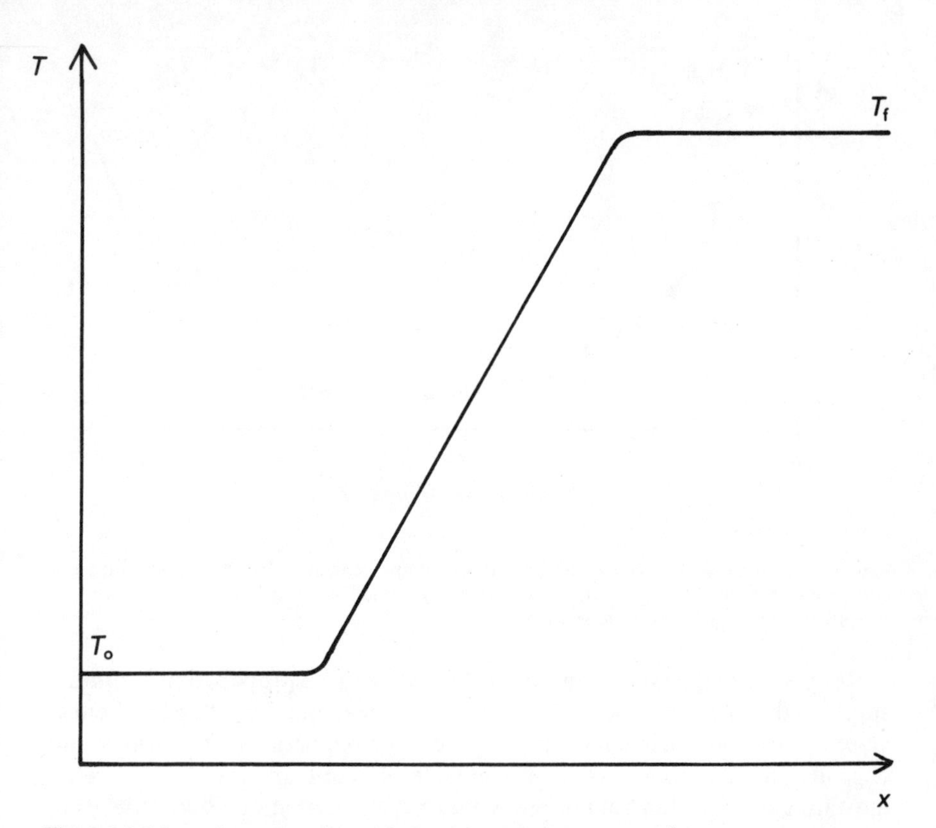

Figure 3.3 Schematic temperature profile in the excess-energy approach

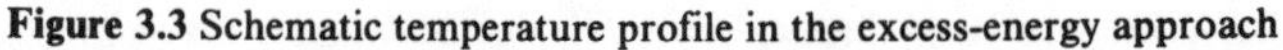

giving a simple link between v_o and H. Integration without the approximation of constant k is possible, provided the variation of k with T is known as an empirical relationship. On the basis of constant k, the excess energy $(e - e_o)$ has a maximum corresponding to a maximum in dT/dx and is nil at $dT/dx = 0$ as noted, so it has the nature of an energy hump or wave, and is often referred to in those terms.

3.3.2 The relationship of excess energy to flammability limits

For methane oxidation:

$$CH_4 + 2O_2 \rightarrow CO_2 + 2H_2O(\text{vap.}),$$

there is no change of mole number, hence equations 3.13 and 3.14 are exact. For various percentages of methane by volume, v_o and T_f are known from experiment and calculation respectively and k is also known. Hence the value of H as a function of mixture composition across the entire range bridging the lower and higher limits of flammability can be deduced. Figure 3.4 shows the results and it is noteworthy that H rises very steeply at the limits but varies only gently for most of the intervening composition region. The upper and lower limits occur at almost equal values of H.

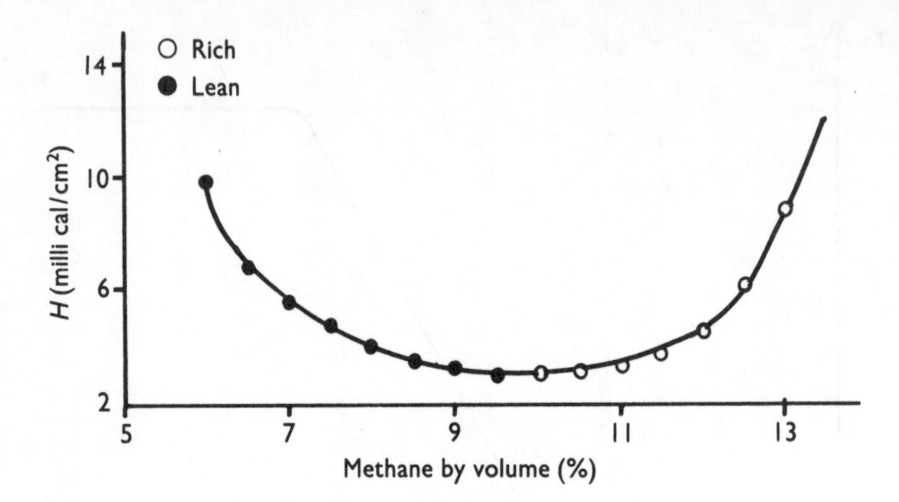

Figure 3.4 Excess energy H against percentage of methane for a methane–air mixture at atmospheric pressure and room temperature (Burgoyne and Weinberg [3])
(Reproduced with the permission of Butterworth Heinemann.)

Results of the type shown in Figure 3.4 are for flame **propagation** in tubes of specified diameter, at speeds of about 6 cm s^{-1} near either limit up to 25 cm s^{-1} close to stoichiometric conditions. The energy requirement for propagation depends on H. In experiments just outside the limit, *spark kernels* develop, indicating that **ignition** but not propagation has occurred. Ignition depends on energy concentration, which in turn depends on H', defined as:

$$H' = \frac{\sigma_0 T_0 k}{m'a} \ln \frac{T_f}{T_0} \qquad \text{J m}^{-3}, \tag{3.16}$$

where a = the length in metres over which the temperature rises to T_f.

Equation 3.16 enables the behavior of flames close to the limits to be understood semi-quantitatively. These have low values of T_f because of the diluent effect of excess air or of fuel and, assuming ln T_f to be the only function of T_f in equation 3.16, lean or rich mixtures will require less energy for ignition than those close to stoichiometric conditions.

3.3.3 Heat loss and minimum ignition energy

Even a mixture well within the flammability limits of composition will not propagate in a tube of diameter below a critical value termed the quenching diameter. These have values of 1–2 mm for many hydrocarbon gases in air at room temperature. This is due to heat losses from the flame to the tube, and clearly these losses are from the excess energy as well as from the post-combustion gases. If losses to the tube were **solely** from the post-combustion gases, there would be no effect of tube diameter on the flame itself.

Heat losses to the tube are manifest as a reduced value of T_f, termed T_f' and given by:

$$T'_f = T_f - \frac{h}{m'C} \qquad \text{K}, \tag{3.17}$$

where h = heat lost to the tube per unit area of the flame, and dependent on the tube dimensions (W m^{-2})
C = heat capacity (J kg^{-1} K^{-1}).

The value of h which just precludes propagation, h_{max}, corresponds to the quenching diameter and the lowest flame temperature T_e for propagation is:

$$T_e = T_f - \frac{h_{max}}{m'C} \qquad \text{K}. \tag{3.18}$$

Figure 3.5 represents spark ignition of a spherical volume of gas of radius r, as discussed in Section 3.2.1. At the flame front, the excess energy is $4\pi r^2 H$

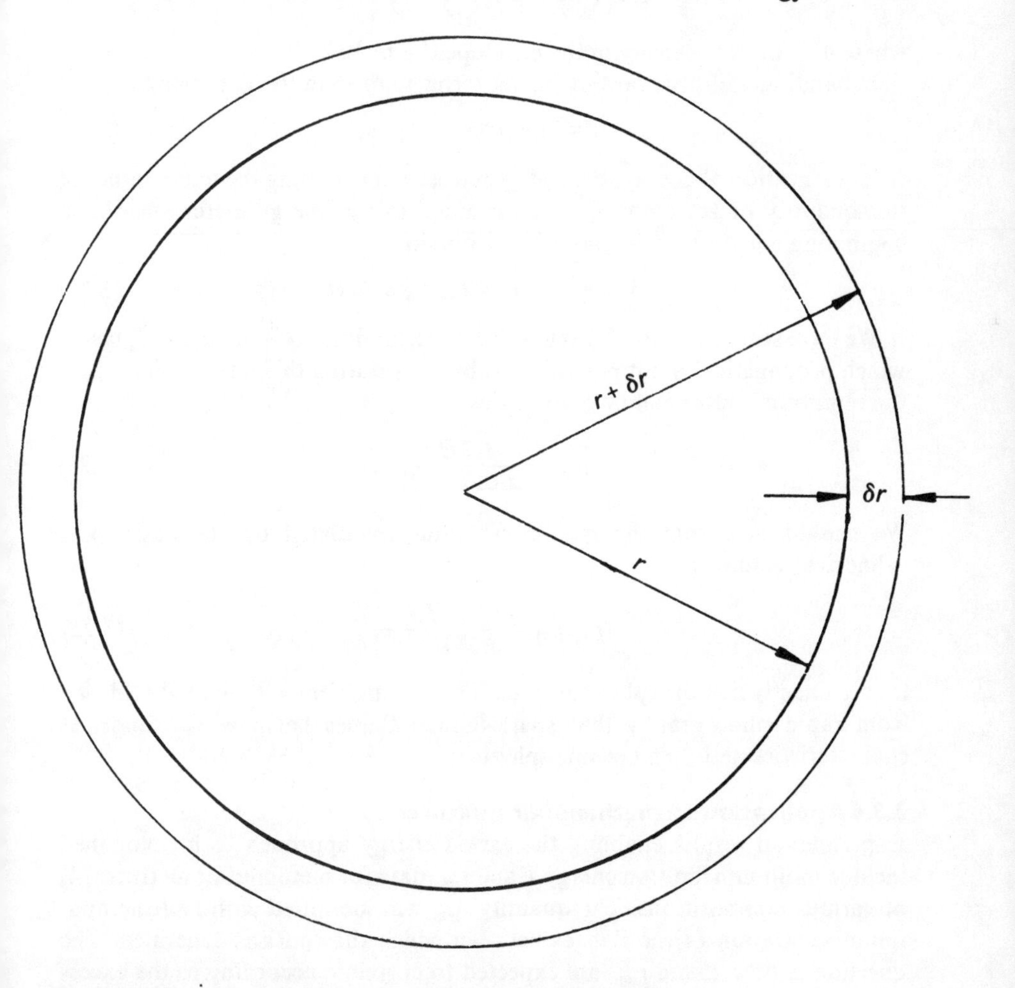

Figure 3.5 Schematic diagram of spark ignition of a spherical volume of gas

joules. Propagation and gas expansion causes $r \rightarrow (r + \delta r)$ and the energy increment δE is:

$$\delta E = H\,[4\pi(r + \delta r)^2 - 4\pi r^2] \qquad \text{J.} \tag{3.19}$$

Expanding and disregarding terms in δr higher than the first gives:

$$\delta E = 8\pi r \delta r H \qquad \text{J.} \tag{3.20}$$

This energy must come from gas passing through the flame front (spherical gas envelope of width δr) which will accordingly cause the flame temperature to be T'_f, lower than T_f as discussed above. The quantity of gas w occupying the shell is given by:

$$w = \sigma_f \left[\frac{4}{3}\,\pi(r + \delta r)^3 - \frac{4}{3}\,\pi r^3\right] \qquad \text{kg,} \tag{3.21}$$

where σ_f is the gas density in the envelope (kg m^{-3}).

Expanding and disregarding higher terms in δr than the first gives:

$$w = 4\pi r^2 \sigma_f \delta r \qquad \text{kg.} \tag{3.22}$$

By conservation of energy, the energy released in dropping the temperature of this quantity of gas from T_f to T'_f is equal to the energy increment δE, so combining equations 3.20 and 3.22 we obtain:

$$4\pi r^2 \sigma_f C \delta r (T_f - T'_f) = 8\pi r \delta r H. \tag{3.23}$$

We have seen (equation 3.18) that there is a minimum temperature T_e below which propagation is not possible and by substituting this into equation 3.23 the minimum radius can be obtained as:

$$r_{min} = \frac{2H}{\sigma_f C(T_f - T_e)}\,. \tag{3.24}$$

We should note that the minimum radius calculated on the basis of a cylindrical volume is:

$$(r_{min})_{cyl.} = \frac{H}{\sigma_f C(T_f - T_e)}\,, \tag{3.25}$$

that is, exactly half the spherical value. This is consistent with limited evidence from rapid photography that spark-ignited flames begin with cylindrical characteristics and then become spherical.

3.3.4 Application to methane/air mixtures

Experimental results enabling the excess energy approach to be examined include minimum ignition energy E and r_{min} data for methane/air mixtures [4] of various compositions. The quantity r_{min} was identified with half the minimum separation of the flanges between which the spark is generated. The question is how E and r_{min} are expected to correlate according to the excess energy approach, and the key is energy concentration H' (equation 3.16).

We expect that for a mixture characterized by E, and r_{min}:

$$\frac{E}{\frac{4}{3}\pi r_{min}^3} = H' = \frac{\sigma_o T_o}{m'a}\int_{T_o}^{T_f} \frac{k}{T}\,\mathrm{d}T, \tag{3.26}$$

where the simplification of constant k made in equation 3.16 has not been made. We expect a graph of:

$$\frac{E}{d^3} \text{ against } \int_{T_o}^{T_f} \frac{k}{T}\,\mathrm{d}T,$$

where d = flange separation = $2r_{min}$, to be linear. The integral can be calculated for the various compositions (methane/air ratios) for which E and r_{min} are known. The graph is shown in Figure 3.6 and points at the lean and rich side of stoichiometric conditions are seen to lie on it without significant scatter. The eventual deviation from linearity is due to drift in the value of the coefficient of the integral in equation 3.26 across a wide composition range.

3.3.5 Critique of the excess energy approach

The conceptual point upon which the excess energy approach stands or falls is that for the energy hump postulated to exist, there will need to be a considerable difference between heat and mass transfer rates in the flame. For the hump to exist, the heat transfer rate by conduction will need to exceed that due to diffusion. We have seen from consideration of equation 3.14 how on this basis the excess energy depends on $\mathrm{d}T/\mathrm{d}x$, which is zero at T_o and T_f and non-zero between them, giving a maximum in the temperature excess and hence a wave or hump. Critics of the approach [5] would include a diffusion component in equation 3.14 which would remove this simple dependence of the excess energy on $\mathrm{d}T/\mathrm{d}x$ and possibly eliminate all significance of the excess energy approach.

Relative rates of heat and mass transfer depend on the Lewis number:

$$\mathrm{Le} = \frac{\alpha}{D}, \tag{3.27}$$

where α = thermal diffusivity = $\frac{k}{c\sigma}$ $\quad m^2\ s^{-1}$

c = heat capacity (J $kg^{-1}\ K^{-1}$)
σ = density (kg m^{-3})
D = diffusion coefficient ($m^2\ s^{-1}$).

Le is often close to unity in gases and flames, a fact that works against the excess energy approach. However, there is a counter view that for mixtures of gases there is scope for D and α to differ sufficiently for Le to be significantly greater than unity, leading to the simple dependence of the excess energy of $\mathrm{d}T/\mathrm{d}x$ outlined.

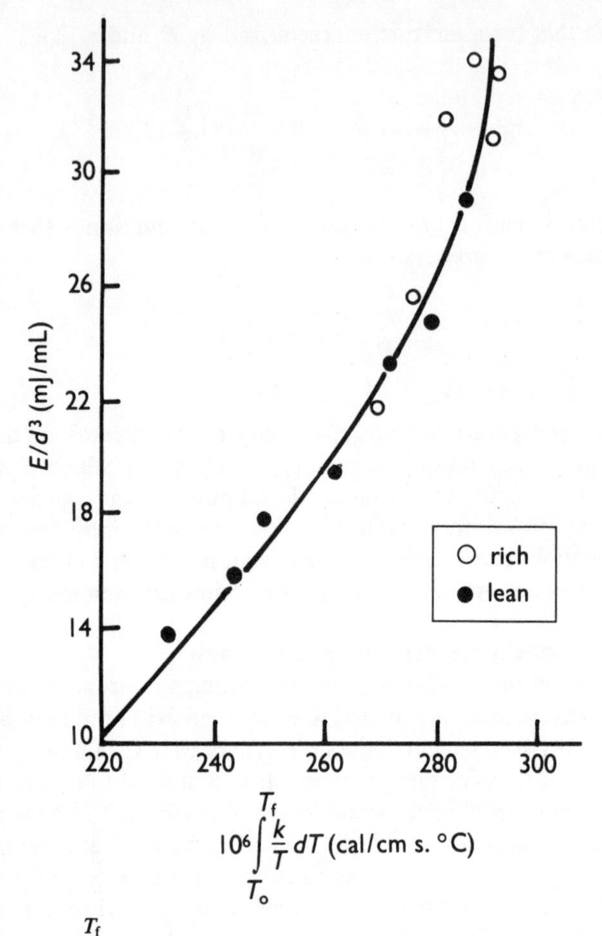

Figure 3.6 $\frac{E}{d^3}$ versus $\int_{T_o}^{T_f} \frac{k}{T}\,dT$ for methane/air mixtures of various composition (Burgoyne and Weinberg [3])
(Reproduced with the permission of Butterworth Heinemann.)

3.4 Electrical effects

The above two descriptions of spark ignition are essentially thermal and involve heat balance and heat transfer equations, and there are some points of contact with the thermal treatment of hot spots discussed in Section 3.7.

While for many spark ignitions the role of the spark is simply the degradation of its electrical energy, there has long been evidence [6] that in some examples of spark ignition we have to look to the nature of the electrical discharge itself for a full understanding. As a simple example, consider otherwise identical spark ignition tests performed with (a) tungsten electrodes or (b) silver electrodes. Silver, unlike tungsten, is a *sputtering electrode*, that is, it releases metal atoms which can accelerate the gas-phase oxidation and exert a strong effect on the kinetics and hence ignition conditions. If a non-sputtering material is used, possible catalytic effects can be eliminated by water cooling.

Also relevant is the ability of the discharge to create ions. This was examined for *electrolytic gas* (hydrogen and oxygen prepared by electrolysis of acidified water) in classical experiments by Finch and Cowen [7], and results were expressed in terms of total initial pressure of electrolytic gas and discharge current necessary for ignition. Ignition was always instantaneous; there was no lag. If a mixture failed to ignite on very brief passage of current, prolonged passage would not make it do so.

Families of ignition limits were obtained for different electrode separations. An example of one of the limits is given in Figure 3.7 and for much of its length it fits the hyperbola:

$$P_0(i + 1.5) = 105, \tag{3.28}$$

where P_0 is expressed in cm of mercury and i is the minimum discharge current required for ignition (mA). The total absence of lag is consistent with the view

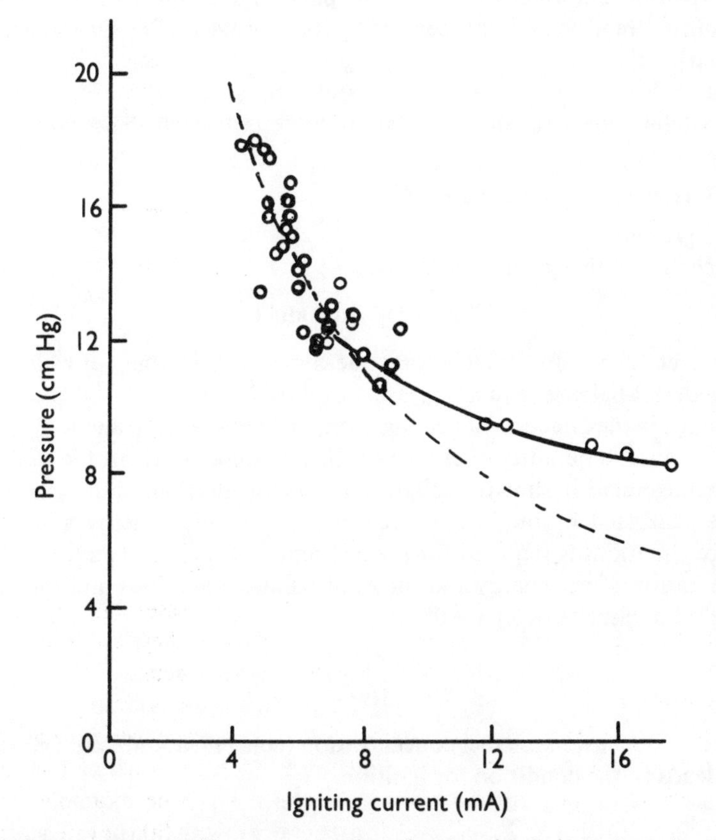

Figure 3.7 Pressure against ignition current for *electrolytic gas* with platinum electrodes separated by 3.8 mm (Finch and Cowen [7]). Dashed line = equation 3.28

that the action of the electrical discharge leading to ignition is ionization and not simple degradation of energy. Mechanistic speculations of how ions generate the chain carriers necessary for ignition of the hydrogen/oxygen mixture (see Section 2.6) must include:

$$H^+ + H_2O \rightarrow H_3O^+$$

$$H_3O^+ + e \rightarrow H + H + OH$$

$$H_3O^+ + e \rightarrow H_2O + H.$$

3.5 More advanced treatments of spark ignition

3.5.1 Introduction

We have so far considered spark ignition in the following terms:

(a) A thermal treatment to some extent custom built to permit comparison with related experimental results.
(b) A treatment requiring the hypothesis of excess energy of the flame front.
(c) A treatment in which the role of the spark is not simply energy degradation but the creation of active centers by ions formed under the impact of the spark.

A spark releases energy very rapidly and a broader and more fundamental approach by Yang [8] enables ignitions to be described which use longer energy release times.

3.5.2 The treatment by Yang [8]

- **Background**

Yang considers the generalized process:

$$\text{fuel} + O_2 \rightarrow \text{products},$$

and his treatment is formulated for a one-dimensional flame, for which energy and material-balance equations are developed. At $x = 0$, where x is the coordinate in the direction of propagation, and between $t = 0$ and $t = t_o$, the heat source releases a quantity of energy causing a temperature profile to develop. The arrangement is shown in Figure 3.8. Incorporation of the temperature profile generated by the heat source into the coupled energy and material balance equations is required for the deduction of ignition conditions.

Integration of the energy and material-balance equations and the requirement that at either side of $x = 0$:

$$\frac{d\cap}{dt} < 0,$$

where $\cap$ = dimensionless fuel concentration (concentration/initial concentration), leads to the condition for ignition:

$$\int_0^\infty \Phi(\cap, T)dx - G(1 - \cap_o)q > 0,$$

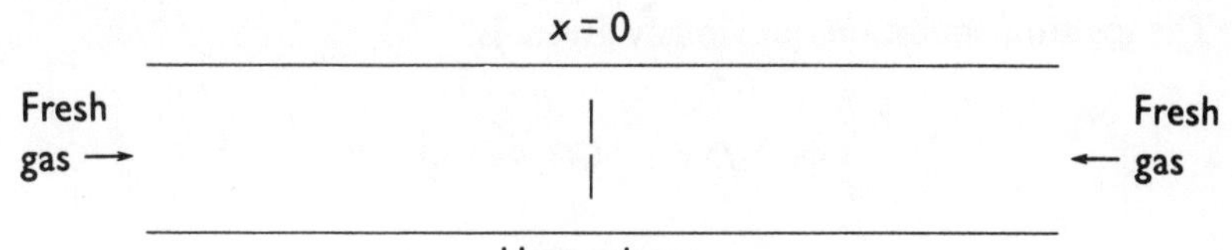

Figure 3.8 Schematic diagram of the system analyzed by Yang [8]

where $\cap_o$ = dimensionless fuel concentration at $x = 0$
G = mass flow rate (kg m-2 s^{-1})
q = heat of reaction (J kg^{-1})
$\Phi(\cap, T)$ = reaction heat release rate (W m^{-3}).

Clearly, $(1 - \cap_o)$ represents the concentration gradient enabling influx of fresh gas.

So the integrated form of the above ignition condition requires knowledge of the initial temperature distribution due to the source, as previously discussed, and this will depend on whether the source is a plane, a line or a point. We first consider a plane source, for example, a heated grid.

- **Plane heating source**

The temperature profile at $t = t_o$ is:

$$T = T_s \operatorname{erfc}\left[\frac{x}{2(kt_o)^{0.5}}\right], \tag{3.29}$$

where T_s = average temperature of the plane (K)
k = thermal conductivity (W m^{-1} K^{-1}),

and the error function complement (erfc) is obtainable from tables. We will encounter error functions again in Chapter 11.

The energy ϵ released by the heater plane in time t_o is:

$$\epsilon = \int_{-\infty}^{\infty} \sigma_o C T S \, dx = 4\sigma_o C T_s S \left(\frac{kt_o}{\pi}\right)^{0.5} \quad \text{J}, \tag{3.30}$$

where σ_o = initial density (kg m^{-3})
S = cross-section of tube (m^2)
C = heat capacity (J kg^{-1} K^{-1}).

The ignition condition, previously given, is:

$$\int_0^\infty \Phi(\cap,T)\,dx - G(1-\cap_o)q \geq 0,$$

and the Arrhenius form of the temperature dependence of reaction (and hence heat release) rate can be approximated by:

$$\Phi(\cap,T) = K(\cap\sigma_o)^m\,T^n \qquad \text{W m}^{-3}, \tag{3.31}$$

where K is a constant incorporating the heat of reaction, with units dependent on m and n which are respectively order of reaction and temperature exponent.

Substituting from equations 3.29 and 3.31 into the inequality gives as the ignition condition:

$$\int_0^\infty K(\cap\sigma_o)^m\,T_s^n\left[\text{erfc}\,\frac{x}{2(kt_o)^{0.5}}\right]^n dx \geq G(1-\cap_o)q.$$

The ignition condition, whereupon the equality is fulfilled, is obtained in terms of a critical source temperature T_{sc} by rearrangement as follows:

$$T_{sc} = \frac{[qG(1-\cap_o]^{1/n}}{\left[K(\cap\sigma_o)^m\int_0^\infty\left[\text{erfc}\,\frac{x}{2(kt_o)^{0.5}}\right]^n dx\right]^{1/n}} \qquad \text{K}, \tag{3.32}$$

and the term $(\cap\sigma_o)^m$ in the denominator is taken to be constant as $\cap \equiv 1$ for $t \leq t_o$.

Substituting from equation 3.30 we obtain the minimum ignition energy ϵ^* as:

$$\epsilon^* = \frac{4S\sigma_oC\,[Gq(1-\cap_o)\,]^{1/n}\,(kt_o/\pi)^{0.5}}{\left[K(\cap\sigma_o)^m\int_0^\infty\left[\text{erfc}\,\frac{x}{2(kt_o)^{0.5}}\right]^n dx\right]^{1/n}} \qquad \text{J}. \tag{3.33}$$

Evaluation of the integral in the denominator gives:

$$\epsilon^* = \frac{4S\sigma_oC\,[\,n^{0.5}Gq(1-\cap_o)\,]^{1/n}(kt_o/\pi)^{0.5(1-1/n)}}{[2K(\cap\sigma_o)^m]^{1/n}} \qquad \text{J}. \tag{3.34}$$

- **Line and point heat sources**

The above equation has been derived for a plane heat source. The corresponding expressions for line and point sources are respectively:

$$[\epsilon^*]_{\text{line}} = 16L\sigma_oC\left[\frac{qG(1-\cap_o)}{16K(\cap\sigma_o)^m}\right]^{1/n}\left[\frac{kt_o}{\pi}\right]^{1-1/n} \qquad \text{J}, \tag{3.35}$$

where L = length of the line (m),
and:

$$[\epsilon^*]_{\text{point}} = 64\sigma_o C\left[\frac{qG(1-\cap_o)n^{3/2}}{64K(\cap\sigma_o)^m}\right]^{1/n}\left[\frac{kt_o}{\pi}\right]^{1.5(1-1/n)} \text{ J}, \quad (3.36)$$

The following trends are predicted by equations 3.34–3.36. First, the minimum ignition energy is for all three types of heat source directly proportional to the heat capacity of the fuel/oxidant mixture. This is also true in the thermal treatment outlined in Section 3.2. In the excess enthalpy approach the occurrence of ignition depends on energy **concentration**, and this, from equations 3.24 and 3.26, appears to have a strong direct dependence on the heat capacity. Direct dependence of ϵ^* on q and k is also indicated in Yang's treatment.

- **Experimental examination**

The time of application of the heat, t_o, is present in each expression for ϵ^*. The dependence is quite strong; $1/n$ in the exponent will usually be much less than unity, in which case the approximate proportionalities hold:

$$(\epsilon^*)_{\text{plane}} \propto t_o^{0.5}$$
$$(\epsilon^*)_{\text{line}} \propto t_o$$
$$(\epsilon^*)_{\text{point}} \propto t_o^{1.5}.$$

These proportionalities can be checked experimentally: a current can be passed along a wire through a gaseous fuel/oxygen mixture and the minimum duration of current application necessary for ignition (t_o) recorded. Clearly:

$$\text{energy supplied} = \mathfrak{R}^2 i t_o \text{ J}$$
$$= E_s \text{ J}, \quad (3.37)$$

where $\mathfrak{R}$ = resistance (ohms)
i = current in amperes.

The system can be designed so that the temperature distribution along the wire has a steep maximum at the center and therefore serves satisfactorily as a point source, as in experiments by Stout and Jones [9] with hydrogen–air mixtures. These experimental results, which predate the Yang treatment of ignition by a heat source, were used by Yang as a test of his theory.

Figure 3.9 shows E_s against $t_o^{1.5}$ for 20% hydrogen/air mixtures ignited by a point source originating in wires of different diameters. It should perhaps be emphasized that Figure 3.9 does not represent *power* application for various times but expresses the experimental observation, supported by Yang's treatment, that a small amount of energy can bring about ignition only if released rapidly. For longer release times, larger amounts of energy are needed. However, it is clear that E_s cannot be equated to ϵ^* for two reasons: no dependence of ϵ^* on wire diameter is expected, and a graph of ϵ^* against t_o should pass through the origin. Moreover, the E_s values are much higher than measured ϵ^* values (see Table 3.1 and Section 3.6).

The explanation is that E_s is the sum of two components, ϵ^* and the heat ϵ_w

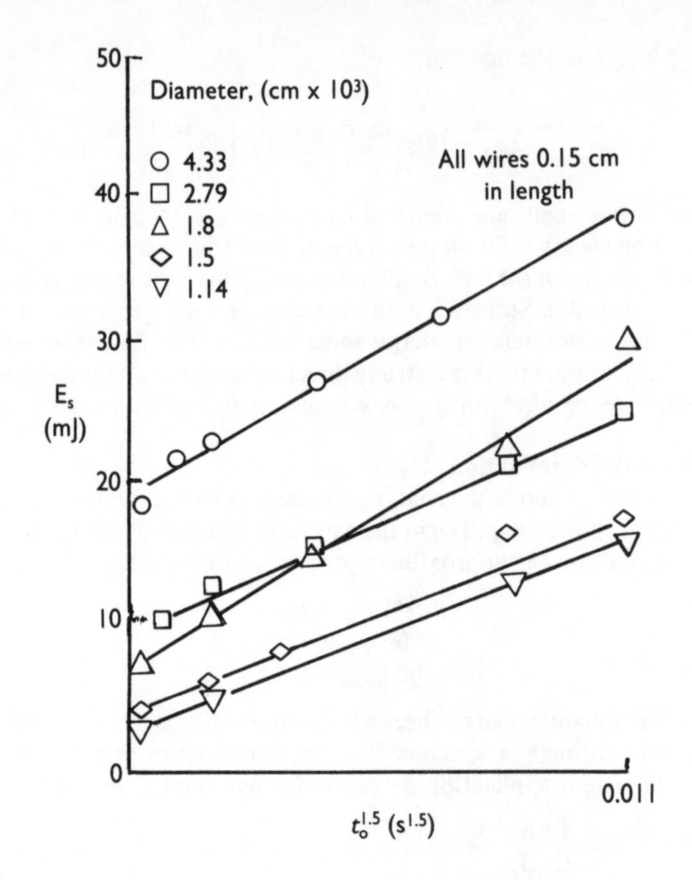

Figure 3.9 E_s versus $t_o^{1.5}$ for a 20% H_2/air flame (Stout and Jones [9])
(Reproduced with the permission of the Combustion Institute.)

entering the wire and raising its temperature. This will clearly depend on wire size and will explain the dependence of E_s on diameter. We can write:

$$E_s = \epsilon^* + \epsilon_w \quad \text{J} \tag{3.38}$$
$$= 64\sigma_o C\left[\frac{qG(1-\cap_o)n^{3/2}}{64K(\cap\sigma_o)^m}\right]^{1/n}\left[\frac{kt_o}{\pi}\right]^{1.5(1-1/n)} + \epsilon_w \quad \text{J.}$$

Hence for large n, as previously discussed, linear graphs of E_s against $t_o^{1.5}$ as shown in Figure 3.9 are quite consistent with the Yang model for a point source, and the intercept is ϵ_w.

The prediction from Yang's model that the source energy required for ignition depends on t_o is wholly consistent with what was previously widely known and schematized by Lewis and Von Elbe [4] as shown in Figure 3.10. A difference is that in Figure 3.10 there is a finite (though small) value $\epsilon^*(0)$ of the source energy at hypothetical *instant release* ($t_o = 0$), whereas equations 3.34 to 3.36 all predict that this value will be zero. The key to the difference is that

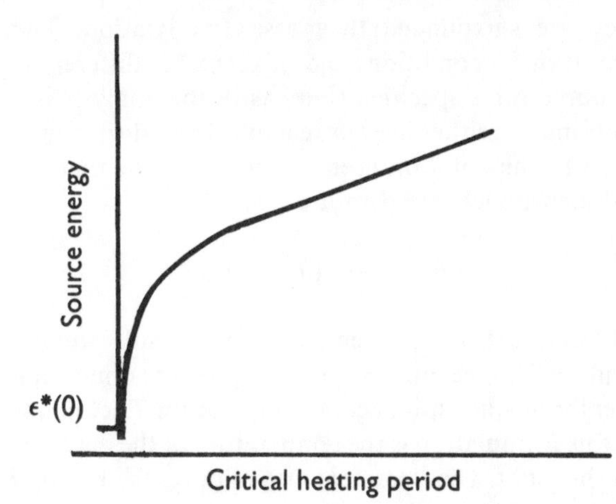

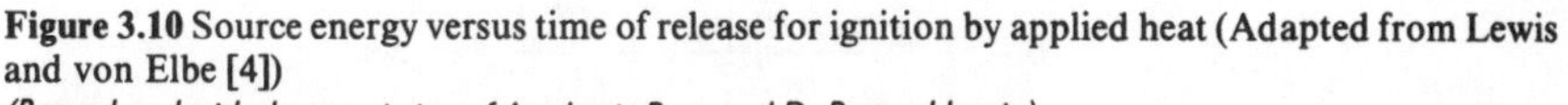
Figure 3.10 Source energy versus time of release for ignition by applied heat (Adapted from Lewis and von Elbe [4])
(Reproduced with the permission of Academic Press and Dr Bernard Lewis.)

whereas the temperature distribution expression in equation 3.29 is normally bell-shaped, for extremely short t_0 it is a spike at $x = 0$ and the resulting high temperature leads to a very rapid reaction, causing local fuel consumption effects to be significant. Hence, while according to equations 3.34 to 3.36, ϵ^* increases with t_0, at very short values of t_0 a reverse effect is also expected, since shorter times mean less fuel consumption. There are therefore two opposing effects and a graph such as Figure 3.10 on the basis of the Yang model is expected to have a short, almost horizontal, region as t_0 approaches zero. Fuel consumption effects are insignificant beyond this horizontal region.

3.6 Measured ϵ^* values

Order-of-magnitude values for organic gases and vapors in air are a fraction of 1 mJ as shown for two examples in Table 3.1. This figure also applies to vapors of petroleum products. Gases containing molecular hydrogen have lower values of the minimum ignition energy and this is a point of some importance as regards the relative merits of manufactured town gas and natural gas. In pure oxygen, hydrocarbons have ϵ^* values of typically 2μJ.

3.7 Hot spot theory

3.7.1 Introduction

We have examined situations in which a localized source of heat is introduced deliberately into a system to ignite it. It is also possible for a localized area of heat to develop in a reactive medium by surface frictional heating, viscous heating or compression. The same principles would apply to a hot piece of metal shaving accidentally finding its way into a combustible solid such as sawdust. These are termed *hot spots* and in transferring heat to the medium with which they are surrounded they assist its ignition. They lose heat according to heat transfer conditions and this is rather different from application of a heat source for a specified time, as in the analysis in the previous section. Hot spots may be either inert or reactive. In analyzing ignition by these sources we return to the notation used in Chapter 1 for thermal ignition, in particular, the dimensionless temperature:

$$\theta = \frac{E}{RT_o^2}(T - T_o), \tag{1.5}$$

and in hot-spot treatment the spot temperature T_i is substituted for T_o in the denominator and for T in the numerator. This gives the dimensionless temperature θ_i at the center of a hot spot, actual temperature T_i, containing E, T_i and T_o, where T_o in this formulation is the temperature of the medium at all points other than at the hot spot. The treatment also requires the Frank–Kamenetskii parameter (equation 1.31):

$$\delta = \frac{r_o^2\, Q\sigma AE \exp[-E/RT_o]}{kRT_o^2},$$

and in hot-spot treatment, T_i is substituted for T_o in the exponential and in the denominator of δ.

The ignition condition is:

$$\delta \geq \delta_{crit}.$$

3.7.2 Numerical treatment of reactive spherical hot spots

Thomas [10] considered a spherical reactive hot spot in an explosive material generating heat according to an Arrhenius dependence. The boundary conditions are:

$$T = T_i \text{ for } r < r_o, \tag{3.39}$$
$$T = T_o \text{ for } r \geq r_o, \tag{3.40}$$

where r = distance coordinate
r_o = sphere (hot spot) radius.

The rate of heat generation q(W m^{-3}) is uniform in view of the single temperature within the spot and the heat balance between the hot spot and the remainder of the material is:

$$T_i - T_o = \frac{qr_o^2}{2k}. \tag{3.41}$$

If $q(T_i)$ has an Arrhenius temperature dependence, this leads [10] to the critical condition:

$$\delta_{crit} = 2\theta_i. \tag{3.42}$$

Combining the expression for δ and that for θ_i (modified versions of equations 1.31 and 1.5 respectively), equation 3.42 can be rearranged to:

$$[r_o]_{crit} = \left[\frac{2\theta_i k R T_i^2 \exp(E/RT_i)}{Q\sigma A E}\right]^{0.5}, \tag{3.43}$$

and this can be tested using experimental data from the literature for liquid explosives. Typically, a hot spot in a liquid explosive with $T_i \approx 350°C$ will have $(r_o)_{crit}$ in the range 10^{-6} to 10^{-5} m.

For other geometries of hot spot, equations of the form of (3.42) can be used:

$$\delta_{crit} = \mathfrak{M}\theta_i, \tag{3.44}$$

where $\mathfrak{M}$ is a constant whose values for spheres are in the range 2, according to Thomas [9], to 4.7, according to Friedman [11]. The difference reflects the different approaches. Thomas' treatment considers a balance of heat released by the reactive hot spot, and heat transferred by conduction, in fulfillment of the conditions expressed in equations 3.39 and 3.40. Friedman's [11] treatment considers the maximum value of δ (recalling that δ is a dimensionless heat release rate) for $T_i(t)$ to take on a negative slope at any value of t. The dependence of $(r_o)_{crit}$ on $\mathfrak{M}$ is not strong. For hot spots of slab or cylindrical shape, Friedman estimates values of $\mathfrak{M}$ of 1 and 2.6 respectively.

3.7.3 Inert spherical hot spots

A critical condition due to Merzhanov et al. [12] for an inert spherical hot spot ($q = 0$) is:

$$\delta_{crit} = 12.1(\ln \theta_i)^{0.6}, \tag{3.45}$$

and we will use this to examine the case of a hot ball bearing dropped into sawdust. Combining equation 3.44 with the expression defining δ, and rearranging:

$$[r_o]_{crit} = \left[\frac{12.1(\ln \theta_i)^{0.6} k \exp(E/RT_i) R T_i^2}{Q\sigma A E}\right]^{0.5}. \tag{3.46}$$

Consider a ball bearing heated to 673K. How large will it need to be to cause ignition if dropped into a pile of sawdust at 298K? Anticipating combustion kinetic data for woody materials given in Chapter 7, let $E = 120$ kJ mol^{-1}, then substitution in the expression for θ for a hot spot gives:

$$\theta_i = 12.0.$$

Using the following data (see Chapter 7):

$$A = 10^6 \text{ s}^{-1}$$
$$\sigma = 500 \text{ kg m}^{-3}$$
$$Q = 17\times10^6 \text{ J kg}^{-1}$$
$$k = 0.15 \text{ W m}^{-1} \text{ K}^{-1},$$

and substitution in equation 3.46 gives:

$$[r_o]_{crit} = 0.5 \text{ cm}.$$

So on the basis of the Merzhanov treatment we expect a ball bearing of this radius to ignite a pile of sawdust of the specifications given if dropped into it. A smaller ball bearing at the same temperature will fail to ignite it.

3.8 Summary

Table 3.2 outlines the approaches in this chapter to ignition by a heat source.

Table 3.2 Comparison of treatments of ignition by a heat source

Method	*Features*	*Ignition condition*	*Application example*
Ignition theory Fenn [2]	Critical radius of gas volume from T_o to T_f by applied heat. E proportional to T_f at the lean limit	H_{min} (J) = $Zx_o^{-1.5}\sigma^{-2}(T_f - T_o)\exp\frac{3E}{2RT_f}$ (Eq. 3.7)	Tested with data sets of H_{min}, T_o or P_o for many simple organics (Figs 3.1 and 3.2)
Excess energy Burgoyne and Weinberg [3]	Increase in energy of flame front at the expense of temperature of propagating gas. Critical temperature T_c below which there is no propagation	H'(J m^{-3}) = $\frac{\sigma_o T_o}{m'a}\int_{T_o}^{T_f}\frac{k}{T}\,dT$ (Eq. 3.26)	Applied *inter alia* to mixtures of methane and oxygen (Fig. 3.6)
Ionization, Finch and Cowen [7]	Electrical discharge creates ions which react to form chain carriers	$P_o(i + 1.5)$ = constant (Eq. 3.28)	Applied to H_2/O_2 mixtures (Fig. 3.7)
Coupled energy and material balance, Yang [8]	Quantity of energy released during a period t_o, and resulting temperature profile incorporated into the energy condition for ignition	ϵ^* (J) $\propto t_o^{0.5}$ (plane source) ϵ^* (J) $\propto t_o$ (line source) ϵ^* (J) $\propto t_o^{1.5}$ (point source) $\epsilon^* = f[C,q,k]$ in each case (Eqs 3.34–3.36)	Applied to H_2/air mixtures (Fig 3.9)
Reactive spherical hot spot, Thomas [10]; nonspherical, Friedman [11]	Element of small radius raised in temperature by friction or viscous heating	$\delta_{crit} = 2\theta_i$ (Eq. 3.42) Similar expressions for nonspherical hot spots (Eq. 3.44)	Applied to liquid explosives
Inert spherical hot spot, Merzhanov [12]	Energy supply by a nonreacting preheated body	$\delta_{crit} = 12.1[\ln\theta_i]^{0.6}$ (Eq. 3.45)	Ball bearing dropped into sawdust (Section 3.7.3.)

References

[1] Brame J. S. S., King J. G., *Fuel-Solid, Liquid and Gaseous*, 5th Edition, London: Arnold (1955)

[2] Fenn J. B. 'Lean flammability limit and minimum spark ignition energy', *Industrial and Engineering Chemistry*, **43** 2865 (1951)

[3] Burgoyne J. H., Weinberg F. J., 'Excess energy hypothesis of flame behaviour', *Fuel* **33** 436 (1954)

[4] Lewis B., von Elbe G., *Combustion, Flames and Explosions of Gases*, New York: Academic Press (1951)

[5] Friedman R., Burke E., cited in: Burgoyne J. H., Weinberg F. J., 'Excess energy hypothesis of flame behaviour — discussion of the basic assumptions', *Fuel* **34** 351 (1955)
[6] Bradford B. W., Finch G. I., 'The mechanism of ignition by electric discharges', *Chemical Reviews* **21** 221 (1937)
[7] Finch G. I., Cowen L. G., 'Gaseous combustion in electrical discharges, II, The ignition of electrolytic gas by d.c. discharges', *Proceedings of the Royal Society of London,* **A116** 529 (1927)
[8] Yang C. H., 'Theory of ignition and auto-ignition', *Combustion and Flame,* **6** 215 (1962)
[9] Stout H. P., Jones E., 'The ignition of gaseous explosive media by hot wires', *Third Symposium (International) on Combustion,* 329, Baltimore: Williams and Wilkins (1949)
[10] Thomas P. H., 'A comparison of some hot-spot theories' *Combustion and Flame,* **9** 369 (1965)
[11] Friedman M. H., 'A correlation of impact sensitivities by means of the hot-spot model', *Ninth Symposium (International) on Combustion,* 294, New York: Academic Press (1963)
[12] Merzhanov et al., cited in: Gray P., Lee P. R., 'Thermal explosion theory', *Oxidation and Combustion Reviews,* **2** 3 (1967)

CHAPTER 4

Flame Structure and Propagation

Abstract

The terms premixed flame, diffusion flame, laminar flame *and* turbulent flame *are explained. A detailed model of a laminar flame is presented by means of which expressions for temperature profiles and heat release rates are obtained. These are applied respectively to propane/air and carbon monoxide/oxygen flames. A derivation of flame velocities is also presented and factors influencing these velocities discussed, with special reference to methane. Some results of velocity measurements of turbulent propane/air flames are discussed. The principles of detonation are also given to clarify the distinction between a flame and a detonation.*

4.1 Introduction

A flame is a combustion reaction which is self-propagating through space and has a luminous zone. Dependence of propagation on heat transfer distinguishes a flame from a detonation, as outlined separately later in the chapter. Flame classification is partly according to whether it is a premixed or diffusion flame. In a premixed flame, fuel and oxidant are brought together before entry to the reaction zone, whereas in a diffusion flame the two must diffuse together in order to react. Further classification is on the basis of the fluid mechanics of the reaction zone. If the gas flow is laminar, all mixing is at the molecular level. If the gas flow is turbulent, there is in addition to molecular mixing, macroscopic mixing due to eddy currents. The threshold distinguishing the two is expressible in terms of a dimensionless group known as the Reynolds number, defined as:

$$\text{Re} = \frac{vx\sigma}{\mu}, \tag{4.1}$$

where v = velocity (m s^{-1})
x = distance (m)
σ = density (kg m^{-3})
μ = dynamic viscosity (kg m^{-1} s^{-1}).

Values of Re in flames between 10^3 and 10^4 signal the transition from laminar to turbulent flow. Eddy currents may be of small diameter, in which case it is possible for separate islands of combustion to develop, as often happens in gas-fired furnaces. If the eddy currents have a large diameter, the effect of turbulence will be to distort the shape of the flame, which will otherwise retain many of the features of a laminar flame. Turbulence can be created by the burnt gas.

Whether a particular flame is premixed or diffusion type depends on the ratio t_D/t_R, where the numerator and denominator are respectively diffusion time and residence time in the flame. If the ratio is very small the flame is premixed, or if very large it is a diffusion flame, and this simple criterion provides a basis for identifying factors causing a flame to be one type or the other. The above ratio will depend inversely on the diffusion coefficient and on the distance travelled by the gases before reaching the reaction zone, and each of these depends inversely on pressure. The ratio t_D/t_R therefore depends directly (and strongly) on pressure, a low pressure favoring a low value of t_D/t_R and hence premixedness. This point can be understood by reference to a candle [1], which is often discussed as a good example of a diffusion flame, as indeed it is at atmospheric pressure. At an air pressure of about 8 percent of atmospheric ($\approx$ 8000 Pa) however, a candle has the features of a premixed flame.

4.2 Simplified treatment of a one-dimensional laminar flame [1]

We will develop equations relevant to a one-dimensional laminar flame and examine ways in which these equations relate to experimental measurements. Consider a steady-state flame propagating in a direction described by coordinate z, as shown in Figure 4.1. Properties such as temperature and density vary

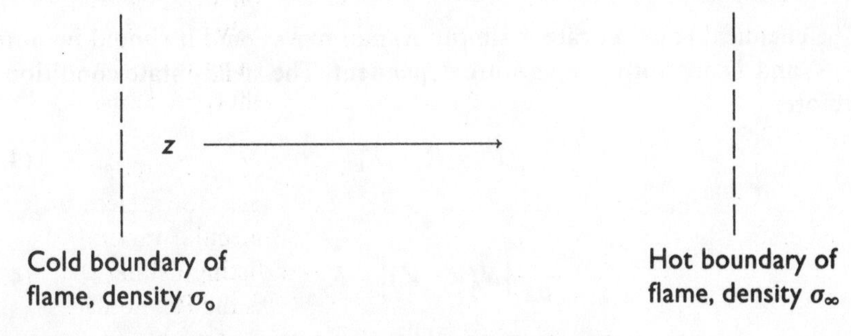

Figure 4.1 Schematic diagram of the flame discussed in Section 4.2

only in the z-direction and not in any direction orthogonal to z. This assumption is also used in analysis of nonreacting fluid flow, for example hot oil in a tube, where averaged values of the properties across a section of the tube are used, this average being different for different values of the coordinate in the direction of flow [2].

For any value of z in Figure 4.1, we have the continuity condition:

$$\sigma v a = \text{a constant}, \tag{4.2}$$

where a = cross-sectional area (m^2)
$\sigma = \sigma_0$, $v = v_0$ and $a = a_0$ at $z = 0$.
$a/a_0 = A$, set to unity if broadening of the flame is neglected, whereupon we have:

$$\sigma_0 v_0 = \sigma v = m, \tag{4.3}$$

where m = mass flow rate (kg m^{-2} s^{-1}).

Turning to individual chemical species, since the flame is in a steady state, the rate of removal of species i by gas flow and diffusion must equal its rate of formation by chemical reaction. The rate of removal by gas flow R_v is:

$$R_v = \frac{d}{dz}(N_i v) \qquad \text{mol m}^{-3}\text{ s}^{-1}, \tag{4.4}$$

where N_i = mole number per m^3.

The rate of removal by diffusion, R_d, is:

$$R_d = \frac{d}{dz} N_i \left(-\frac{D}{X_i}\frac{dX_i}{dz}\right) \qquad \text{mol m}^{-3}\text{ s}^{-1}, \tag{4.5}$$

where D = diffusion coefficient (m^2 s^{-1})
X_i = mole fraction.

Setting $V_i = -\dfrac{D}{X_i}\dfrac{dX_i}{dz}$, we obtain:

$$R_d = \frac{d}{dz}(N_i V_i) \quad \text{mol m}^{-3}\text{ s}^{-1}. \tag{4.6}$$

The chemical reaction rate is simply K_i mol m^{-3} s^{-1} and it should be noted that K_i and D are both temperature dependent. The steady state condition is therefore:

$$R_v + R_d = K_i \tag{4.7}$$

or

$$\frac{d}{dz}\left\{N_i(v + V_i)\right\} = K_i. \tag{4.8}$$

When considering energy conservation we make use of the fact that for any value of z the rate of enthalpy influx by gas flow or diffusion is equal to its rate of transfer by conduction. The rate R_1 of enthalpy influx by gas flow is:

$$R_1 = \sigma v H' \quad \text{W m}^{-2}$$

$$= \sigma_0 v_0 H' \quad \text{W m}^{-2}, \tag{4.9}$$

$$\text{where } H' = \frac{1}{\sigma} \sum H_i N_i \quad \text{J kg}^{-1}, \tag{4.10}$$

where H_i = molar enthalpy (J mol^{-1})

The diffusion enthalpy influx R_2 is:

$$R_2 = \sum N_i H_i V_i \qquad \text{W m}^{-2}, \tag{4.11}$$

where V_i denotes diffusion velocity (see equation 4.5), units m s^{-1}.

The heat transfer by conduction R_3 is:

$$R_3 = -k \frac{\mathrm{d}T}{\mathrm{d}z} \qquad \text{W m}^{-2}. \tag{4.12}$$

As we have seen, under steady state conditions the sum of all the R terms is constant, therefore:

$$\sigma_0 v_0 H' + \sum N_i H_i V_i - k \frac{\mathrm{d}T}{\mathrm{d}z} = \text{constant}. \tag{4.13}$$

The transport terms both vanish at the hot boundary of the flame, denoted by the subscript ∞, therefore the constant is $\sigma_0 v_0 H'_\infty$.

Equations 4.2, 4.8 and 4.13 form the basic equations of a laminar flame. A point deserving mention is that conductive and diffusional heat transfer are within the flame only and any heat transferred to the surroundings is from the post-combustion gases beyond the hot boundary of the flame itself. We have encountered this idea previously in the discussion of spark ignition. In this respect we should note that for the excess energy approach discussed in Chapter 3 to be valid, the diffusional heat transfer term in equation 4.13 would need to be negligible.

The following points need to be addressed with regard to the equations developed:

(a) To what extent is the assignment of a single velocity justified for experimental flames?
(b) What information can be deduced about temperature profiles and heat release rates?

4.3 One-dimensional and quasi-one-dimensional laminar flames

4.3.1 Structure

The natural way for a flame to behave is to move through space so stabilization of a flame requires a device known as a burner in which propagation and gas flow offset each other. The part of the flame coincident with primary reaction is called the luminous zone, succeeded by a secondary zone called the post-luminous zone (though showing weak luminosity). Primary luminosity in hydrocarbon flames is due to radical reactions, whereas secondary luminosity is due largely to CO oxidation.

There are two noteworthy examples of flames that actually are one-dimensional. One is the flat flame, stabilized on a grid burner in such a way that the dimension in the direction of propagation is very much smaller than either of the others. It is easy to visualize how the condition of heat transfer to the surroundings from the post-combustion gases only would hold for a flat flame. The second is the spherical flame, with a radial distance coordinate, readily achievable experimentally. The question of the extent to which flames of other shapes, for example V-shaped and conical, can be regarded as one-dimensional has been addressed by measurements on propane–air flames of various shapes [3].

Laminar premixed stoichiometric propane–air flames at 0.25 atm pressure (assumed constant throughout the flame) were obtained on burners affording conical, button-shaped and V-shaped flames. Gas velocity measurements normal to the flame front (identified visually by reason of the luminous zone) were made at various distances with respect to the center of the luminous zone by particle track techniques. For these flame shapes Figure 4.2 shows the gas velocity as a fraction of the influx cold gas velocity (v/v_o) against distance from

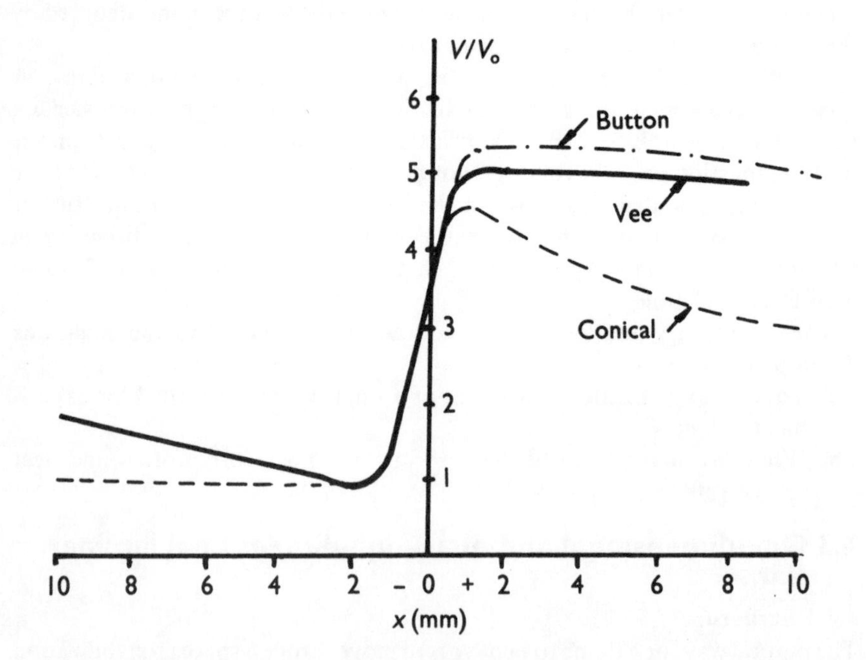

Figure 4.2 Velocity profile for a propane/air flame (Fristrom [3])
(Reproduced with the permission of the American Institute of Physics.)

the center of the luminous zone. The three curves are seen to be coincident in regions close to the luminous zone at either side of it. Moreover, the velocities were found to be independent of the orthogonal coordinate along the flame front. These facts are viewed as strong evidence of the quasi-one-dimensional nature of even these relatively complex shapes of flame. The density profile in the flame is obtainable from the velocity measurements and the continuity condition:

$$\sigma_0 v_0 a_0 = \sigma v a.$$

At the center of the luminous zone, we have, from Figure 4.2:

$$\frac{v}{v_0} = 4,$$

and this leads to a density in the center of the luminous zone of about one-quarter that of the influx. The temperature profile in the flame can be estimated in a similar way. We combine equation 4.2 with the Ideal Gas Law, which gives an expression for the flame pressure P:

$$P = \frac{\sigma RT}{M} \quad \text{N m}^{-2}, \tag{4.14}$$

where M = molecular weight (kg mol^{-1}), to be estimated as an average from a knowledge of the flame composition at various z-values.

So we have:

$$\frac{\sigma_0 T_0}{M_0} = \frac{\sigma T}{M}, \tag{4.15}$$

and combining this with equation 4.2 gives:

$$T = T_0 \frac{Mva}{M_0 v_0 a_0} \quad \text{K} \tag{4.16}$$

For the propane–air flames examined, with $T_0 \approx 300$K, this gives $T \approx 1300$K in the center of the luminous zone, rising to 2000K in the post-luminous mantle. The quantity $\mathrm{d}T/\mathrm{d}z$ is always positive and in quasi-one-dimensional flames the profile with respect to the orthogonal coordinates is expected to be almost flat.

4.3.2 Heat release rates [1]

- **Calculation of the heat release rate from the flame equations**

The heat release rate is:

$$Q = \sum H_i K_i, \tag{4.17}$$

and in obtaining heat release rates we have to return to the species continuity and energy-balance equations 4.8 and 4.13. We define a fractional mass flux G_i of species i due to both gas flow and diffusion as:

$$G_i = \frac{N_i M_i (v + V_i)}{\sigma v}. \tag{4.18}$$

Clearly $$\sum G_i = 1. \tag{4.19}$$

By means of this substitution the species continuity equation 4.8 becomes:

$$\frac{\sigma_o v_o}{M_i}\frac{\mathrm{d}G_i}{\mathrm{d}z} = K_i, \tag{4.20}$$

and the heat-balance equation 4.13 becomes:

$$\sigma_o v_o \sum \frac{H_i G_i}{M_i} - k\frac{\mathrm{d}T}{\mathrm{d}z} = \sigma_o v_o H'_\infty. \tag{4.21}$$

We can now differentiate equation 4.21 to give:

$$\frac{\mathrm{d}}{\mathrm{d}z}\left\{\sigma_o v_o \sum \frac{H_i G_i}{M_i} - k\frac{\mathrm{d}T}{\mathrm{d}z}\right\} = 0, \tag{4.22}$$

since the term on the right of equation 4.21 does not depend on z. We can apply the product rule for differentiation to give:

$$\sigma_o v_o \left\{\sum \frac{H_i}{M_i}\frac{\mathrm{d}G_i}{\mathrm{d}z} + \sum \frac{G_i}{M_i}\frac{\mathrm{d}H_i}{\mathrm{d}z}\right\} - \frac{\mathrm{d}}{\mathrm{d}z}\left(k\frac{\mathrm{d}T}{\mathrm{d}z}\right) = 0. \tag{4.23}$$

Now: $$\frac{\mathrm{d}H_i}{\mathrm{d}z} = C_i\frac{\mathrm{d}T}{\mathrm{d}z}, \tag{4.24}$$

where C_i = heat capacity (J mol^{-1} K^{-1}).

Substituting equations 4.20 and 4.24 into 4.23 gives:

$$\sum H_i K_i + \sigma_o v_o \frac{\mathrm{d}T}{\mathrm{d}z}\sum \frac{G_i C_i}{M_i} - \frac{\mathrm{d}}{\mathrm{d}z}\left(k\frac{\mathrm{d}T}{\mathrm{d}z}\right) = 0. \tag{4.25}$$

The first term on the left of equation 4.25 is simply the heat release rate Q (see equation 4.17), hence we have:

$$Q = -\sigma_o v_o \frac{\mathrm{d}T}{\mathrm{d}z}\sum \frac{G_i C_i}{M_i} + \frac{\mathrm{d}}{\mathrm{d}z}\left(\mathrm{k}\frac{\mathrm{d}T}{\mathrm{d}z}\right) \qquad \mathrm{W\ m^{-3}}. \tag{4.26}$$

• Comparison with experiment

In the above treatment, $\sum \frac{G_i C_i}{M_i}$ can often be approximated to a single value of the heat capacity C of the mixture (J kg^{-1} K^{-1}) and equation 4.26 then provides a route to experimental determination of heat release rates in one-dimensional or quasi-one-dimensional flames. In very lean flames, C and k can be taken to be those of oxygen, having a dependence on T but not on z. This was the approach taken [4] in the measurement of heat release rates in premixed CO–O_2 flat flames with about 80% (molar basis) oxygen. Thermocouple temperature measurements at various z, and C and k as a function of T from the literature, are input to numerical solution of equation 4.26, giving Q at different z-values. This was found to be ≈ 0.4 MW m^{-3}, close to the onset of luminosity.

4.3.3 Flame velocities

We have so far mainly considered flames stabilized on burners but, as we have already noted, in the absence of a suitable burner a flame will propagate through space. The flame velocity v_o is the speed of the influx cold gases normal to the flame front. For a flat flame stabilized on a burner, v_o will be numerically equal to v_g, the influx speed of the gas. For a conical flame, v_o and v_g will be linked by a trigonometric function of the cone angle.

A simple treatment of flame velocities applicable to flat flames was given over one hundred years ago by Mallard and Le Chatelier and its important conclusions have been confirmed by more recent and more advanced investigations.

If the temperature of ignition is T_{ig}, the linearized temperature gradient along the flame is, with other symbols as previously defined:

$$\text{gradient} = \frac{T_\infty - T_{ig}}{v_o t_R} \qquad \text{K m}^{-1}, \tag{4.27}$$

where t_R has units s. By energy balance we have:

conductive heat flux (W m^{-2}) = rate of enthalpy gain by the following mass (W m^{-2}),

that is:

$$k\frac{T_\infty - T_{ig}}{v_o t_R} = mC_p\,(T_{ig} - T_e), \tag{4.28}$$

where C_p = heat capacity at constant pressure (J kg^{-1} K^{-1})
T_e = gas entry temperature (K).

Making use of equation 4.3 and rearranging gives:

$$v_o = \left[\frac{k}{C_p\sigma_o}\,\frac{T_\infty - T_{ig}}{T_{ig} - T_e}\,\frac{1}{t_R}\right]^{0.5}. \tag{4.29}$$

The group $k/C_p\sigma_o$ is recognizable as the thermal diffusivity (usual symbol α, units m^2 s^{-1}) and it is an extremely useful rule of thumb that flame velocity is directly proportional to the square root of this quantity. If we identify t_R with reaction time, its reciprocal will depend on reaction **rate**, leading to a direct relationship between reaction rate and flame velocity.

4.3.4 Factors affecting flame velocities

It is a common observation that a flame in a cylindrical tube moves more rapidly the larger the tube diameter. Equation 4.29 cannot, without enlargement, account for this. The reason is that the measured rate v' of progression of a flame along a cylindrical tube will equal v_o only if the flame surface across the tube is perfectly flat, a condition rarely even approximated. Physical factors cause the flame surface to be curved and v' and v_o are linked by the relationship:

$$v_o a = v'\pi r^2, \tag{4.30}$$

where r = tube radius (m)
a = flame front area (m^2).

The relationship can be tested experimentally either by use of geometrical shapes (for example, hemispherical, paraboloidal) to describe a surface or by direct determination of a by photographic techniques. Coward and Hartwell [5], with methane–air mixtures and photographic determinations of a, examined equation 4.30 and found it to hold well. Examples of their results are given in Table 4.1.

Table 4.1 Flame speeds of 10% CH_4/air mixtures in horizontal cylindrical tubes (Coward and Hartwell [5])

r (m)	v' (m s^{-1})	a (m^2)	$\frac{\pi r^2 v'}{a}$ (m s^{-1})
0.05	1.11	0.03	0.29
0.025	0.92	0.0066	0.27
0.0125	0.715	0.00126	0.28

The quantity in the fourth column is v_o, seen to be constant within experimental error. Together with the findings described in Section 4.3.1 with respect to flames other than flat ones (for example, conical) this provides support for the view that v_o is a fundamental, shape-independent quantity for flames of particular mixtures.

In considering the effect of pressure on flame velocity it must be remembered that while k and C_p are only weak functions of pressure for gases, σ varies strongly and the reaction rate depends on pressure in ways not easy to predict without kinetic information. Moreover, the presence of a diluent affects the pressure dependence of flame velocity. Equation 4.29 cannot therefore be used to predict general trends of v_o with pressure and our purpose is best served by examining representative experimental results.

In experimental determinations of velocities of laminar methane flames [6] the burner method of determination was used in preference to measurement of propagation rate along a tube. This method requires simply a knowledge of the volumetric gas flow rate and area of a flame stabilized on a burner. Measurements were made for stoichiometric methane–**air** mixtures across the pressure range 0.1 to 40 atm. Across this pressure range the flame velocity declines about tenfold from $\approx$ 50 to $\approx$ 5 cm s^{-1} and this is broadly indicative of the effect of pressure on thermal diffusivity and, in turn (equation 4.29), on flame velocity. Across the pressure range 0.1 to 2 atm, the flame velocity drops by about 40 percent. However, when stoichiometric methane–**oxygen** flames were investigated over this pressure range it was found that there was no pressure dependence of the flame velocity; it was steady at $\approx$ 400 cm s^{-1}. The higher values than with air are of course easily explained in terms of the effect of nitrogen as a diluent. The different pressure dependence of flame velocity is due to the different effects of the released heat in the two flames. With air, almost all the heat released simply raises the temperature of the post-combustion gases (that is, it becomes sensible heat in the gases):

combustion products/nitrogen → combustion products/nitrogen at T_f,

where T_f is the flame temperature. On the other hand, with oxygen, a substantial proportion of the heat released is involved in dissociation processes to form radicals from the products, for example:

$$H_2O \rightarrow OH + H,$$

and the extent of dissociation is pressure dependent. The remainder of the heat released by the reaction becomes sensible heat. Pressure is therefore relevant to thermal diffusivity, reaction rate and extent of dissociation, and hence distribution of the released heat between product heating and dissociation reactions. In the methane–oxygen flames studied [6] the resultant of these influences is an overall independence of flame velocity on pressure.

4.4 Turbulent flames

4.4.1 Introductory principles

Many industrial combustion processes use turbulent conditions and in natural gas utilization the difference between domestic- and industrial-scale use is often that between laminar and turbulent conditions. We have seen how the two are distinguished by the Reynolds number and Figure 4.3 shows the classical results of Damkohler [7] on propane–air mixtures in a tube at various Re numbers. The ratio S_T/S_L, turbulent to laminar flame velocity at the same mixture composition, pressure and tube diameter, is seen to be constant at unity up to Re $\approx$ 2000, normally considered the threshold distinguishing laminar and turbulent flow in a tube. At higher Re values the ratio rises, a high degree of turbulence favoring high velocities. To describe turbulent combustion we need to define a scale and a velocity of turbulence.

Turbulent flow is characterized by flow velocities $u_o \pm u'$ where u' is root mean square (r.m.s.) fluctuation velocity, and by a scale ℓ of the turbulence, units m. These lead to an alternative form of the Reynolds number for use in analyzing turbulent systems, the turbulent Reynolds number Re′:

$$\mathrm{Re}' = \frac{u' \ell \sigma}{\mu}. \tag{4.31}$$

The scale of turbulence reflects eddy diameters. We will first refer to the case where $\ell \ll$ flame dimensions. We have already seen that in laminar flow the flame velocity is proportional to the square root of the thermal diffusivity α (equation 4.29) and if conditions are turbulent the dependence is modified in the following way:

$$\frac{S_T}{S_L} = \frac{\alpha + u'\ell}{\alpha}. \tag{4.32}$$

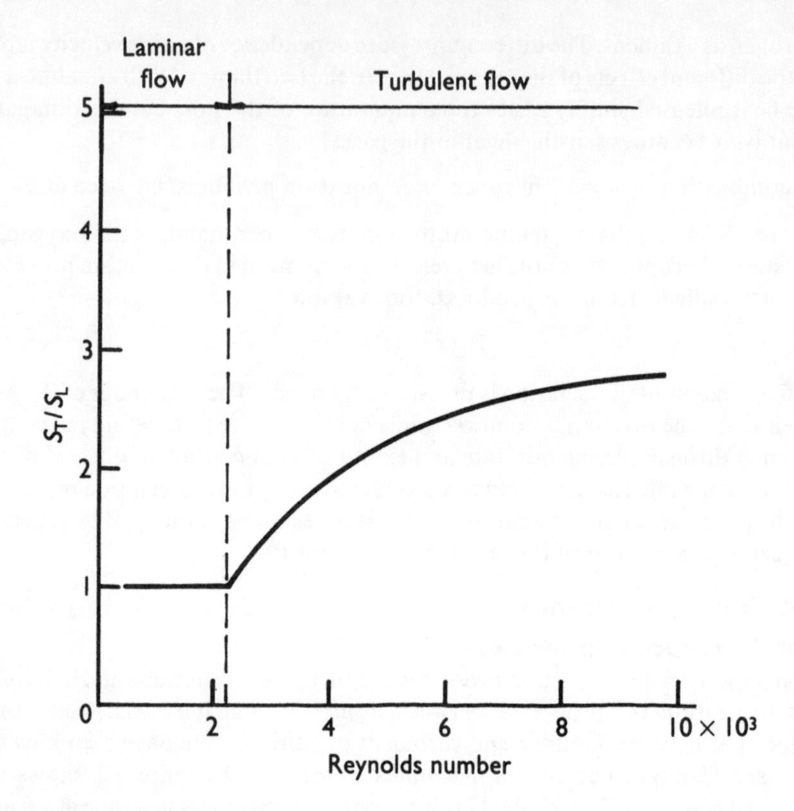

Figure 4.3 Effect of Reynolds number on flame velocity for a propane–air mixture. After Damkohler [7]
(Reproduced with the permission of NASA.)

The effect of small-scale turbulence is to enhance transport of heat and reactive intermediates, hence its positive effect on velocity. Equation 4.32 is one of a number of correlations linking S_T with S_L via u' for turbulent conditions of various specifications. The simplest is:

$$S_T = S_L + u', \tag{4.33}$$

for larger scale turbulence, indicating that under circumstances where this correlation is applicable, $S_T \approx u'$ at high turbulence velocities. Degree of turbulence is often expressed as a percentage:

$$\% \text{ turbulence} = \frac{u'}{u_o} \times 100. \tag{4.34}$$

Another relevant quantity is eddy diffusivity D', defined as:

$$D' = u'\ell \qquad \text{m}^2\ \text{s}^{-1}. \tag{4.35}$$

4.4.2 Experimental measurements on turbulent premixed flames

Experimental methods that can be used in the examination of turbulent flames include:

(a) Creation of turbulence by an assembly of fans in a gas mixture. Grids can also be used to promote turbulence.
(b) Estimation of eddy dimension by photographic techniques.
(c) Measurement of flow rates, turbulence scale and percentages of turbulence by anemometry.
(d) Anchorage of a turbulent flame on a burner and calculation of the flame velocity from the influx gas velocity and the angle between the flame surface and the axis of the burner.

Techniques of this type were brought to bear on premixed stoichiometric propane–air flames in seminal work by Lefebvre et al. [8a, 8b] which aimed to examine the separate factors of percentage turbulence and turbulence scale on S_T, where S_L is 0.448 m s^{-1}. It is instructive to examine these results in detail for the principles of turbulent combustion that they convey. Findings could be grouped into three categories according to experimental conditions, as follows:

(i) $u' < 2S_L$ (actually ≈ 0.27 or ≈ 0.63 m s^{-1}, giving u'/S_L of 0.6 or 1.4); $\ell = (2.8$ to $7.1) \times 10^{-4}$ m; u_o = 9–30 m s^{-1}, 2–9% turbulence.

For either value of u'/S_L, S_T/S_L rises linearly with ℓ under these conditions up to a value of ≈ 6. This effect is attributed to increases in D' with ℓ, causing flame wrinkling and enhanced heat transfer to the unburnt gas. However, since u' is of the same order as S_L, there is, under these conditions, no laceration of the flame. This is evident from photographic records of the flames.

(ii) $u' \approx 2S_L$ (actually ≈ 0.9 or ≈ 1.2 m s^{-1}, giving u'/S_L of 2.0 or 2.7); $\ell = (2.8$ to $25.4) \times 10^{-4}$ m; u_o = 6–43 m s^{-1}, 2–14% turbulence.

In this region S_T/S_L is almost independent of ℓ, being 5 ± 1. Photographic records for flames in this category with ℓ-values differing by a factor of five show no differences in flame structure.

(iii) $u' > 2S_L$ (actually ≈ 2.2, ≈ 2.9 or ≈ 3.7 m s^{-1}, giving u'/S_L of 5.0, 6.5 or 8.2); $\ell = (4$ to $25) \times 10^{-4}$ m; u_o = 13–44 m s^{-1}, 5–14% turbulence.

Under these conditions, S_T was found to bear an inverse relationship to u'.

Correlations were fitted to the experimental results for regions (i), (ii) and (iii) as follows.

In region (i):

$$\left(\frac{S_T}{S_L}\right)^2 = 1 + 2.15(u'\ell)^2. \qquad (4.36)$$

In region (ii): $S_T = 2u'$ (that is, independent of ℓ). (4.37)

In region (iii): $$\left(\frac{S_T}{S_L}\right)^2 = 10^{-4}\,\frac{u'^2}{\ell}, \tag{4.38}$$

that is, in this region, S_T depends inversely on ℓ.

We will revisit turbulent flames in our discussions of natural gas combustion.

4.5 Diffusion flames

4.5.1 Simplified treatment of a laminar diffusion flame [9]

A simple treatment of a laminar diffusion flame, leading to very useful semi-quantitative results, can be developed around the model shown in Figure 4.4 [9]. The air and fuel gas flows have the velocity u_o, which has no r-dependence, and diffusion takes place only perpendicularly to the flow. This treatment describes, for example, a luminous Bunsen burner flame.

The flame height h can be estimated with the supposition that the oxygen profile in the y-direction will vary from 21% at $y = 0$ to zero at $y = h$. Since oxygen supply is by diffusion, at $y = h$ the average distance travelled into the flame by diffusion must be approximately r. We can apply the Einstein relation for diffusion to give:

$$r^2 = 2Dt \quad \text{at } y = h, \tag{4.39}$$

and

$$t = \frac{y}{u_o}, \tag{4.40}$$

that is,

$$y = \frac{u_o r^2}{2D}. \tag{4.41}$$

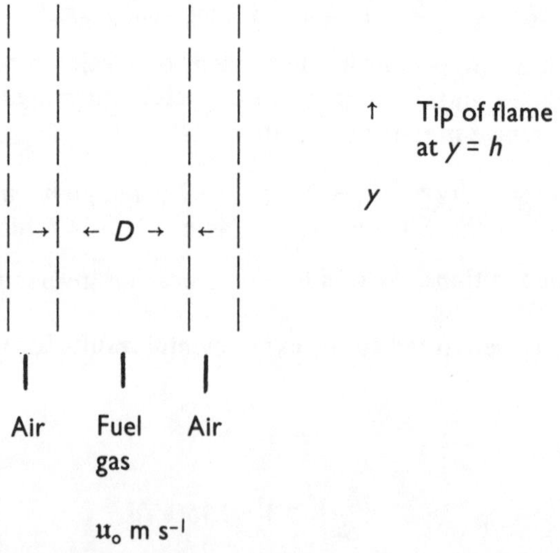

Figure 4.4 Schematic diagram of a co-flow diffusion flame

Now $u_o r^2 = V/\pi$, where V = volumetric flow rate. Therefore:

$$y = \frac{V}{2\pi D}. \tag{4.42}$$

4.5.2 Flame heights

Several useful conclusions can be drawn from equations 4.41 and 4.42. The flame height depends directly on u_o and V, but for given V does not depend on burner diameter. The quantities y and D change in opposite directions and D depends inversely on mean molecular weight. From the theory of diffusion in a binary gas mixture [10] we have the equality:

$$D_{\text{fuel-oxygen}} = D_{\text{oxygen-fuel}}. \tag{4.43}$$

We therefore expect heavier fuel gases to have higher flames. This is borne out by the observation that, other things being equal, a carbon monoxide diffusion flame is about 2.5 times higher than a hydrogen flame.

Testing of this treatment by measurements on a methane–air diffusion flame [11] led to satisfactory agreement between theory and experiment, of which the following is an example. With a burner of radius 1.27 cm and with $u_o = 1.55$ cm s^{-1}, a flame height of 3.3 cm was recorded and D was estimated as 0.5 cm^2 s^{-1}. Putting r, u_o and D into equation 4.41 gives $y = 2.5$ cm, in satisfactory agreement with experiment.

The treatment of diffusion flames summarized in equations 4.41 and 4.42 has been for laminar conditions only. This is likely to be true for a small burner with $u_o \approx 1$ cm s^{-1}, but less likely to apply to industrial scale combustion as the following example illustrates [9]. For $r = 5$ cm and $u_o = 1000$ cm s^{-1}, according to equation 4.41, $y = 250$ m and furnaces are not designed to accommodate flames of this size. The message of this calculation is that in industrial situations the flow must be supplemented by turbulence.

The effect of turbulence on heights of diffusion flames can be appreciated by reference to work on city gas (35% H_2) diffusion flames in air [12]. For a 0.4 inch tube diameter, the volumetric flow rate V was incrementally increased from 0 to ≈ 1400 cm^3 s^{-1}, and the flame height y was measured. Hence at the lower flow rates the flow is laminar, becoming turbulent at the higher flow rates. Since the burner diameter is constant, each increment in the gas flow represents a proportional increment in Reynolds number (equation 4.1). The results are shown in Figure 4.5 where in the low Re region the flame height rises fairly rapidly with V, and at higher Re the dependence of y on V is seen to be much weaker.

4.6 The distinction between flames and detonations

4.6.1 Phenomenological distinction

Our earlier description of a flame as being a combustion reaction self-propagating through space would also be true of a detonation, so in concluding the chapter it is appropriate to explain how to distinguish them. In purely

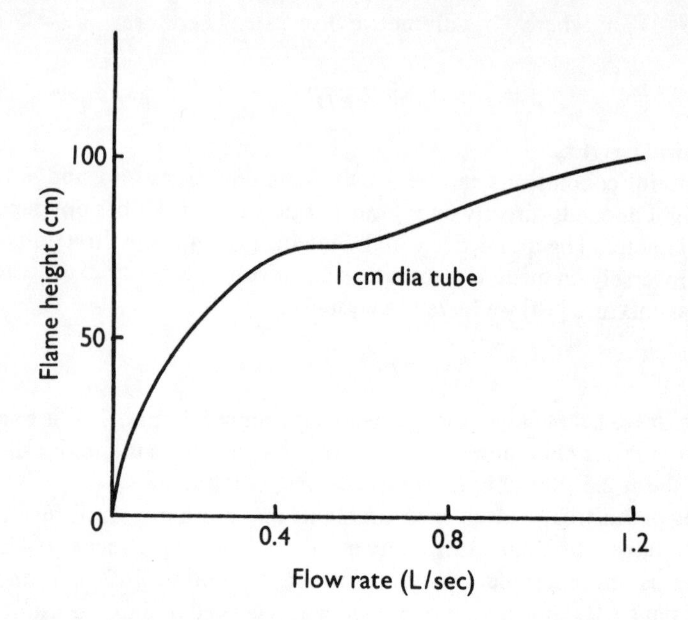

Figure 4.5 Flame height versus volume flow rate for city gas in a 0.4 inch diameter burner (Wohl, Gazley and Kapp [12])
(Reproduced with the permission of the Combustion Institute.)

phenomenological terms it is not difficult since speeds associated with detonations are much higher than those associated with flames, being in excess of 1000 m s^{-1}, that is, supersonic. Detonations are associated with shock waves of gas and their behavior is governed by energetics and the principles of rapid gas compression, but not by rates of heat transfer by conduction or diffusion as in a flame, propagation of which is subsonic.

4.6.2 Outline of the fundamentals of detonations

When a nonreacting gas is rapidly compressed, mass, energy and momentum conservation lead to the relationship [13]:

$$E_1 - E_o = \frac{1}{2}(V_o - V_1)(P_o + P_1), \qquad (4.44)$$

where E = energy (J kg^{-1})
V = specific volume (m^3 kg^{-1})
P = pressure (N m^{-2}),

and the subscripts o and 1 denote, respectively, the gas before and after compression.

Conservation laws also give the speed C of propagation as:

$$C = V_o \left[\frac{P_1 - P_o}{V_o - V_1}\right]^{0.5} \quad \text{m s}^{-1}. \qquad (4.45)$$

There is no energy transfer to the surroundings in this model, so conditions are adiabatic and the equation is known as the Hugoniot adiabatic. If the gas experiencing compression is exothermically reacting, equation 4.44 becomes:

$$E_1 - E_o = \frac{1}{2}(V_o - V_1)(P_o + P_1) + Q, \tag{4.46}$$

where Q = energy released by the reaction (J kg^{-1}).

The change of P and T may lead either to a flame or to a detonation, the latter being by far the more vigorous, with speeds in excess of 1000 m s^{-1}. The Hugoniot adiabatic provides a basis for deducing whether a flame or a detonation will result from the compression. Since the change is adiabatic, we can put:

$$E_1 - E_o = \frac{P_1V_1 - P_oV_o}{\Phi - 1}, \tag{4.47}$$

where $\Phi = \dfrac{\text{specific heat at constant pressure}}{\text{specific heat at constant volume}}$.

Then equation 4.44, for a nonreacting gas undergoing compression, becomes:

$$\cap = \frac{\epsilon(\Phi + 1) - (\Phi - 1)}{(\Phi + 1) - \epsilon(\Phi - 1)}, \tag{4.48}$$

where $\cap = \dfrac{P_1}{P_o}$ and $\epsilon = \dfrac{V_o}{V_1}$, the dimensionless pressure rise and the compression ratio respectively. Since a shock wave is characterized by an extremely sharp pressure rise, an approximate criterion for its occurrence can be obtained by setting $\cap = \infty$, requiring:

$$(\Phi + 1) - \epsilon(\Phi - 1) = 0. \tag{4.49}$$

For an ideal gas, $\Phi = 1.67$, so equation 4.48 yields $\epsilon = 4$ as the criterion for a shock wave. The equation corresponding to (4.48) for the reacting gas contains a second term due to the exothermicity:

$$\cap = \frac{\epsilon(\Phi + 1) - (\Phi - 1)}{(\Phi + 1) - \epsilon(\Phi - 1)} + Q\frac{2\epsilon(\Phi - 1)}{P_oV_o(\Phi + 1) - \epsilon(\Phi - 1)}, \tag{4.50}$$

which can also be written:

$$\frac{P_1V_1 - P_oV_o}{\Phi - 1} = \frac{1}{2}(V_o - V_1)(P_o + P_1) + Q, \tag{4.51}$$

and equation 4.45 holds for the reacting gas as it does for the nonreacting gas. These equations provide a framework within which Figure 4.6, the Hugoniot adiabatic diagram for the reacting gas, can be understood.

Point A on Figure 4.6 is P_o, V_o and the case where the product gas falls at or below point F on the Hugoniot adiabatic will first be considered. This

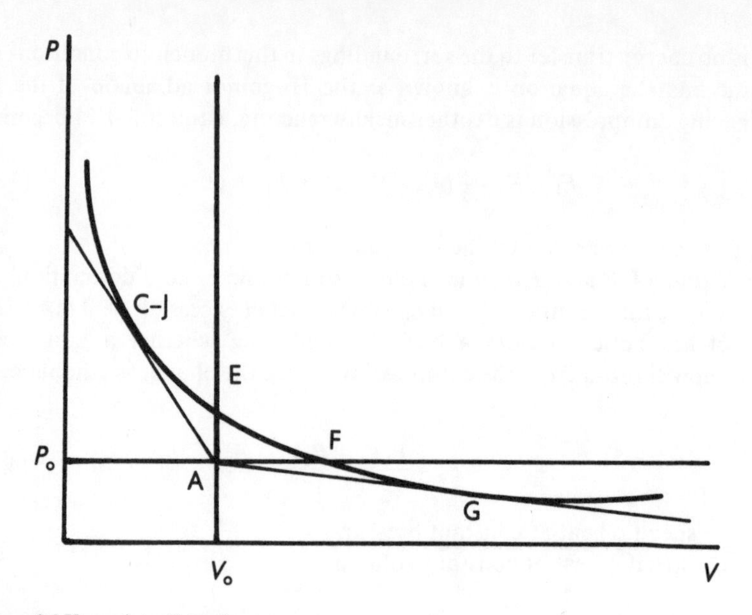

Figure 4.6 Hugoniot adiabatic

operation has resulted in increased volume at reduced pressure. A line from A to any point on the adiabatic below F has a slope of:

$$\frac{P_1 - P_o}{V_o - V_1} < -\frac{dP}{dV},$$

where the subscript 1 denotes the point of intersection of the adiabatic and the differential on the right is the slope of the adiabatic at the point of intersection.

It follows from the above inequality that:

$$V_o\left[\frac{P_1 - P_o}{V_o - V_1}\right]^{0.5} < V_o\left[-\frac{dP}{dV}\right]^{0.5}.$$

Now, from equation 4.45 the left-hand side is the propagation speed. The right-hand side is **by definition** the speed of sound under the prevailing conditions. It therefore follows that compression from P_o, V_o to a point on the adiabatic at or below F will result in subsonic propagation, an ordinary flame whose behavior is subject to restrictions by heat transfer. This propagation is sometimes called deflagration.

A hypothetical compression operation taking the system from point A to a point on the adiabatic between E and F would result in increased pressure **and** increased volume. This is physically impossible and would lead to imaginary solutions to equation 4.45. This part of the adiabatic is therefore not experimentally achievable.

We now consider cases where the product gas falls on a point in part E to C–J of the adiabatic; C–J, the Chapman–Jouguet point, is named after the original investigators. Any line from A to this part of the adiabatic will have a slope greater than the adiabatic and therefore by arguments analogous to that above, it follows that the propagation is supersonic. The limiting case at C–J is where the line joining A and C–J and the adiabatic are tangential and the propagation speed is equal to the speed of sound. Any experiment taking the system to a point on the part of the adiabatic between E and C–J will lead to a stable detonation. A notional operation taking the system from initial state A to any point to the left of E fulfills the requirement of volume drop in response to pressure rise and is physically realistic in that respect, but only the region E to C–J signifies stable detonations so the C–J point represents the minimum propagation speed of detonation for the system under consideration.

The Chapman–Jouguet condition can be expressed as:

$$C \geq U + U_s,$$

where C = speed of the detonation (see Equation 4.45)
U = speed of the product gases
U_s = speed of sound in the product gases,

and the equality is the tangency condition. This condition is not fulfilled at points on the Hugoniot adiabatic to the left of C–J.

4.6.3 Comparison with experiment

In interpreting experimental results according to the Hugoniot adiabatic we need to take the theory further in the following way [13]. The compression is adiabatic and if it is therefore taken to be approximately isentropic, the following relation holds:

$$\frac{dP}{dV} = -\frac{\Phi P_1}{V_1}. \tag{4.52}$$

Applying this to the C–J point and combining with equation 4.45 gives:

$$C = V_o \left[\frac{\Phi P_1}{V_1}\right]^{0.5}. \tag{4.53}$$

Utilizing the fact that the Ideal Gas Equation applies to the gas in its reacted state, and recalling that $\epsilon = V_o / V_1$, gives:

$$C = \epsilon [\Phi n_1 R T_1]^{0.5}, \tag{4.54}$$

where n_1 denotes mole number.

Equation 4.52 at the C–J (tangency) point is equivalent to:

$$\frac{P_1 - P_o}{P_1} = \Phi \frac{V_o - V_1}{V_1}, \tag{4.55}$$

which rearranges to:

$$1 - \frac{P_o}{P_1} = \Phi(\epsilon - 1). \tag{4.56}$$

Application of the Ideal Gas Law to the system before and after reaction, and rearranging, leads to:

$$\Phi\epsilon^2 - \epsilon(\Phi + 1) + \frac{n_o R T_o}{n_1 R T_1} = 0. \tag{4.57}$$

Now that T_o and T_1 have been introduced into the analysis via the ideal gas substitutions, equation 4.46 can be rewritten:

$$\begin{aligned} c(T_1 - T_o) &= \frac{1}{2}(V_o - V_1)(P_o + P_1) + Q \\ &= \frac{1}{2}(\epsilon - 1)[n_1 R T_1 + (n_o R T_o/\epsilon)] + Q, \end{aligned} \tag{4.58}$$

where c = heat capacity of the product gases (J kg^{-1} K^{-1}). Equations 4.57 and 4.58 provide a route to calculation of the quantities ϵ, n_1 and T_1, and hence, via equation 4.54, the detonation propagation speed. This can then be compared with experimental values. Accurate measurement of the detonation propagation speeds has been possible for over a century [14] by electrical timing devices; photographic techniques have complemented them in more recent times.

Jouguet [15] calculated the detonation propagation rate for stoichiometric hydrogen/oxygen mixtures by first omitting terms containing ϵ in equation 4.58 and putting c as a power series in T_1. This simplified form of equation 4.58 was then solved for T_1, and the value for T_1 inserted into equation 4.57, Φ being straightforward to estimate reliably. The resulting value of ϵ was put back into equation 4.58, together with the power series for c, and an improved T_1 was obtained, which by means of substitution in equation 4.57, led to an improved ϵ, and so on, until two successive iterations produced indistinguishable results. This led to a value of 2629 m s^{-1} for the propagation velocity, against an experimental value of 2820 m s^{-1}, which is very encouraging agreement.

In the development of the equations for calculating the detonation propagation speed we have taken the tangency condition, that is, the C–J point. This is because of the experimental fact that a particular mixture at particular initial conditions has a single value of the propagation speed of stable detonation, often referred to as the C–J speed.

4.6.4 Conclusion

In summary, flames and detonations are distinguished by subsonic speeds in the former, and supersonic in the latter, and whereas a particular mixture at particular initial conditions has a single detonation propagation rate, such rates for flames depend on conditions affecting heat transfer. Density drops on deflagration but rises on detonation, as became apparent in our examination of Figure 4.6. The fundamental distinction between the two can be understood by reference to the Hugoniot adiabatic, and regions of the adiabatic for one or the other are separated by a region with no physical fulfillment. Flames and detonations are sometimes referred to as *combustion waves* and a third type of combustion wave is *smoldering* which will be considered in Chapter 11. We will return to detonations when discussing high explosives.

References

[1] Fristrom R. M., Westenberg A. A., *Flame Structure*, New York: McGraw-Hill (1965)
[2] Holman J. P., *Heat Transfer*, SI Metric Edition, New York: McGraw-Hill (1989)
[3] Fristrom R. M., 'Flame zone studies. II, Applicability of one-dimensional models to three-dimensional laminar Bunsen flames'. *Journal of Chemical Physics* **24** 888 (1956)
[4] Friedman R., Nugent R. G., 'Flame structure studies. IV Premixed carbon monoxide combustion', *Seventh Symposium (International) on Combustion*, 311, London: Butterworths (1959)
[5] Coward H. F., Hartwell F. J., 'Studies in the mechanism of flame movement. I, The uniform movement of flame in mixtures of methane and air in relation to the tube diameter', *J. Chem. Soc.* 1996 (1932).
[6] Diederichsen J., Wolfhard H. G., 'The burning velocity of methane flames at high pressure', *Transactions of the Faraday Society* **52** 1103 (1956)
[7] Damkohler G. (1939), cited in: Beer J. M., Chigier N. A., *Combustion Aerodynamics*, London: Applied Science Publishers (1972)
[8] (a) Lefebvre A. H., Reid R., 'The influence of turbulence on the structure and propagation of enclosed flames', *Combustion and Flame* **10** 355 (1966). (b) Ballal D. R., Lefebvre A. H., 'Turbulence effects on enclosed flames', *Acta Astronautica* **1** 471 (1974)
[9] Jost W., *Explosion and Combustion Processes in Gases*, (Translated by H. O. Croft), New York: McGraw-Hill (1946)
[10] Geankoplis C. J., *Transport Processes—Momentum, Heat and Mass*, Boston: Allyn and Bacon (1983)
[11] Burke S. P., Schumann T. E. W., 'Diffusion flames', *Industrial and Engineering Chemistry* **20** 998 (1928). Reprinted 1964 in the retrospectively published *Proceedings of the First Symposium on Combustion*, Pittsburgh: The Combustion Institute.
[12] Wohl K., Gazley C., Kapp N., 'Diffusion Flames', *Third Symposium (International) on Combustion*, 288, Baltimore: Williams and Wilkins (1949)
[13] Sokolik A. S., *Self-ignition, Flame and Detonation in Gases*, Translated by N. Kaner, Jerusalem: Israel Program for Scientific Publications (1963)
[14] Berthelot M., Vielle P., cited in: Lewis B., von Elbe G., *Combustion, Flames and Explosions in Gases*, New York: Academic Press (1961)
[15] Jouguet E., *Méchanique des Explosifs*, Paris, 1917 (cited by Sokolik [13])

CHAPTER 5

Gaseous Fuels

Abstract

Several fuel gases of practical importance are discussed in terms of their combustion properties. Producer gas is the first to be described and detailed accounts of its use in transport and in process heating are given, as well as a quantitative discussion of the feasibility of its use in electricity generation. Coke-oven gas, which has about 4 times the heat value of producer gas, is described, as well as a number of other manufactured gases which have about the same heat value as coke-oven gas. The discussion of natural gas is used as a background against which to present some important principles of gas burner operation, including the possible interchangeability of gases on a particular burner.

5.1 Introduction

Large-scale use of **manufactured** fuel gas dates from early nineteenth-century England when gas from the carbonization of coal was used in lighting. Coal carbonization plant had been used in the previous century, but the primary product was coke, of which gas was only a by-product. A milestone in the development of industrial fuel gases was the operation of the first gas producer in Germany in about 1840. Incentives for conversion of solid fuels to gaseous fuels include:

(a) Relative ease of transportation of gas in a pipe.

(b) Freedom of gaseous fuels from ash.

Significant **natural** gas usage dates from about 1870. Gaseous fuels are classified mainly on the basis of their heat values from producer gas (low heat value) through town gas to natural gas; natural gas has a heat release per kg substantially exceeding that of coal or petroleum products. Even higher heat values are obtainable from refinery gas or LPG. Nevertheless, low-heat-value gases such as producer gas have served industry well for several generations and deserve attention. In this summary of gaseous fuels we treat them in ascending order of heat value, starting with producer gas.

5.2 Producer gas

5.2.1 Composition and heat value

Producer gas is made by passing air or (more commonly) air and steam through a bed of hot coal or coke, or possibly cellulosic material, for example, coconut waste. Table 5.1 gives typical properties of producer gas manufactured from coke [1].

Table 5.1 Properties of producer gas

Composition (% molar)		*Density* (air = 1)	*Heat value*
CO_2	4–6	0.9	135 BTU ft^{-3}
O_2	0.4		≈ 5 MJ m^{-3}
CO	28–30		≈ 4.5 MJ kg^{-1}
H_2	10–12		
CH_4	0.4–0.6		
N_2	53–56		

Note: Unit volume in the heat value is at 30 inches of mercury (101.4 kPa) pressure and 60° F (15.6° C).

The effect of the nitrogen is to cause a low heat value, although the adiabatic temperature of a producer-gas flame is not greatly lower than that of a natural-gas flame. Whereas the adiabatic flame temperature of a natural gas/air mixture is about 2000° C, values for producer gas, depending on precise composition, are also of this order. In practical systems, producer gas/air mixtures can achieve flame temperatures in excess of 1500° C. At first consideration this is surprising in view of the low heat value. In fact, up to about 10% inert gases, many fuel gases have almost the same theoretical flame temperature when burnt in air. The reason can be understood by reference to the quotient:

$$\frac{\text{heat released}}{\Sigma(\text{mole number} \times \text{specific heat of products})},$$

on which the flame temperature depends. In simple terms, increased heat release is offset by increased thermal capacity of the post-combustion gases. The effect of the diluent nitrogen in producer gas is to increase the heat capacity of the post-combustion gases by only about 35% with respect to the combustion products of the equivalent nitrogen-free gas, hence its effect on temperature is correspondingly small.

Users of producer gas include metallurgical and ceramics industries. The proportion used in power generation is small. The exigencies of World War 2 led to considerable research and development into the use of producer gas for land transport. Some of the important features of producer-gas combustion can be understood by reference to this research program.

5.2.2 Exploratory use in land transport [2]

Trials were made of a small gas producer mounted on a motor lorry, whose performance was tested over 107-mile journeys. Gasification of the solid fuel by steam and air provided a gaseous fuel for the engine. Several fuels—anthracite coals, cokes, briquettes—were tested for suitability as feedstock and a number of gasifier designs were investigated. Performance criteria included:

(a) Calorific value of the gas from the mobile producer.
(b) Average speed over the 107-mile test course.
(c) Water usage rate.
(d) Quantity of ash remaining.

Table 5.2 gives data for one of the fuels:

Table 5.2 Performance of a one-ton lorry using producer gas [2]

Gasifier feedstock	Coal carbonized at 600° C*
Heat value of the gas	5 MJ m^{-3}
Fuel (feedstock) consumption rate	0.91 lb $mile^{-1}$ (≡ 3.9 km kg^{-1})
Average speed	34 km h^{-1}
Water consumption	0.17 kg per kg fuel
Weight of ash after 107-mile trial	0.113 kg

*Semi-coke in modern terminology (see Chapter 6)

The feedstock fuel had a heat value of ≈ 30 MJ kg^{-1}, so the figure in the third row of the table converts to 3.9 km per 30 MJ or 0.13 km MJ^{-1}. If we estimate that a one-ton lorry will achieve 15 miles per gallon (≈ 3.3 kg) of a petroleum-based fuel (heat value ≈ 46 MJ kg^{-1}), the corresponding figure for the more conventional fuel is 0.16 km MJ^{-1}. Efficiency in these terms is superior for gasoline and is expected to be because gasification is largely partial combustion and therefore inevitably involves loss of part of the heating value. The primary flammable constituent of producer gas is carbon monoxide, formed by:

$$C + \frac{1}{2}O_2 \rightarrow CO,$$

which is exothermic and accounts for ≈ 9 MJ kg^{-1} of the total 30 MJ kg^{-1} in principle obtainable from the coke. This exothermicity keeps the fuel bed at the required temperature for gasification but represents energy lost in terms of utilization of the gasification product.

5.2.3 Notes on producer gas combustion and utilization

Producer gas has a density close to that of air (Table 5.1). This, and the high heat capacity, cause producer gas flame propagation to be fairly slow (see equation 4.29). Laminar producer gas/air flames in a 1-inch tube [3] have a maximum flame velocity depending on fuel-to-air ratio of 2 ft s^{-1}. Coal-derived gases abundant in hydrogen (see Section 5.3) are capable of velocities of 5 ft s^{-1} under these conditions, and hydrogen itself, a velocity of 16 ft s^{-1}.

Producer gas of the composition given in Table 5.1 requires ≈ 9 mol oxygen for combustion of 1 m^3, which is close to the number of moles of oxygen present in 1 m^3 of air at the same temperature and pressure. Stoichiometric producer gas/air combustion therefore requires roughly equal volumes of air and gas. A common design of producer gas burner [1] admits gas through a central port and an equivalent volume of air through multiple smaller ports. The combustion rate of gas in these burners is typically 200 000 ft^3 (5660 m^3) h^{-1}. We can estimate the heat release rate q from a burner of this specification using producer gas of 5 MJ m^{-3} as:

$$q = \frac{5 \times 5660}{3600} \text{ MW} \tag{5.1}$$
$$= 8 \text{ MW}.$$

Loss of part of the energy of the coal due to the exothermicity of the gasification process itself has already been noted and for simple air-blown producer gas from coke, likely to contain little hydrogen and methane from devolatilization, the ratio ϕ of MJ in the gas to MJ in the coke feedstock is simply:

$$\phi_p = \frac{\text{heat of reaction of } CO + \frac{1}{2}O_2 \rightarrow CO_2}{\text{heat of reaction of } C + O_2 \rightarrow CO_2} \tag{5.2}$$
$$= \frac{282 \text{ kJ mol}^{-1}}{393 \text{ kJ mol}^{-1}} = 0.72.$$

Modern gas producers utilizing coal and air/steam are capable of ϕ_p values of about 0.77 and this was the value used in studies by the United Aircraft Corporation [4] into the feasibility of 1000 MW power generation with producer gas. Electrical efficiencies ϕ_e, defined as:

$$\phi_e = \frac{\text{electrical energy generated}}{\text{energy content of the fuel}}, \tag{5.3}$$

depend on the generating plant, and that proposed to be used with the producer gas had values on the basis of the producer gas of 45 percent, so on the basis of the original coal, 0.77 × 45 = 35 per cent. In fact, electricity generation direct from coal usually has a ϕ_e value in the 35–38 percent range, so improved plant capable of obtaining 45 percent makes electricity from producer gas quite viable. Even with conventional plant, ϕ_e using producer gas approaches 30 percent provided that good values of ϕ_p are obtained.

While 5 MJ m^{-3} is the expected heat value of producer gas, it should be remembered that this refers to clean gas, free from soot and tar. In applications where it is not necessary to remove these contaminants they obviously contribute to the heat value, and values of ≈ 8 MJ m^{-3} are then possible. Such a gas will have a slight yellow color.

Blast-furnace gas is a low-heat-value gas sometimes regarded as a special case of producer gas, with heat values 3.5 – 4 MJ kg^{-1}. It is a by-product of iron production, whereas ordinary producer gas is a primary product and this is clearly economically important. The heat value is certainly low and performance improvement in combustion plant using blast-furnace gas can be effected by:

(a) pre-heating influx gases to the burner; or
(b) enrichment of the gas with organic material, for example, creosote vapor, yielding a gas of up to 9 MJ m^{-3}.

5.3 Utilization of manufactured gases of $\approx$ 20 MJ m^{-3} heat value

5.3.1 Introduction

Several gases fall into this category and they are outlined in Table 5.3. *Town gas* is of course a broad term for gas supplied to domestic and commercial users.

5.3.2 Relative combustion properties of these gases and producer gas

This comparison can be understood by reference to a classical investigation of the relative merits of coke-oven gas and producer gas in the steel industry [5]. The important combustion properties are compared point by point.

- **Air requirement and total influx gas**

For producer gas of the composition given in Table 5.1 the oxygen requirement is 1.68 mol per MJ released, or 8.0 mol air. The total influx gas required for release of 1 MJ is given by:

$$\text{influx gas} = \text{fuel} + \text{air}, \tag{5.4}$$

and recalling that 1 m^3 of gas at room temperature contains approximately 42 mol, equation 5.4 becomes:

$$\begin{aligned}\text{influx gas} &= \frac{42}{5}\ \text{mol gas MJ}^{-1} + 8.0\ \text{mol air MJ}^{-1}\\ &= 16.4\ \text{mol MJ}^{-1}\ \text{total gas.}\end{aligned}$$

For coke-oven gas of the composition given in Table 5.3 the oxygen requirement is 1.86 mol per MJ released, or 8.9 mol air. Equation 5.4 in this case becomes:

$$\begin{aligned}\text{influx gas} &= \frac{42}{20}\ \text{mol gas MJ}^{-1} + 8.9\ \text{mol air MJ}^{-1}\\ &= 11.0\ \text{mol MJ}^{-1}\ \text{total gas,}\end{aligned}$$

that is, coke-oven gas combustion requires about one-third less incoming total gas (molar basis) per MJ than producer gas.

Table 5.3 Manufactured gases of ≈ 20 MJ m^{-3} heat value

Gas	*Typical composition (molar basis)*	*Method of manufacture*	*Application example*
Retort coal gas	H_2 50% CO 8–15% Hydrocarbons including methane: 30–35% CO_2 1 - 4% O_2 < 1% N_2 4 - 7%	Carbonization of coal at 950–1150°C	Town gas
Coke-oven gas	H_2 55% CO 5 - 6% CH_4 28% Other hydrocarbons: 2–3% CO_2 2% O_2 < 1% N_2 5–6%	By-product of coke manufacture	Process heating
Carburetted water gas	H_2 35% CO 34% CH_4 15% C_2H_4 13% N_2 2% CO_2 1%	Steam gasification of coke and blending with cracked petroleum material	Town gas
Ammonia (15 MJ m^{-3})	NH_3	Synthesis gas	Land transport

- **Post-combustion emissions**

Using the methods of stoichiometry it is easily shown that the producer gas will give rise to the following quantities of gas, expressed on a dry basis (water assumed to have condensed), on combustion:

CO_2 (from CO combustion + amount initially present)	2.8 mol MJ^{-1}
N_2 (from producer gas + air)	10.8 mol MJ^{-1}
Total *dry* gas	13.6 mol MJ^{-1}

The corresponding figures for coke-oven gas are as follows:

CO_2 (from CO and hydrocarbon combustion* and amount initially present)	0.8 mol MJ^{-1}
N_2 (from gas + air)	7.1 mol MJ^{-1}
Total *dry* gas	7.9 mol MJ^{-1}

Hence much more effluent on a dry basis is produced by the low-heat-value gas. The factor of three in the CO_2 emissions should be noted.

Approximate water yields per MJ of heat released are 2.3 mol for coke-oven gas and 1.1 mol for producer gas. The above calculations are on the basis of:

$$\text{producer gas or coke-oven gas} + \text{air} \rightarrow CO_2 + H_2O\ (\text{liquid}).$$

The heating values incorporate the exothermic process:

$$H_2O\ (\text{vapor}) \rightarrow H_2O\ (\text{liquid}).$$

Such heat values are termed gross values or, in the older literature, higher values. The net or lower values are adjusted to allow for the entire product water in the vapor phase, and clearly the higher the carbon:hydrogen elemental ratio of any fuel, the smaller is the difference between the gross and net values.

In a practical combustion system, heat can be lost owing to removal of water from the system as vapor and the actual heat released may lie between the gross and net values. Returning to our comparison of producer-gas and coke-oven-gas combustion, there is more potential for loss of this type with coke-oven gas.

- **Intensity of heating**

Intense heating, meaning a rapid rate of heat release (**not** an unusually high flame temperature), is required in industrial melting processes, and coke-oven gas has an advantage over producer gas for this purpose. Coke-oven gas is sometimes used to bring an ingot of metal close to its melting temperature quickly, whereupon producer gas is used to continue the process. Even without full quantitative comparison of heat release rates along the lines outlined in Chapter 4, we can identify the properties of coke-oven gas that cause it to display an intense flame.

The thermal conductivity of gases has an inverse relationship to molar weight**, about equal to that of air for producer gas and about a factor of two lower for coke-oven gas. Hence coke-oven gas has a significantly higher thermal conductivity, but the molar heat capacities do not differ as much. This is because the heat capacities of N_2, H_2 and CO (all diatomics) are very close to each other across a wide temperature range. That of CH_4 is somewhat higher and for this reason the coke-oven gas is likely to have a slightly higher molar heat capacity. Perhaps most importantly, coke-oven gas has a faster reaction

*Approximating the C_{2+} hydrocarbons to ethane, C_2H_6.

**Molar weight for gas mixtures such as coke-oven gas means the weight of one Avogadro number of molecules.

rate with oxygen because of its high hydrogen content. All of these factors lead to a high flame velocity for coke-oven gas, which in turn is consistent with a high heat release rate and choice of the gas as fuel in melting processes.

5.3.3 Ammonia as fuel

Ammonia as a fuel has a gross specific energy of approximately 15 MJ m^{-3}, hence it is appropriate to deal with it in this section. It has the advantage that it can be manufactured from coal feedstock via synthesis gas, or the hydrogen required can be obtained from electrolytic gas (see Chapter 3). This explains the use of ammonia as a civilian land transport fuel during World War 2. The combustion reaction is:

$$2NH_3 + \frac{3}{2}O_2 \rightarrow N_2 + 3H_2O(l),$$

for which the exothermicity is 15.3 MJ m^{-3} ammonia.

A clue can be gained from the above stoichiometry to one of the most remarkable features of ammonia combustion: its extremely high minimum ignition energy (ϵ^*). We have seen that ϵ^* values are usually a fraction of 1 mJ for hydrocarbons in air, the precise value depending on the fuel:air ratio and on the shape of the heat source. Ammonia/air mixtures require $\approx$ 10 mJ and the reason is that dissociation to form hydrogen is required:

$$NH_3 \rightarrow \frac{1}{2}N_2 + \frac{3}{2}H_2.$$

This is endothermic to the extent of $\approx$ 2 MJ m^{-3} ammonia and heat must come from the source to create sufficient hydrogen for initiation. This is easily verifiable by partial substitution of nitrogen and hydrogen according to the dissociation stoichiometry for ammonia and measurement of ϵ^* compared with the value for ammonia/air only, under otherwise identical conditions. Table 5.4 summarizes results so obtained using spark ignition.

Table 5.4 ϵ^* for ammonia and partially dissociated ammonia in air [6] under stoichiometric conditions

Percent substitution	ϵ^* (mJ)
0	10
5.6	4
14.1	1
28.0	0.2

The dependence of ϵ^* on hydrogen gas initially present is very strong, in accord with the view that the high value for ammonia/air is due to the need to make hydrogen by dissociation.

Ammonia/air flames are also characterized by low flame velocities, typically 15 cm s^{-1} in laminar tube propagation, whereas, other things being equal, methane/air flames have velocities of about 100 cm s^{-1}. Methane and ammonia have very similar molecular weights, thermal conductivities, densities and heat capacities. Methane has the higher exothermicity but, according to the treatment in Chapter 4, this affects the flame speeds only rather weakly via T_∞. The difference in flame velocities is therefore related to reactivity. Initiation in methane oxidation is via the reaction [7]:

$$CH_4 + O_2 \rightarrow CH_3 + HO_2,$$

followed by steps including:

$$CH_3 + O_2 \rightarrow CH_2O + OH$$

$$CH_2O + O_2 \rightarrow CHO + HO_2$$

$$CHO + O_2 \rightarrow CO + HO_2.$$

However, according to flame spectrometric studies [8] the analog of the methyl in ammonia oxidation, NH_2, instead of reacting with oxygen is further stripped of hydrogen atoms to form first NH and H. This is followed by formation of molecular hydrogen, the primary fuel reactant in the hottest part of the flame, and molecular nitrogen, which is an end product. It is not difficult to understand the different velocities of methane/air and ammonia/air flames, in spite of the physical similarities of the two fuels when these mechanistic differences, and their effect on reaction rate, are noted.

5.4 Utilization of natural gas

5.4.1 Introduction

Natural gas is really a generic term, and although the dominant constituent is almost always* methane, the composition varies considerably from one source to another . A fairly typical example having 90% methane (molar basis), 5% ethane, and the balance non-combustibles, has a gross heat value of 37.5 MJ m^{-3} and a density relative to air of 0.6. In many places, natural gas succeeded town gas of $\approx$ 20 MJ m^{-3} for domestic use and in outlining relative properties of the two it is necessary to introduce a quantity widely used in fuel gas technology, termed the Wobbe index, Wo:

$$Wo = \frac{\text{gross heat value}}{[\text{density relative to air}]^{0.5}} \text{ MJ m}^{-3}. \tag{5.5}$$

The importance of the Wobbe index is that rate of delivery of a gas through an orifice is inversely proportional to the square root of its density. The rate of supply of fuel heat value to burner, sometimes called the *thermal delivery*, is therefore proportional to the Wobbe index. However, if two gases have the same value of Wo, that does not *necessarily* mean that they can be supplied interchangeably to a particular burner, since equality of Wo is no guarantee

*Examples with a preponderance of ethane C_2H_6 are not unknown, for example, a source in Kentucky, USA, has 69.7% ethane and the correspondingly very high heat value of 58 MJ m^{-3} [9].

that the air requirements will be the same. For example, pure carbon monoxide and a 20 percent mixture of ethene (C_2H_4) in nitrogen have precisely equal Wobbe numbers, since nitrogen, carbon monoxide and ethene have the same molar weight, and the exothermicity of ethene combustion per mole is five times that of carbon monoxide combustion. However, air requirements are 10.2 mol MJ^{-1} and 8.5 mol MJ^{-1} for the ethene/nitrogen mixture and pure carbon monoxide respectively. Nor does the Wobbe index relate to flame velocities. In gas supply, attention has to be paid both to Wobbe index values and to flame velocity for consistently satisfactory combustion in appliances [10].

5.4.2 Ignition characteristics

The Wobbe indices of a typical natural gas and a typical coal-derived gas of ≈ 50% hydrogen are given in Table 5.5, together with other combustion properties.

Table 5.5 Comparative combustion properties of natural gas and retort coal gas

	Wo (MJ m^{-3})	v_0* (m s^{-1}) in a 1-inch tube [1]	*Flammability limits (%) in air at room temperature and* 1 atm *pressure*	*Min.** *ignition energy* (mJ) [8]
Coal gas	≈ 30	1.5	5–32	0.03–2
Natural gas	≈ 48	0.6	5–14	0.30–2

*Precise value dependent on fuel:air ratio

A noteworthy feature of natural gas, compared to coal gas, is the much narrower range of flammable compositions (column 4 in Table 5.5). This can be understood in semi-quantitative terms if it is accepted that the flammability limits are related to the peak temperature achievable by a mixture. We have seen (Section 5.2.1) that for stoichiometric fuel gas/air mixtures this is usually about 2000° C. It will be lower for mixtures other than stoichiometric because the excess fuel or air raises the heat capacity. For methane/air or coal gas/air on the lean side, the molar heat capacity of the excess gas will be almost that of air. On the other hand, on the rich side, the heat capacity of the excess gas will still be close to that of air in the case of coal gas, whose most abundant constituents are diatomics (H_2,CO) with molar heat capacities close to those of nitrogen and oxygen. However, moving from stoichiometric to rich mixtures with natural gas causes a sharp increase in the heat capacity of the mixture since the molar heat capacity of methane is substantially higher than that of air and rises much more steeply with temperature. Therefore as the proportion of fuel increases, the heat capacity rises much more rapidly (hence the flame temperature drops more rapidly) with natural gas/air mixtures than with coal gas/air. This is consistent with the narrower limits of flammability.

5.4.3 The performance of natural gas in furnaces

A study of the performance of UK natural gas [11] under turbulent conditions in a vertical experimental furnace enabled the radiant heat flux — a quantity of primary interest in furnace design and use — to be measured and related to combustion conditions. These included the use or otherwise of swirl, that is, the influx gases experiencing recirculation. Swirl causes reacting gases to impart some of their heat to unreacted gases.

The furnace arrangement [11] was a cylinder of 0.6 m diameter with the burner at the top, and thermal and analytical measurements were recorded as a function of X/D, where X and D are respectively the distance from the burner and the diameter. The visible flame extends only to $X/D \approx 1.5$. Radiation flux was measured by pyrometry and Figure 5.1 shows the flux for two trials under identical conditions, except for the use of swirl in the trial denoted with filled-in points, but not in the other. Swirl measurably enhances the heat flux. The flame temperature peaked at about 1675° C and the wall temperature at about 1100° C.

Although these results relate to turbulent conditions, laminar natural-gas

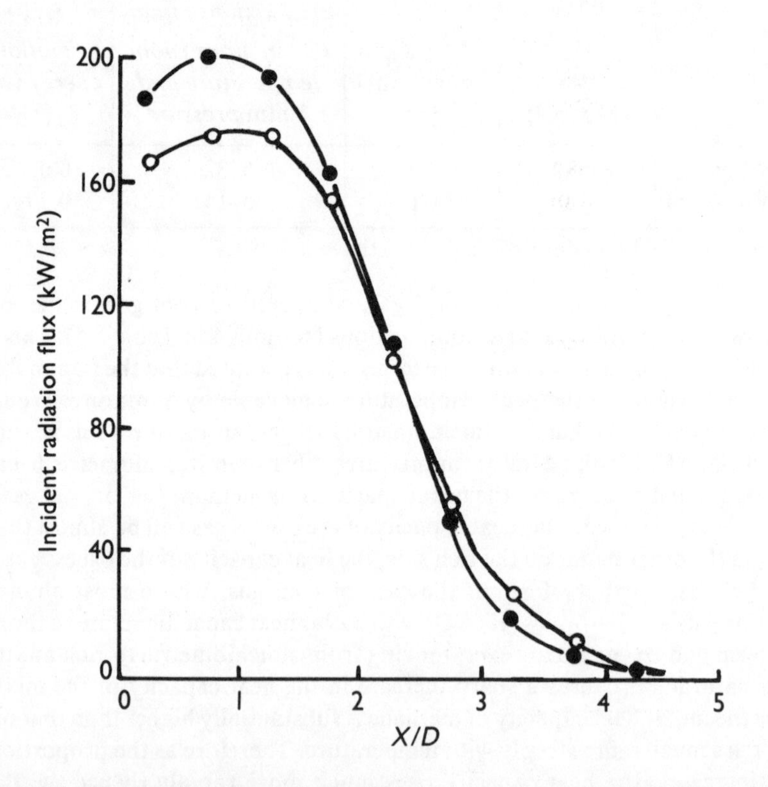

Figure 5.1 Radiant heat flux against distance from the burner for a cylindrical furnace utilizing natural gas (Hassan, Lockwood and Moneib [11])
(Reproduced with the permission of the Combustion Institute.)

flames are important in smaller scale applications. Laminar methane/air flames are discussed in Sections 4.3.4 and 10.1.2.

5.4.4 Burners for natural gas

We will consider the *atmospheric* class of burner, since it accounts for more natural gas usage than any other type. Fuel gas is admitted at low pressure (typically 2000 Pa above atmospheric) and by its momentum entrains atmospheric air, with which it travels to the port and into the combustion area. This air, called primary air, usually needs to be supplemented by air diffusing to the flame, called secondary air.

The function of a burner is to anchor a flame in such a way that the gas velocity and flame velocity balance and the flame is therefore stationary. A burner can do this only within certain limits of the operating conditions. An important consideration in burner design is the thermal delivery, discussed above in the context of the Wobbe index. The maximum thermal delivery for natural-gas burners is about 0.9 kW cm^{-2} port area. Also central is the primary air supply and the proportion of the total (stoichiometric) air requirement provided by the primary air. Figure 5.2 shows the limits of satisfactory burner performance for natural gas/air in a particular configuration of burner [12] and we can see that two factors discussed—thermal delivery and primary air supply as a percentage of theoretical—must be kept within a specified region of the graph. Consequences of attempted use outside this region will be one of the following:

(a) *Flash-back.* This is the name given to the burner instability resulting in flame propagation back down the tube conveying the fuel gas and

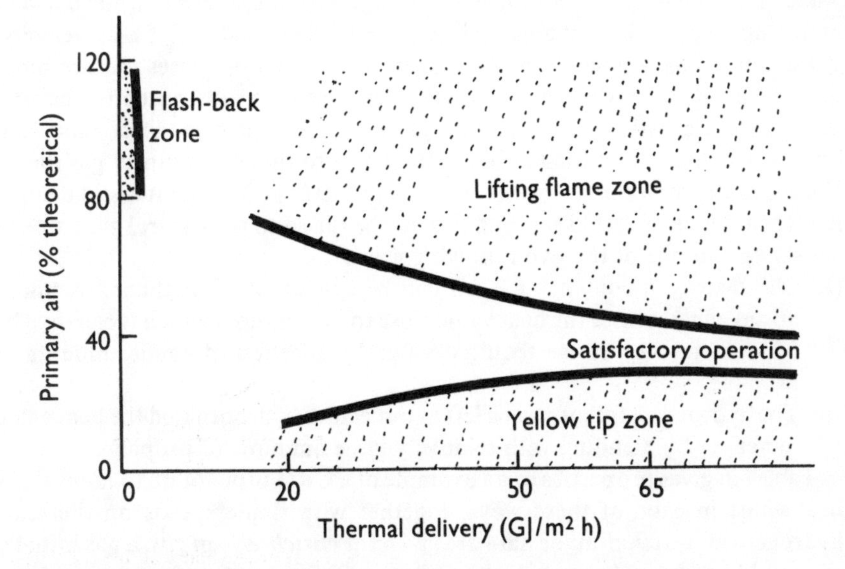

Figure 5.2 Operating limits for a natural gas burner (Weber and Vandaveer [12])
(Reproduced with the permission of Industrial Press.)

entrained air to the port and is seen to occur when the air supply is about 100 percent and at low thermal deliveries. Low thermal deliveries mean low gas velocities, and provision of all of the required air means a high temperature in the combustion zone, hence a high flame velocity which the low gas velocity cannot offset, with the result described. Natural gas in fact is much less susceptible to flash-back than manufactured gases and this is consistent with the fact (Table 5.5) that natural gas has lower flame velocities.

(b) *Flame lift*. This occurs at high thermal deliveries and is caused by a failure of the flame velocity to offset the gas velocity, resulting in propagation in an opposite direction from flash-back. This imposes the 0.9 kW cm^{-2} limitation.

(c) *Yellow tip*. This occurs at low air supply and is due to pyrolysis of the gas to carbon particles whose subsequent combustion causes the yellow color.

5.4.5 Substitutes for natural gas

Gases with equivalent heat values to natural gas are in two categories: 'high-BTU' oil gas, made by cracking petroleum materials, and propane-enriched gases. Oil gas can be made up to 56 MJ m^{-3} heat value and diluted as necessary, for example, with process discharge gas of 80% N_2, 20% CO_2. We have seen that the thermal delivery depends on the Wobbe index and a substitute gas can be matched to the *adjustment gas* (that is, the gas for which the appliances being supplied are adjusted) either in terms of heat value or in terms of Wobbe index and hence thermal delivery, and emergency substitution of suitable oil gas for natural gas is straightforward [12]. It should be noted that neither heat value matching of the substitute with the adjustment gas, nor thermal delivery matching, **ensures** interchangeability, since neither relates to flame velocity.

Relative flame velocities of adjustment and substitute gases require more thought in the case when propane-enriched gases are substituted for natural gas. These are hydrogen-containing gases supplemented with propane, and hydrogen causes rapid flame velocities relative to that of a natural-gas flame. For example, *catalytic rich gas* (CRG), manufactured from light distillate fuel, has a heat value of 25 MJ m^{-3} and this can be raised to the natural gas value of 37 MJ kg^{-1} in one of the two following ways:

(a) The hydrogen content of CRG can be converted to methane, (methanation), which raises the heat value close to the required value. It can then be brought to the precise required value by addition of small amounts of propane.

(b) The hydrogen content can be left as molecular hydrogen and the heat value increased by blending with relatively large amounts of propane.

Table 5.6 gives properties of an example of a CRG brought up to natural gas heat value in each of these ways, together with flame speeds on the scale hydrogen/air = 100. Similar data are given for *enriched lean gas*, a gas initially of ≈ 10 MJ m^{-3} supplemented with propane. Reference data for pure methane are also given. The four gases are close to identical in terms of heat value and

Table 5.6 Properties of selected natural gas substitutes [10]

	CO %	CH_4 %	C_2H_6 %	H_2 %	*Added* C_3H_8 %	*Heat value (MJ m*$^{-3}$*)*	*Density (air = 1)*	Wo (MJ m^{-3})	v_0 (H_2 = 100)
Methanated CRG	nil	95.7	nil	3.5	0.8	37.3	0.546	50	15
CRG, no methanation	0.9	72.3	nil	18.8	8.1	37.3	0.546	50	20
Enriched lean gas	2.6	11.7	0.8	58.3	26.6	37.3	0.544	51	32
Pure methane	-	100	-	-	-	37.8	0.555	51	14

Wobbe index, but show measurable differences in flame speeds. Indeed the value of CRG without methanation is at the low end of the range for hydrogen-containing manufactured gases. Burners adjusted for natural gas can accommodate only quite limited departure from the natural gas value of the flame velocity, and this has to be considered when use of a substitute is proposed.

5.5 Gas-turbine combustion

5.5.1 Introduction

The discussion of industrial-scale use of fuel gases has to this point been chiefly in the realm of process heating, for example, the melting of metal ingots, and this needs to be complemented by reference to use of fuel gases in gas turbines, a major outlet for fuel gases. A point of terminology requires clarification: a *gas turbine* is a device in which product gases from a combustion process are used to create power by impingement on the blades of a turbine. This eliminates the intermediate step of steam raising necessary with a steam turbine. The fuel generating the combustion gases for a gas turbine can be solid, liquid or gaseous, hence the alternative and perhaps preferable term *combustion turbine*.

5.5.2 Choice of fuel

Liquid fuels are the most commonly used fuels in this type of application, while the most frequently employed gaseous fuel is natural gas. However, coal-based gases, including producer gas and even blast-furnace gas have been used in gas-turbine combustion. A combustion characteristic relevant to gas-turbine operation is the amount of post-combustion gas which, as we have seen, differs markedly between gaseous fuels.

In a gas turbine, some of the turbine power is used to compress the post-combustion gases. We saw in section 5.3.2 that a typical producer gas (gross heat value $\approx$ 5 MJ m^{-3}) gives rise to 13.6 mol dry post-combustion gas per MJ heat released, whereas coke-oven gas ($\approx$ 20 MJ m^{-3}) gives rise to 7.9 mol MJ^{-1}. The corresponding figure for natural gas is about 10 mol MJ^{-1}. Where a low-heat-value gas is substituted for a high-heat-value one, and therefore the amount of post-combustion gas is increased, this can lead to operating difficulties [13]. A possible solution is to feed the excess back into the gasification plant. This is favorable in terms of energy economy because of the sensible heat that the excess post-combustion gas will contain.

5.6 Conclusion

We have examined natural gas and manufactured gases up to ≈ 35 MJ m^{-3}. Higher values than this are obtainable from petroleum-derived gases including LPG, the discussion of which in Chapter 13 is chiefly in the context of accidental explosions.

References

[1] Brame J. S. S., King J. G., *Fuel, Solid, Liquid and Gaseous*, Fifth Edition, London: Arnold (1961)
[2] Hurley T. F., Fitton A., 'The emergency use of producer gas for land transport', *Journal of the Institute of Fuel* **21** 283 (1947)
[3] Payman and Wheeler, cited in: Haslam R. T., Russell, R. P., *Fuels and their Combustion*, New York: McGraw-Hill (1926)
[4] Robson F. L. et al., cited in: Hottel H. C., Howard J. B., *New Energy Technology — Some Facts and Assessments*, Cambridge, MA: MIT Press (1971)
[5] Wilson L. M., 'The uses of coke oven gas in the steel industry', *Fuel* **6** 29 (1927)
[6] Verkamp F. J., Hardin M. C., Williams J. R., 'Ammonia combustion properties and performance in gas turbine burners', *11th Symposium (International) on Combustion*, 985, Pittsburgh: The Combustion Institute (1967)
[7] Minkoff G. J., Tipper C. F. H., *Chemistry of Combustion Reactions*, London: Butterworths (1962)
[8] Gaydon A. G., Wolfhard H. G., *Flames: Their Structure, Radiation and Temperature*, 4th Edition, London: Chapman and Hall (1979)
[9] Johnson A. J., Auth G. H., *Fuels and Combustion Handbook*, 1st Edition, New York: McGraw-Hill (1951)
[10] BP Ltd, *Gas Making and Natural Gas*, London (1972)
[11] Hassan M. M., Lockwood F. C., Moneib H. A., 'Measurements in a gas-fired cylindrical furnace', *Combustion and Flame* **51** 249 (1983)
[12] Weber E. J., Vandaveer F. E., *Gas Engineer's Handbook* (C. G. Segeler, Editor-in-Chief) New York: The Industrial Press (1966)
[13] Odgers J., Kretschmer D., *Gas Turbine Fuels and their Influence on Combustion*, Tunbridge Wells: Abacus Press (1986)

CHAPTER 6

Aspects of the Combustion of Coal and Solid Coal Products

Abstract

Three methods of coal combustion—pulverized-fuel (p.f.) combustion, grate (fixed-bed) combustion and fluidized-bed combustion (f.b.c.)—are outlined, using selected reports from the research literature. For p.f. combustion the roles of the coal volatiles and of the residual solid material are discussed fully and a model for a p.f. flame is presented. Factors affecting grate combustion, including radiative heat transfer from flames, are discussed, and a model for coal oxidation in a fixed bed is presented. Fluidized-bed combustion is treated by giving a detailed description of carbon particle combustion under f.b.c. conditions and supplementing this with reports of direct observation of volatile participation in f.b.c. The combustion characteristics of a number of solid coal derivatives, including coke, are also outlined.

6.1 Background on the origin and nature of coal

Coals vary widely in properties and the degree of maturity is a central factor in the utilization of any particular coal. The coal formation sequence, termed the peat–anthracite series, is:

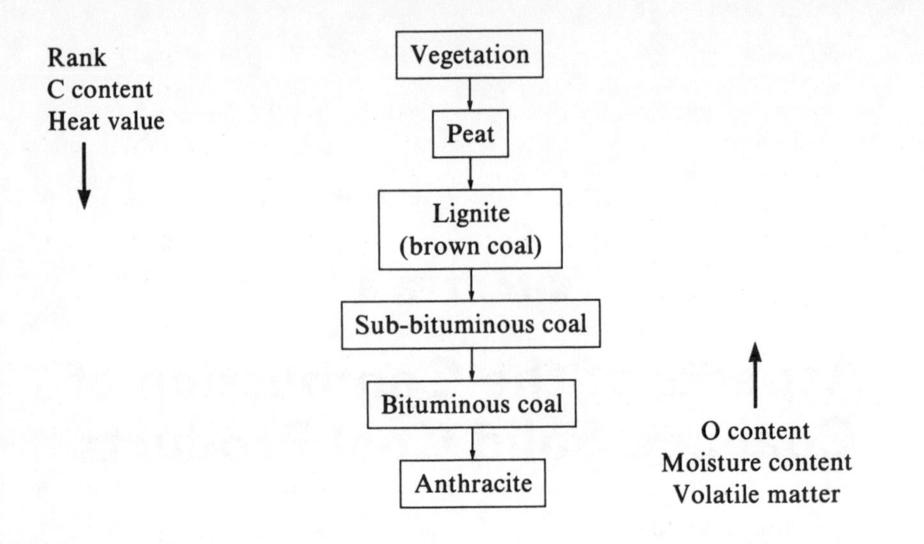

Figure 6.1. Schematic representation of the coalification sequence, showing property trends with maturity

Hence a bituminous coal is of greater maturity or, in the terminology of coal science, of higher rank, than a lignite, and an anthracite is of higher rank than a bituminous coal. Property trends are indicated in Figure 6.1, and partial analysis figures together with heat values for examples of US coals of particular rank are given in Table 6.1. It should be noted that the figures are on an as-received basis, that is, as received by the coal analyst, without any artificial drying or allowance for the ash content. Considerable changes can occur if such figures are re-expressed on a dry coal or dry, ash-free coal basis.

Table 6.1 Analyses of representative US coals

Source and rank	*Moisture* %	*Volatile matter** %	C %	O %	*Heat value* (MJ kg^{-1})
Rhode Island anthracite	4.5	3.0	82.4	1.8	27.1
Washington bituminous	10.3	30.4	61.3	14.4	25.7
Wyoming sub-bituminous	20.8	35.4	57.8	31.1	23.5
N. Dakota lignite	36.9	24.9	37.4	45.0	14.0

*Determined at 950° C

The organic part of the coal other than the volatile matter is called the fixed carbon (in fact a misnomer, as this component of the coal also contains some hydrogen). Volatile loss due to heat is referred to as pyrolysis or devolatilization. Most of the volatile matter is flammable so a detailed study of coal combustion requires examination of devolatilization and of volatile matter and fixed carbon participation in the combustion. Solid material remaining after devolatilization, comprising essentially the fixed carbon and the ash, is called coke or char. These materials are of considerable importance as fuels and chemical feedstocks.

We first consider the respective roles of volatiles and fixed carbon in coal combustion by reference to experimental work on pulverized (finely milled) bituminous coal particles. Pulverized fuel (p.f.) is widely used in industry, and among its advantages is almost complete combustion of the fuel carbon to CO_2, the CO levels being extremely low.

6.2 Pulverized-fuel

6.2.1 An investigation of the role of volatiles in pulverized-fuel (p.f.) combustion of a bituminous coal [1]

This work made use of a Pittsburgh bituminous coal of volatile matter 36.0% (dry basis) at particle sizes up to 200 microns, though the particle size distribution was such that most were in the 0–60 micron range. The coal was burnt in air using a specially designed experimental burner under the conditions summarized below:

(a) The coal/air flame was in effect one-dimensional, as there were no recirculation currents within it.

(b) Heat transfer between the flame and the surrounding gas was entirely radiative, conduction and convection being negligible.

The p.f./air mixture was admitted to the burning zone and small samples of p.f. removed by suction at specified distances from burner entry (convertible into times from a knowledge of the velocity profile) and analyzed for volatile matter content. In this way Figure 6.2, showing the volatile matter of the p.f. as a percentage of the original at various times, was obtained. The fixed carbon (F.C.) of the p.f. at the sampling points is simply:

$$\text{F.C.} = 100 - (\text{volatiles} + \text{moisture} + \text{ash})\ \%. \tag{6.1}$$

Figure 6.3 shows the fixed carbon content, as a fraction of the original, as a function of time from entry to the burner. From Figures 6.2 and 6.3, in each of which the occurrence of the flame front is also shown, it can be seen that devolatilization occurs almost entirely **after** ignition, and that fixed-carbon combustion begins **at** ignition. Given the overlap of devolatilization and solid burnout shown in Figures 6.2 and 6.3 respectively, it is clear that some volatiles will be released and burn in the gas phase, while some undecomposed volatiles will burn in the solid phase. Now, at any time t from burner entry:

$$V(t) = V_o - V_s - V_v, \tag{6.2}$$

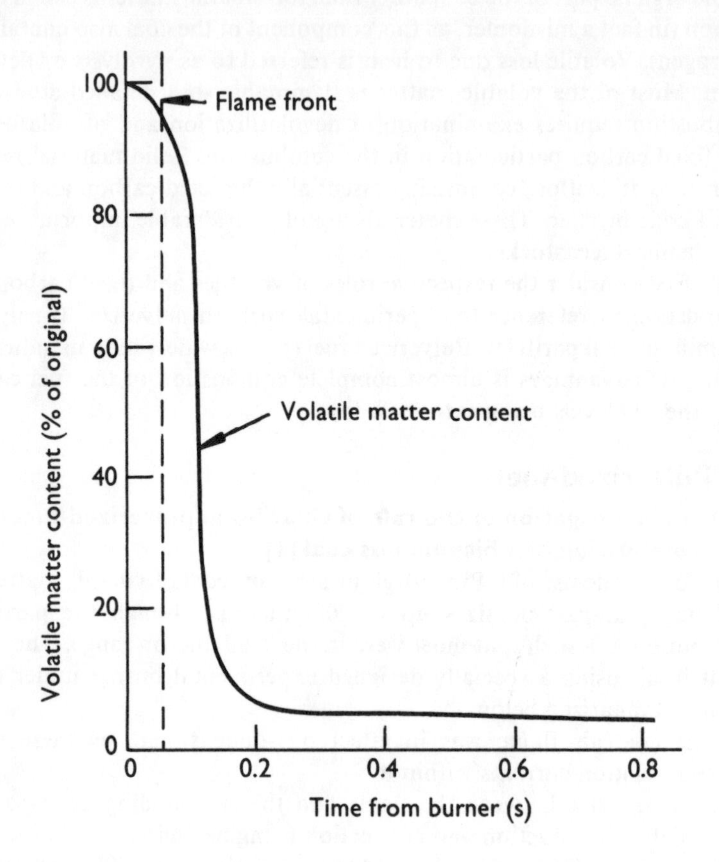

Figure 6.2 Volatile loss from bituminous coal p.f. as a function of time from burner entry (Howard and Essenhigh [1], slightly simplified)
(Reproduced with the permission of the Combustion Institute.)

where $V(t)$ = volatile content of the solid (g per g coal)
V_o = initial volatile content (g per g coal)
V_s = volatile content burnt as solid by time t (g per g coal)
V_v = volatiles released by time t into the vapor phase (g per g coal).

Rearranging:

$$\frac{V(t)}{V_o} = 1 - \frac{V_s}{V_o} - \frac{V_v}{V_o}. \tag{6.3}$$

$V(t)$ is known for various t by measurements of the p.f. samples and V_o is also known. The values of V_s at various t were obtained graphically from a knowledge of $V(t)$ and fixed-carbon contents, so the way is open for calculation of V_v at various times, and hence resolution of the coal volatile content into material burnt undecomposed and material released into the vapor phase. This is shown in Figure 6.4 from which it can be seen that over 20 per cent of the volatile matter burns undecomposed with the fixed carbon.

It is possible that the precise values of V_v, V_s, etc. are subject to errors due to the assumption that the proximate analysis values of the volatile matter of the original and partly burnt coal samples were appropriate to the flame conditions, that is, that:

$$\frac{\text{volatile matter content under rapid-heating p.f. conditions}}{\text{volatile matter content under proximate analysis conditions}} = 1. \quad (6.4)$$

Examination of this point requires comparison of the volatile matter content from proximate analysis with that from a rapid-heating technique such as electrical grid heating under inert conditions. In fact such information is available [2] for a bituminous coal of 36.4% volatile matter, dry basis, for which the ratio in equation 6.4 has the value 1.01, despite the fact that the peak temperature in the rapid-heating experiment is 150°C higher than that in the proximate analysis determination of volatile matter. However, counter-

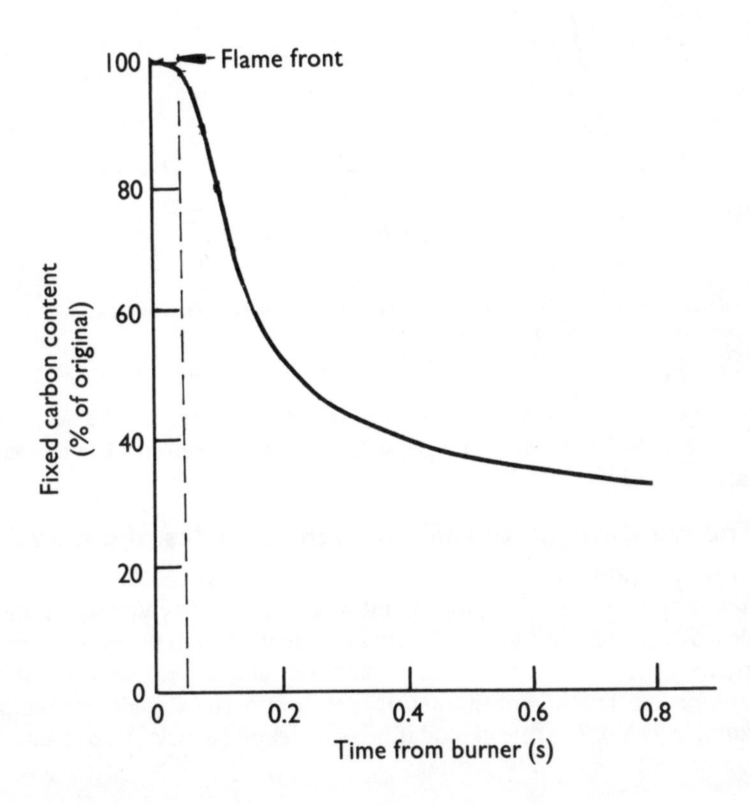

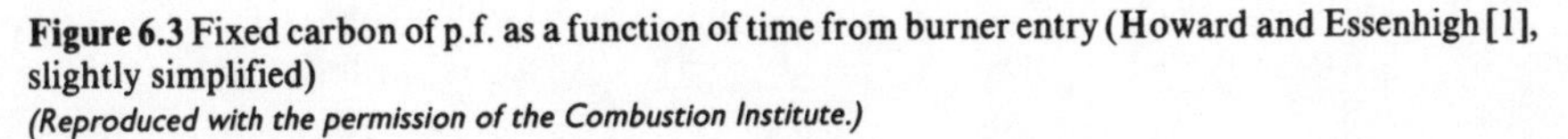
Figure 6.3 Fixed carbon of p.f. as a function of time from burner entry (Howard and Essenhigh [1], slightly simplified)
(Reproduced with the permission of the Combustion Institute.)

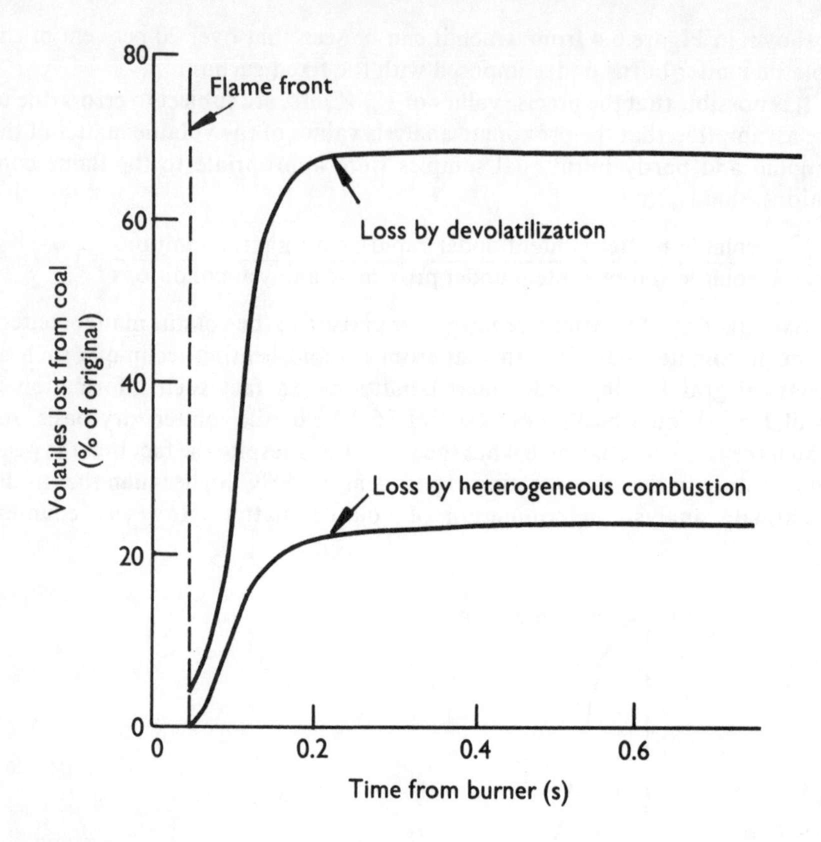

Figure 6.4 Resolution of total volatile content of p.f. into that released and burnt as vapor and that burnt undecomposed (Howard and Essenhigh [1])
(Reproduced with the permission of the Combustion Institute.)

examples could be given and therefore equation 6.4 should not be taken to be generally true. We will return to this question in a later section when discussing devolatilization kinetics.

6.3 The combustion of solid carbon particles of p.f. size

6.3.1 Introduction

In full descriptions of coal combustion it is necessary to include equations for volatile loss and solid burnout. The latter can be studied independently by use of particles with no volatile release, either material inherently low in volatiles (for example, anthracite or semi-anthracite) or material already devolatilized (for example, char). The rate of oxidation of a carbon particle is the resultant of two components and can be written:

$$R_m = \frac{1}{\dfrac{1}{R_d} + \dfrac{1}{R_c}}, \tag{6.5}$$

where R_m = measured rate (g g^{-1} coal s^{-1} atm^{-1} O_2 if first order in oxygen)
R_d = rate that would be observed if the kinetics depended solely on oxygen transport to the particle external surface (same units)
R_c = rate that would be observed if the kinetics depended solely on chemical reaction at the particle surface (same units).

Clearly, a measured rate will depend on these two limiting rates to different degrees according to combustion conditions.

6.3.2 Experimental measurements on low-volatile coal [3]

In this experimental program, particles of semi-anthracite of p.f. size were passed through a tubular flow reactor and gas samples withdrawn at intervals along the furnace axis. Oxide of carbon yields at these sample points were convertible into extent of reaction of the carbon particles at various times, providing a straightforward means of determining R_m from mass changes and the oxygen pressure. Rearranging equation 6.5:

$$R_c = \frac{R_m R_d}{(R_d - R_m)}. \tag{6.6}$$

Hence R_c can be obtained from experimental values of R_m and calculated values of R_d. The chemical components of the rates were also expressed in terms of particle external surface area instead of particle weight:

$$R_{ca} = \frac{R_c d\sigma}{6}, \tag{6.7}$$

where R_{ca} = chemical component of the reaction rate (g s^{-1} cm^{-2} atm^{-1} O_2)
d = median particle diameter (cm)
σ = particle density (g cm^{-3}) estimated experimentally at the sample points and dependent on the degree of burnout (see below).

The degree of burnout u at the various times of sampling was calculated as:

$$m(t) = m_o (1 - u), \tag{6.8}$$

where $m(t)$ = remaining mass of solid at time t
m_o = mass at $t = 0$.

Particle temperatures (T_p) at the sampling points were determined with a radiation pyrometer. Corresponding gas temperatures (T_g) were also known from direct measurement.

Some representative results are given in Table 6.2. In each case the measured rate differs from the limiting rates on the basis of solely diffusional or solely chemical control. The diffusion part is more important for the larger particle size, as would be expected. In this respect it should be noted that even at the high extent of burnout in the example shown, the particle diameter is a substantial proportion of the initial value of 78 microns. Variations in particle size with degree of burnout are discussed further below.

Table 6.2 Reaction rates and related data for semi-anthracite [2]

d (microns)	u	T_p (K)	T_g (K)	R_m*	R_d**	R_c†	R_{ca}††
22	0.650	1635	1620	173.2	3181	183.2	0.049
78	0.965	1993	1715	419.0	1324	613.1	0.219

* Units g g^{-1} coal s^{-1} atm^{-1} O_2, measured
** Same units, calculated
† Same units, calculated from equation 6.6
†† Units g cm^{-2} s^{-1} atm^{-1} O_2, calculated from equation 6.7.

The results for the semi-anthracite at different temperatures and particle sizes (hence external surface areas) were plotted as log R_{ca} against reciprocal temperature on the same axes as results of the same sort for other anthracites or semi-anthracites obtained by independent investigators. The aggregate graph could be drawn as a single line without excessive scatter, which could be fitted by an Arrhenius-type equation:

$$R_{ca} = 20.4 \exp[-79\,800/RT_p] \qquad \text{g s}^{-1}\text{ cm}^{-2}\text{ atm}^{-1}\text{ O}_2, \qquad (6.9)$$

that is, there is an activation energy of 79.8 kJ mol^{-1}.

The diffusional part of the rate relates to diffusion to the particle surface. Since carbon materials are porous in varying degrees, it is to be expected that the chemical rate R_{ca} will be influenced by intraparticle diffusion. Two scenarios are relevant to the results on semi-anthracite:

(a) In a situation where the diffusion was sufficient for uniform concentration of oxidant throughout the particle volume, there would be purely chemical control, and particles would burn at essentially constant size and sharply diminishing density.
(b) In a situation where the rate of reaction at the pore surface was comparable to the diffusion rate into the pores, there would be an overall rate of reaction subject to both diffusional and chemical control. A consequence of this would be more rapid reaction at the outer layers of the particle than in the interior.

In the work on semi-anthracite, sample results from which are given in Table 6.2, examination of particles at various degrees of burnout led to the view that it is the second of the two possibilities outlined above that applies to the 22- and 78-micron particles. However, with 6-micron particles of the same semi-anthracite the first possibility applies and, in such small particles, oxygen penetration was not a factor influencing the rate. Hence the chemical rate R_{ca} is perhaps inappropriately named since it depends on factors other than intrinsic reactivity, including the extent of internal surface and the particle size.

6.3.3 Rates of combustion of low-rank coal chars

The porosity (fraction of the apparent volume accounted for by voids) is only 5 percent or less for high-rank coals, yet this is sufficient for significant reaction inside the particle and rates of oxidation dependent on pore diffusion. Chars have very much higher porosities, so the factor of intraparticle diffusion is even more important. This is evident in results for low-rank coal chars of p.f. size [4] burnt at temperatures between 900K and 2200K in a tubular-flow reactor and also at temperatures between 630K and 760K in a small fixed-bed reactor. In the higher temperature range, 49- and 89-micron fractions of the char react at a rate influenced both by intrinsic reactivity and by pore diffusion, so they shrink as they burn at constant density (scenario (b)). In the lower temperature range, where reaction rates are lower, oxygen penetration is sufficient for uniform burning across the particle volume with reduction in particle density with burnout (scenario (a)).

Apparent activation energies in the high and low temperature ranges were respectively 68 and 134 kJ mol^{-1}. Carbon oxidation influenced both by chemistry and by pore diffusion has significantly lower activation energies than oxidation which is purely chemically controlled and usually has values in excess of 120 kJ mol^{-1}.

Carbonized solid fuels such as chars are of practical importance in combustion and gasification (in which case they undergo partial oxidation). However, information regarding the kinetics of their oxidation is valuable as input to modelling of p.f. combustion. Such modelling also requires devolatilization kinetics and this has been the subject of considerable experimental investigation. Devolatilization kinetics and burnout kinetics jointly provide much of the input for numerical simulation of p.f. combustion.

6.4 Devolatilization kinetics

Heat transfer conditions prevailing in p.f. plant are such that heating rates of the order of 10^4 K s^{-1} or higher are experienced by the particles, which therefore reach ceiling temperature extremely rapidly. At the most fundamental level a differential equation for volatile release can be written:

$$\frac{dX}{dt} = \frac{H_p}{h_v}, \tag{6.10}$$

where X = fraction of total volatiles released at time t

H_p = rate of heat uptake by a particle (itself the subject of a heat balance equation, units W kg^{-1})

h_v = enthalpy of devolatilization (J kg^{-1}).

As an example of a value of h_v, for a particular British bituminous coal it was measured [5] as endothermic to a degree of 0.6 MJ kg^{-1}. It therefore has about 2 percent of the magnitude of the heat value of the coal, and an opposite sign in the thermochemical sense.

A more common way of expressing the kinetics of volatile release is with an equation of the form:

$$\frac{dV}{dt} = k(V_m - V), \tag{6.11}$$

where V = volatile yield at time t (kg kg^{-1} coal)
V_m = volatile yield at infinite time, that is, the maximum volatile release of which the sample is capable (same units)
k = rate constant, with an Arrhenius temperature dependence (s^{-1}).

The value of V_m is, strictly speaking, temperature-dependent but is often approximated to temperature invariance in numerical work; we discussed in Section 6.2.1 the fact that the volatile matter content by proximate analysis is often an adequate estimate of the quantity of volatile matter released in a p.f. flame, even though the peak temperatures are significantly different. Values of k can be obtained experimentally in the following way [6]. Pulverized fuel is passed along a tubular reactor with nitrogen as the gas, so the weight loss on furnace transit is clearly due to devolatilization. Experiments can be performed at different transit times (in the range 30 to 110 ms) and temperatures, and devolatilization time is taken to be the transit time minus 20 ms for heating to temperature. Volatile matter can be determined for particles having undergone furnace passage and the difference between these and the volatile content of the original material gives $V(t)$.

Integration of equation 6.10 gives:

$$\frac{V(t)}{V_m} = \{1 - \exp(-kt)\}. \tag{6.12}$$

Although V_m is taken to be temperature independent, the volatile matter content by proximate analysis, and the quantity of volatiles released under particular combustion conditions, are considered in this approach to differ because of the heating rates* according to the following relationship:

$$V_m = fg\,VM_o, \tag{6.13}$$

where VM_o is the proximate analysis value, g has the temperature-independent value of 0.85, except for particles strongly swelling on devolatilization, and f has values in the range 1.2 to 1.5. Hence the ratio in equation 6.4 has values 1.02 to 1.28. Putting together equations 6.12 and 6.13 and taking logs gives:

$$-kt = \ln\left\{1 - \frac{V(t)}{fVM_o g}\right\}, \tag{6.14}$$

hence k at a particular temperature can be determined by plotting the logarithm against transit time t, and the temperature dependence of k can be checked by plotting $\ln k$ against $1/T$, from which Arrhenius parameters can be deduced.

As a sample result [6] we consider a British bituminous coal examined in this way at temperatures in the range 923K to 1223K and transit times between 30

*A dependence of the quantity of volatiles on heating rate is not **always** evident and at least in the case of certain low-rank coals, values obtained at ≈ 10 K s^{-1} and ≈ 10^3 K s^{-1}, peak temperature ≈ 1000°C in each case, are indistinguishable [7]. Where the equality or otherwise of values of the volatile matter at different temperatures and/or heating rates for a particular coal is important, there is clearly a need to examine it experimentally.

and 110 ms. The median particle diameter was 20 microns. With $\mathfrak{g} = 0.85$ and $\mathfrak{f} = 1.276$, the rate constants obtained fitted the expression:

$$\mathfrak{k} = A \exp[-E/RT], \tag{6.15}$$

where A = pre-exponential factor
$= 1.15 \times 10^5\ s^{-1}$
E = activation energy
$= 73\ kJ\ mol^{-1}$.

Hence for this particular coal, substitution in equation 6.15 at 800° C gives:

$$\mathfrak{k} = 1.15 \times 10^5 \exp[-73\ 000/(8.314 \times 1073)]$$
$$= 32.1\ s^{-1}.$$

For a transit time of 80 ms, equivalent to a devolatilization time of 60 ms, the volatile yield is obtainable from equations 6.12 and 6.13, which combine to give:

$$V(t) = \mathfrak{f}\mathfrak{g}VM_o\{1-\exp(-\mathfrak{k}t)\}. \tag{6.16}$$

Putting $\mathfrak{f} = 1.276$, $VM_o = 0.35$ kg kg^{-1} coal and $\mathfrak{g} = 0.85$ gives for the volatile yield:

$$V(60\ ms, 800°C) = 0.32\ kg\ kg^{-1}\ coal.$$

Hence 90 percent of the slow heating value of the volatile yield is achieved in 60 ms devolatilization of 20-micron particles.

This approach is a good basis for understanding the role of devolatilization in p.f. combustion. There are, however, two ways in which it can be expanded for a more detailed treatment:

(a) Separate equations similar to (6.11) can be used for each of the volatile products (CO, H_2, tar, etc.).
(b) Two-step pyrolysis can be considered, where a primary product breaks down further to form secondary ones. For example, tars may crack to form simple hydrocarbons.

6.5 Pulverized-fuel flames

6.5.1 Propagation of p.f. flames

The simple treatment of flame velocities in Chapter 4 made use of purely conductive heat flux within the flame, likely to be valid for gaseous fuels but not for solid dust cloud flames, for example, p.f. In these there is also significant radiative transfer of heat from the flame to the particles; a description of p.f. flames by Essenhigh and Csaba [8], with some of the features of an earlier one by Nusselt [9], has the essential features shown in Figure 6.5.

(a) The dust (p.f.) cloud has coordinates $x = 0$ at the plane flame front, and $x = L_i$ at the burner port. Properties are invariant in the y- and z-planes for any x.
(b) The particles are assumed to be spherical and all of the same diameter.

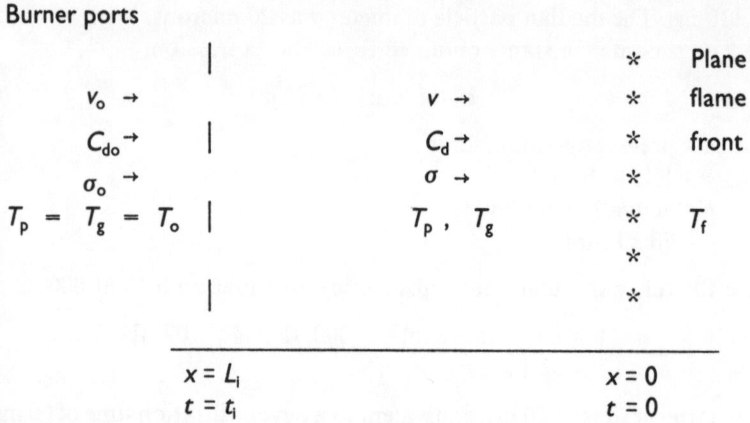

Burner ports

$v_o \rightarrow$	$v \rightarrow$	*	Plane
$C_{do} \rightarrow$	$C_d \rightarrow$	*	flame
$\sigma_o \rightarrow$	$\sigma \rightarrow$	*	front
$T_p = T_g = T_o$	T_p , T_g	*	T_f
$x = L_i$		$x = 0$	
$t = t_i$		$t = 0$	

Figure 6.5 Model of a p.f. plane flame (Adapted from Essenhigh and Csaba [8]) *(Reproduced with the permission of the Combustion Institute.)*

(c) Motion of the particles and gas relative to each other is assumed negligible, making the immediate environment of any one particle effectively quiescent.

Such a system requires heat balances for both radiation and conduction heat transfer. Radiation of intensity I (W m^{-2}) incident upon the particles obeys the expression:

$$\frac{dI}{dx} = + \Phi I, \tag{6.17}$$

where the coefficient Φ has units m^{-1}.

The positive sign arises because of the choices of coordinate system whereby x decreases in the direction from the burner port to the flame front. The quantity Φ is estimated as the ratio of the apparent particle area when they are viewed as two-dimensional to the total cloud volume, that is:

$$\Phi = \frac{n\pi a^2}{V} = \pi a^2 n', \tag{6.18}$$

where n = number of particles
V = cloud volume (m^3)
a = particle radius (m)
$n' = n/V$ = number of particles/unit volume (m^{-3}).

The dust-cloud concentration C_d (kg m^{-3}) is given by:

$$C_d = \frac{4}{3}\pi a^3 \sigma' n', \tag{6.19}$$

where σ' = solid density of the particles. From equations 6.18 and 6.19 it follows that:

$$\Phi = \frac{3C_d}{4a\sigma'}. \tag{6.20}$$

The equation for radiation heat transfer from flame to particles therefore becomes:

$$\frac{dI}{dx} = \frac{3C_d}{4a\sigma'} I, \tag{6.21}$$

and this can be expressed in terms of the particle and gas temperatures, respectively T_p and T_g (K), to give:

$$\frac{dI}{dx} = C_d c_p \frac{dT_p}{dt} + \sigma_g c_g \frac{dT_g}{dt}$$

$$= \Phi I, \tag{6.22}$$

where c_p = heat capacity of particle (J kg^{-1} K^{-1})
c_g = heat capacity of gas (J kg^{-1} K^{-1})
σ_g = density of gas (kg m^{-3}).

Now
$$\frac{d}{dx} = \frac{1}{v} \frac{d}{dt},$$

where v = velocity (m s^{-1}),
and from continuity:

$$vC_d = v_o C_{do} \tag{6.23}$$

$$v\sigma = v_o \sigma_o. \tag{6.24}$$

Substitution in equation 6.22 gives:

$$\frac{dI}{dt} = C_{do} v_o c_p \frac{dT_p}{dt} + \sigma_o v_o c_g \frac{dT_g}{dt} \tag{6.25}$$

$$\frac{1}{v_o} \frac{dI}{dt} = C_{do} c_p \frac{dT_p}{dt} + \sigma_o c_g \frac{dT_g}{dt}$$

$$= \Phi I = \frac{\mu}{v_o} I = \frac{\mu}{v_o} I_f \exp(\mu t), \tag{6.26}$$

where μ (s^{-1}) = coefficient re-expressed on a time basis.
$I = I_f$ at $t = 0$.

The formulation of the model is such that time, like distance, decreases along the direction from the flame front to the burner. For long ignition distances this leads to the condition that $\exp(\mu t_i) \rightarrow 0$, where $t = t_i$ at the burner, and this provides a simplified analysis for sufficiently long ignition distances (see below).

For conduction from particles to gas, the heat balance equation is obtained simply from Fourier's Law as:

$$c_g \sigma_g V \frac{dT_g}{dt} = 4\pi a^2 nk \frac{T_p - T_g}{a}, \tag{6.27}$$

where k = thermal conductivity of the gas (W m^{-1} K^{-1}).

This simplifies to:

$$\frac{\mathrm{d}T_g}{\mathrm{d}t} = \frac{4\pi a n' k}{\sigma_g c_g}(T_p - T_g). \tag{6.28}$$

Now $k/c_g\sigma_g = \alpha$, the thermal diffusivity, hence:

$$\frac{\mathrm{d}T_g}{\mathrm{d}t} = 4\pi a n'\alpha\,(T_p - T_g). \tag{6.29}$$

Substitution from equation 6.18 gives:

$$\frac{\mathrm{d}T_g}{\mathrm{d}t} = \frac{4\Phi\alpha}{a}(T_p - T_g)$$

$$= \frac{4\mu\alpha}{a v_o}(T_p - T_g)$$

$$= K(T_p - T_g). \tag{6.30}$$

The system of equations 6.26 and 6.30 (each of which contains v_o, by definition the flame velocity) provides a means of predicting this quantity for p.f. flames when solved subject to the conditions:

$$T_p = T_g = T_o \text{ at } t = t_i, \tag{6.31}$$

where $t = t_i$ at $x = L_i$ (see Figure 6.5). Full analytical solutions in the form of $(T_p - T_o)$ and $(T_g - T_o)$ as a function of t were obtained [9] but are unwieldy and not very transparent. However, simplified solutions for various combustion conditions were also obtained and we will examine a selection of them and use them to draw conclusions about the effect of combustion conditions on flame velocity. These simplified solutions fall into two classes:

(a) Those for which the ignition distance is very long, so that $I(t_i) = 0$, as discussed.
(b) Those for which the ignition distance is short and the above approximation is not valid.

In practice the difference is that between an ignition distance of the order of a metre (case (a)) and an ignition distance of only a fraction of a metre (case (b)). For the effectively infinite case, if there is also efficient heat transfer between the gas and the particles so that $T_p \approx T_g = T$, the solution to equations 6.26 and 6.30 is:

$$T - T_o = \frac{I_f \exp(\mu t)}{v_o[C_{do}c_p + \sigma_o c_g]}. \tag{6.32}$$

Recalling (Figure 6.5) that at $t = 0$, $T = T_f$:

$$v_o = \frac{I_f}{(T_f - T_o)(C_{do}c_p + \sigma_o c_g)}. \tag{6.33}$$

Now, by Stefan's law:

$$I_f = \mathfrak{C}\Omega T_f^4, \tag{6.34}$$

where $\mathfrak{C}$ = emissivity
Ω = Stefan's constant,

therefore v_o can be estimated from measurement of T_f together with the constants in the denominator of equation 6.33.

Alternatively, for the effectively infinite case the approximation T_g = constant may hold, signifying a gas phase of high heat capacity, that is, the gas acts as a heat sink for the burning particles. A simplified solution of equations 6.26 and 6.30 is then:

$$T_p - T_o = \frac{I_f \exp(\mu t)}{v_o\{C_{do}c_p + [\sigma_o c_g/(\mu/K)]\}}. \tag{6.35}$$

Solving for v_o at the flame front ($t = 0$) gives:

$$v_o = \frac{I_f}{(T_i - T_o)\{C_{do}c_p + [\sigma_o c_g/(\mu/K)]\}}, \tag{6.36}$$

where T_i = temperature of the particles at the flame front.

For the finite case, the approximate solutions contain t_i, as with the expression for a short flame under the condition that $T_p = T_g$:

$$v_o = \frac{I_f[1 - \exp(\mu t_i)]}{(T_f - T_o)(C_{do}c_p + \sigma_o c_g)}, \tag{6.37}$$

which becomes equation 6.33 as $t_i \rightarrow -\infty$. Combining equations 6.33 and 6.37 we can write:

$$v_o = (v_o)_\infty[1 - \exp(\mu t_i)], \tag{6.38}$$

where $(v_o)_\infty$ is for the infinite case.

In making use of the equations derived above to predict trends of p.f. flame velocities with combustion conditions, we recognize that T_i will have its maximum value for the value of C_{do} corresponding to stoichiometric conditions, and that on the fuel-rich side of stoichiometric conditions it will remain close to the maximum value. This is because the increased heat capacity of the mixture is likely, in a real system where the particles display a size distribution, to be offset by preferential ignition of fines. Hence, on the rich side of stoichiometric conditions (but not on the lean side) we take T_i, the temperature of the particles of the flame front, as independent of C_{do} and so can conclude:

(a) From equation 6.36, the flame velocity decreases with increasing C_{do}.
(b) Also from equation 6.36, flame velocity increases with increasing T_o, so the effect of preheating the gases on their way to the burner ports is to increase the flame velocity.

Whereas p.f. *plane* flame velocities are usually in the range 0.1 to 0.5 m s^{-1}, p.f. *jet* flame velocities are an order of magnitude higher and the reason for this can be understood in terms of T_0. In a jet flame, recirculation of hot combustion products markedly raises the effective value of T_0, with corresponding increases in the flame velocity. There is, of course, no recirculation in a plane flame.

6.6 Combustion of coal on grates

6.6.1 Introduction

In p.f. combustion, the coal and primary air are admitted together into the combustion system. When coal is burnt in a bed, air is supplied separately. Coal burnt in beds may be run-of-mine or it may have been crushed to a desired size. Crushed coal has particle sizes of ≈ 3 mm upwards. Clearly run-of-mine or crushed coal will have higher thermal inertia* than p.f. The combustion of coal in beds was the subject of a landmark symposium in London in 1937 [10] and reports were given of laboratory simulations of fuel beds which enabled factors influencing such combustion to be assessed. These factors include air-flow rate, particle size and heating by radiation from above the bed. In industrial fuel beds, air supplied at the grate is called primary air and this is supplemented by secondary air which is involved in the combustion of volatiles away from the grate.

6.6.2 Experiments to examine the effects of combustion conditions

A common way of studying fuel beds is by means of a pot furnace comprising a grate at its base through which primary air can be admitted, and a small (about one-foot high) cylindrical steel pot to hold the fuel. Results from such an arrangement were presented at the 1937 symposium. The pot furnace had the additional features of an electric radiator at the top to simulate radiation heat transfer from a flame to fuel closer to the grate, and a pilot flame to assist the volatile matter in igniting.

The effect of air-flow rate was first examined. Since this is admitted at ambient temperature it will have a cooling effect, working against ignition. For different air-flow rates the radiator output required to cause ignition of the coal was recorded and the expected positive correlation for a particular British bituminous coal is shown in Figure 6.6.

For a number of coals the maximum air-flow rate at which ignition could occur for a given setting of the radiator was measured and these values showed clearly the effect of coal rank. Higher rank coals were more difficult to ignite. For example, at a radiation heat supply of 3 kW, a coal of 80.9% carbon (dry, ash-free basis) could ignite at air-flow rates of up to ≈ 500 kg m^{-2} h^{-1}, whereas a coal of 88.9% carbon and of the same particle size could not ignite at air-flow rates above ≈ 250 kg m^{-2} h^{-1}.

The effects of particle size were examined in a similar way. A bed composed of large particles was found to ignite more easily than one composed of smaller

*The concept of thermal inertia is fully discussed in Chapter 12.

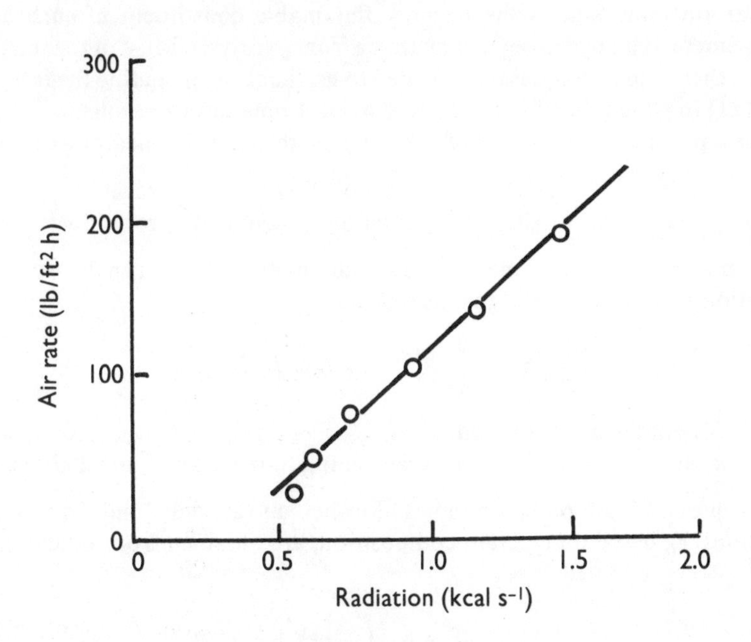

Figure 6.6 Air flow rate against radiative power required for coal ignition ([10])
(Reproduced with the permission of the Institute of Energy.)

particles up to sizes of about 2.5 cm. This is attributed to the smaller surface-to-volume ratio of the larger particles and, for fixed radiation heat transfer rate, there is an approximately linear correlation between maximum air-flow rate for ignition and particle size. Above particle sizes of 2.5 cm the increased voidage of the bed exacerbates cooling and ignition becomes more difficult. Added surface water increases the ease of ignition. A particular coal at a particular radiation heat transfer rate had a maximum air-flow rate for ignition of 490 kg m^{-2} h^{-1} with 5% added moisture and a 25% higher value with 15% added moisture. One of the factors responsible for this effect is coalescence of particles owing to the presence of water and its influence on convective cooling by the influx air.

With charges of coal smaller than those in the pot-furnace experiments it is possible for ignition to be brought about by feeding preheated air from below. In these experiments there will be an apparent critical temperature of preheated gas below which coal ignition does not occur. For example [10], a 2 g sample of Malayan lignite failed to ignite at an influx air temperature of 490K (though the center of the charge rose to 550K) but ignited with an air influx temperature of 495K.

6.6.3 A model of solid fuel oxidation in a bed

As well as outright combustion to yield combustion products, beds of coal or coke are also used for partial oxidation to make a fuel gas, as we saw in Chapter

5. Carbon monoxide is the primary flammable constituent of such a gas, supplemented by hydrogen and methane from pyrolysis. Most of the relatively recent theoretical analyses have related to gasification, including the following one [11] in which theory was checked against pot-furnace results.

Air is passed through a bed of coke and the relevant chemical equations are:

$2C + O_2 \rightarrow 2CO$	Reaction 1, combustion reaction.
$CO_2 + C \rightarrow 2CO$	Reaction 2, gasification reaction.

By arguments based on kinetics and stoichiometry [11], we can develop a rate equation for a bed of solid fuel as follows:

$$\frac{d}{dt}\left\{N_{O_2} + \frac{N_{CO_2}}{2}\right\} = -k_1 N_{O_2} - k_2(N_{CO_2}/2), \tag{6.39}$$

where N denotes mole fraction

k_1 and k_2 are rate constants pertaining to reactions 1 and 2 above (s^{-1}).

The values of N are based on moles of influx gas (air only) and the true mole fractions Y_i, based on reactant composition, are obtained from stoichiometry as:

$$Y_i = N_i\left(1 - \frac{Y_{CO}}{2}\right) \tag{6.40}$$

and

$$N_i^o = Y_i^o, \tag{6.41}$$

where superscript o denotes $t = 0$.

We first consider the case where $k_1 \approx k_2 = k^*$. This will be so under conditions where reactions 1 and 2 are both governed by diffusion. Reactions 1 and 2 then differ only according to the different diffusion coefficients in air of oxygen and carbon dioxide, a difference of about 20 percent. Integrating equation 6.39 for $k_1 = k_2 = k^*$ and substituting from equations 6.40 and 6.41 gives:

$$\frac{Y_{O_2} + Y_{CO_2}/2}{1 - Y_{CO}/2} = Y_{O_2}^o \exp(-k_1 t). \tag{6.42}$$

Therefore the hypothesis that reactions 1 and 2 are both diffusion controlled is amenable to experimental check. A log/linear plot of the left-hand side of equation 6.42 against t, where the mole fractions are obtained from gas analysis during a pot-furnace experiment, will yield a straight line if the hypothesis is correct. For different k_1 and k_2 a simple analytical treatment is not possible but approximate solution indicates that the above graph will have marked curvature. The mole fractions as a function of time can be found by measurement of gas composition at various heights above the grate and knowledge of the velocities once the pot combustion is in a quasi-steady state. This was done for charges of coke at several air-flow rates and the results are shown in Figure 6.7

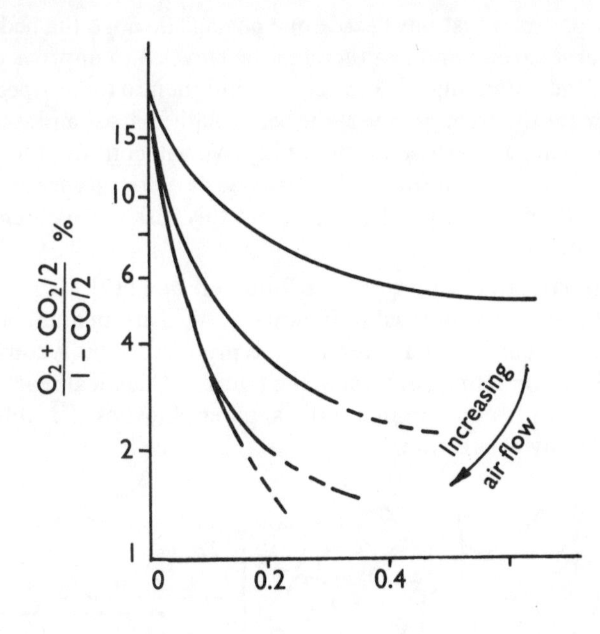

Figure 6.7 Log/linear plot of $\frac{O_2 + CO_2/2}{1 - CO/2}$ percent against time for 1.6 cm particles of coke in a pot furnace at various air-flow rates (Eapen, Blackadar and Essenhigh [11])
(Reproduced with the permission of the Combustion Institute.)

where, for each flow rate, the graph is curved, indicating that it cannot be assumed that reactions 1 and 2 are both diffusion controlled.

Deduction of k_2 at various temperatures from the approximate solution to equations 6.39 to 6.41 led to an activation energy for reaction 2 of $\approx$ 200 kJ mol^{-1}, a value much too high for a diffusion-controlled process. The gasification reaction is therefore chemically controlled.

The combustion reaction 1, however, is believed to be controlled by diffusion as well as by chemical rates. Support for this view is indicated by experimental results in which partially burnt out particles were examined and found to display constant density but smaller diameters in the hotter regions of the pot.

6.7 Fluidized bed combustion (f.b.c.)

6.7.1 Introduction

Fluidized-bed combustion (f.b.c.) utilizes a bed of inert particles (for example, sand, coal ash) supported by a grate through which air passes sufficiently rapidly for the bed to become fluidized. After preheating of the bed at startup, coal particles of size intermediate between those used as p.f. and those used in grates are injected and their throughput will account for 1–5 percent of the bed weight at any time. Bubbles will travel vertically through the bed. Ash generated from particles that have made one passage through the bed is likely to contain unburnt carbon and can therefore be recycled to improve combustion efficiency. Understanding of f.b.c. requires insight into three aspects: mass transfer between the different phases of the bed; volatile release and subsequent combustion; and carbon particle burnout [12]. We will consider the first and third aspects in the context of some laboratory-scale work on carbon–particle combustion in a fluidized bed and complement this with some comments on the role of volatiles.

6.7.2 Carbon particle combustion in a fluidized bed [12]

The approach taken is summarized in Figure 6.8, the inert medium being ash. At the surface of the particle (diameter *d*) CO is produced. This is converted to CO_2 in a thin diffusion flame enveloping the particle. Chemical reaction rates are very rapid and diffusion dominates the apparent kinetics. The diffusion is not supplemented by convection.

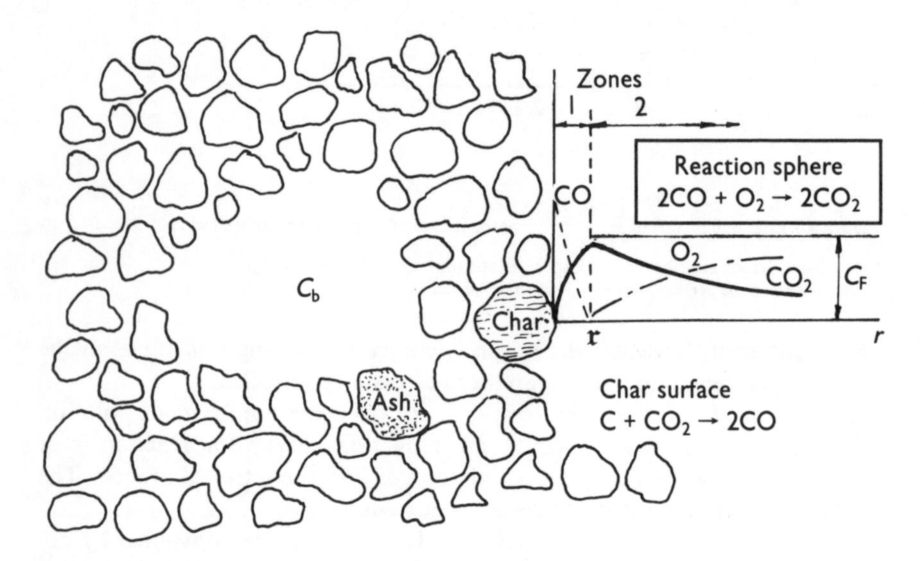

Figure 6.8 Carbon particle in f.b.c. (Avedesian and Davidson [12])
(Reproduced with the permission of the Institution of Chemical Engineers.)

Gas concentration profiles around a particle are shown in Figure 6.8 and the gaseous reaction zone is defined by $r = \mathfrak{r}$, where r = spherical coordinate = 0 at the center of the particle. At $r = \mathfrak{r}$, the O_2 and CO concentrations are both zero; CO drops to zero from the particle side and O_2 from the other side. The CO_2 formed in this zone diffuses partly back to the carbon surface where it is converted to CO, and partly to the gas stream where it is carried away. Application of Fick's Law to the diffusion of oxygen at $r = \mathfrak{r}$ gives:

$$n_o = 4\pi\, \mathfrak{r}^2 D' \frac{d[O_2]}{dr}, \qquad (6.43)$$

where n_o = oxygen flux (mol s^{-1})
D' = effective diffusion coefficient.

$D' < D$, the diffusion coefficient, because of the effect of the ash particles.

By making use of boundary conditions and the dimensionless Sherwood number, defined by $\mathrm{Sh} = k'd/D$, where k' = mass transfer coefficient (m s^{-1}) and d is the particle diameter, the flux equation becomes:

$$n_o = 2\pi \mathrm{Sh} D d c_f, \tag{6.44}$$

where c_f = concentration of oxygen in the particle phase. $\mathrm{Sh} \approx 2\epsilon_o$ for the char particles of interest, where ϵ_o is the fractional voidage of the ash particles*.

Now, if the fluidizing gas velocity is U and the velocity at incipient fluidization is U_o, the difference between them is manifest as bubbles, therefore:

$$\text{total gas flow through the bed} = UA^\bullet \ \mathrm{m^3\ s^{-1}} \tag{6.45}$$
$$\text{flow through the particle phase} = U_oA^\bullet \ \mathrm{m^3\ s^{-1}} \tag{6.46}$$
$$\text{flow as bubbles} = (U - U_o)A^\bullet \ \mathrm{m^3\ s^{-1}}, \tag{6.47}$$

where $A^\bullet$ = cross-sectional area of the bed. Now, for a bed with a single charge of coal of m kg, the number of particles N in the charge is:

$$N = \frac{6m}{\sigma_c \pi d_i^3}, \tag{6.48}$$

where $d = d_i$ at $t = 0$
σ_c = char density.

So, combining equations 6.44 and 6.48, oxygen flux to all the char particles is given by:

$$Nn_o = \frac{12m\mathrm{Sh}Ddc_f}{\sigma_c d_i^3}, \tag{6.49}$$

and this can be equated with the consumption due to oxidation, given by:

$$\mathrm{O_2}\ \text{consumption} = KA^\bullet H_o c_f, \tag{6.50}$$

where H_o = bed height at incipient fluidization (m)
K = specific rate (s^{-1}).

Combining equations 6.49 and 6.50:

$$K = \frac{12m\mathrm{Sh}Dd}{\sigma_c d_i^3 A^\bullet H_o}. \tag{6.51}$$

This expression gives the correct limiting case that $K = 0$ at $d = 0$ (nil reaction after total burnout) and clearly provides a way to estimate burnout times if d

*Sh = 2 for an isolated spherical particle in a convection-free (diffusion only) environment. The lower value $2\epsilon_o$ is due to the ash particles and their effect on gas transport to the char particles.

can be expressed in terms of t, as it can from oxygen balance in the following way:

Rate of oxygen exit from the bed as bubbles

$$= (U - U_o)c_b A^\bullet \qquad \text{mol s}^{-1}, \tag{6.52}$$

where c_b = concentration of oxygen in the bubble phase at the top of the bed (mol m^{-3}).

Rate of oxygen exit from the bed other than as bubbles

$$= U_o c_f A^\bullet \qquad \text{mol s}^{-1}. \tag{6.53}$$

Hence the total rate of exit is obtained by summing equations 6.52 and 6.53 and c_b is given by the expression [14]:

$$c_b = c_f + (c_o - c_f)\exp(-X), \tag{6.54}$$

where c_o = entry concentration and X is the number of times a bubble is flushed out in passing through the bed.

The rate of oxygen entry to the bed = $Uc_o A^\bullet$. (6.55)

Clearly the difference between the entry rate (equation 6.55) and the total exit rate (equations 6.52 and 6.53) is the rate of oxygen reaction, by stoichiometry equal to the rate of carbon combustion, given by:

$$\text{rate of carbon combustion} = -\frac{N}{M_R}\frac{dm'}{dt}, \tag{6.56}$$

where M_R = 0.012 kg mol^{-1} and m' is the mass of one particle of fuel.

Combining equations 6.53 to 6.56 in the way indicated, we obtain:

$$(c_o - c_f)A^\bullet [U - (U - U_o)\exp(-X)] = -\frac{N}{M_R}\frac{dm'}{dt}. \tag{6.57}$$

Now

$$\frac{dm'}{dt} = \frac{d}{dt}\left[\frac{4}{3}\pi(d/2)^3\sigma_c\right] = \frac{\pi\sigma_c}{6}\frac{d}{dt}(d^3), \tag{6.58}$$

and by the chain rule in differentiation:

$$\frac{d}{dt}(d^3) = 3d^2\frac{dd}{dt}.$$

Equations 6.57 and 6.58 can therefore be combined to give:

$$(c_o - c_f)A^\bullet [U - (U - U_o)\exp(-X)] = -\frac{\pi\sigma_c d^2 N}{2M_R}\frac{dd}{dt}. \tag{6.59}$$

Clearly the right-hand side of the above equation, when divided by N, is the oxygen consumption rate of one particle n_o which is given in equation 6.44. This can be used to eliminate c_f from equation 6.59, which when subsequently integrated, subject to the initial condition $d = d_i$ at $t = 0$, gives:

$$t = \frac{m[1 - (d/d_i)^3]}{12c_o A^\bullet [(U - U_o)\{1 - \exp(-X)\} + U_o]} + \frac{\sigma_c d_i^2[1 - (d/d_i)^2]}{96\text{Sh}Dc_o}. \tag{6.60}$$

At $t = t_B$, the burnout time, $d = 0$, therefore:

$$t_B = \frac{m}{12c_o A^\bullet [(U - U_o)\{1 - \exp(-X)\} + U_o]} + \frac{\sigma_c d_i^2}{96 \mathrm{Sh} D c_o}. \tag{6.61}$$

Equation 6.61 indicates that large charges (m) and large particle diameters will cause long burnout times and that low velocities also cause extended burnouts. Graphs of t_B against m for fixed d_i and U **or** of t_B against d_i^2 for fixed m and U will yield a straight line from which Sh and X are obtainable from the slopes and intercepts. In a small laboratory f.b.c. unit, results from experiments with coal char were processed in this way [12]. The bed was preheated to 900° C by natural gas before admitting solid fuel. The inert solid was coal ash of average diameter 0.39 mm and t_B was determined visually by means of a stopwatch. The relationships expected were found to hold and Figure 6.9 is a graph of t_B against m for other conditions as indicated and Figure 6.10 shows t_B against d_i^2. From these and other similar experiments on the char, values of Sh in the range 1.1 to 1.5 and values of X of about 0.6 were calculated.

6.7.3 Participation of volatiles in f.b.c.

Because of the larger particles, volatile release is slower in f.b.c. than in p.f. combustion. In section 6.4 it was shown that a p.f. particle will approach maximum devolatilization in a time of the order of 60 ms. By contrast, particles of f.b.c. size are likely to require seconds, or even tens of seconds. Subsequent distribution of the volatiles in the bed depends on the mixing, and ignition of volatiles is subject to a delay.

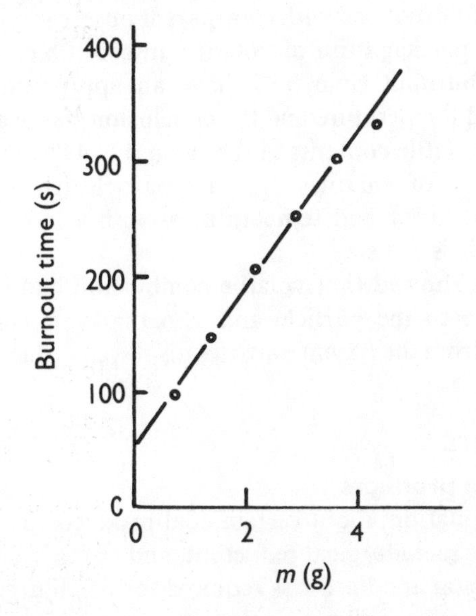

Figure 6.9 Graph of t_B against m for char particles in a small f.b.c. unit. d_i = 1.87 mm, U = 0.214 m s^{-1}, U_o = 0.046 m s^{-1}, σ_c = 720 kg m^{-3}. (Avedesian and Davidson [12])
(Reproduced with the permission of the Institution of Chemical Engineers.)

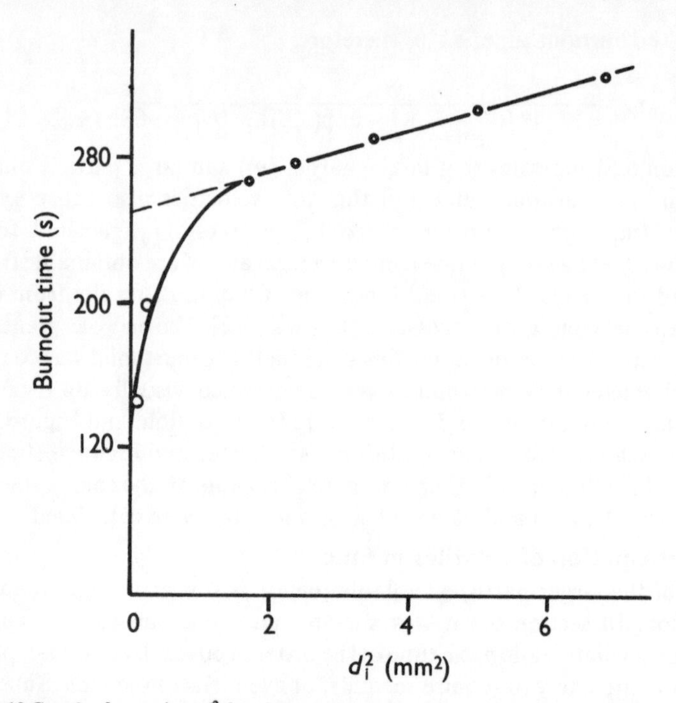

Figure 6.10 Graph of t_B against d_i^2 for char particles (m = 3.6 g, other conditions as for Figure 6.9.) *(Reproduced with the permission of the Institution of Chemical Engineers.)*

Prins [13] used a fluidized bed with transparent observation ports and video techniques to study participation of volatiles in coal f.b.c. Volatile ignition delay and volatile burnout time both have an approximately hyperbolic relationship with bed temperature and the conclusion was drawn for the suite of coals examined (volatile contents in the range 11–48%) that at higher bed temperatures, ignition of volatiles precedes particle ignition and possibly assists it, whereas at lower bed temperatures, with a less rich envelope of volatiles released, this is not so.

The video records showed that volatile combustion had the features of a diffusion flame close to the particle and also that volatiles dispersed and burning in isolation from the parent particles displayed a blue flame with some premixed character.

6.8 Coal products

6.8.1 Carbonization products

Table 6.3 gives information about certain coal products in fuel use. Coke is produced largely as a metallurgical reductant, and cokes from weakly coking coals, lacking the fusion and hardness required for metallurgical application, tend to be more reactive in combustion than cokes from strongly coking coals. If a change is to be made from bituminous coal to coke in a fuel bed, the following points have to be considered.

Table 6.3 Coal products used in combustion

Product	*Method of maufacture*	*Heat value* ($MJ\ kg^{-1}$)	*Examples of use as fuel*
Coke	Carbonization of a coking coal at 1200K or higher	≈ 33*	Steam raising Heating
Coke breeze	Accompanies coke production	≈ 26	Steam raising
Semi-coke	Low-temperature (e.g. 900K) carbonization	Similar to coke	Smokeless fuels
Briquettes	Low- or high-rank coal, ground and pressed into regular shapes (e.g., ovoid). May involve a binder (e.g., pitch, starch) or may be binderless	Equivalent to the parent coal or better	Can be substituted for unprocessed coal in many applications
Coal–water slurry	Coal milled and mixed with water to form a suspension which can be pumped. Dewatering before combustion to form *coal cake*	Lower than dry coal	Steam raising
Coal–water mixture (CWM)	Coal (≈ 70%)/water suspension, burnt as a liquid fuel. Suspension stability crucial	≈ 24	Process heating Substitute for fuel oil
Coal–oil mixtures	p.f.-size coal particles (≈ 40% by weight) mixed with heavy fuel oil	≈ 40	Shipping

*Depending on the ash content

(a) Carbonization to form coke involves swelling, therefore, although bituminous coals and cokes have similar heat values on a weight basis, cokes have lower values on a particle volume basis. This might necessitate the use of thicker fuel beds.
(b) Cokes are extremely low in volatiles, so combustion in a bed is mainly at or close to the grate. Hence there is greater reliance on primary air and less on secondary air. Not only are cokes low in volatiles but such volatiles as are present are largely CO and H_2. Cokes consequently burn without significant smoke.

Coke breeze comprises particles of ≈ 2 cm or lower and has a higher concentration of ash-forming constituents than the coke of which it is a by-product and hence has a greater tendency to foul combustion plant and a lower heat value. Semi-cokes (sometimes referred to as chars) are also obtained by carbonization but at lower temperatures than coke. Ease of ignition of cokes and chars has an inverse relationship to carbonization temperature. Semi-cokes were developed to provide smokeless fuel for domestic use.

6.8.2 Briquettes

Benefits accruing from the use of briquettes include the potential to convert coal fines into lumps suitable for transportation and general purpose use.

Enhancement of the heat value is especially marked in briquettes made from low-rank coal, since the briquettes are much less moist than the coal as won. Very large amounts of coal are purchased as briquettes. One of the world's leading briquette making countries is Germany which has an annual production of the order of 50 million tonnes.

Necessary for satisfactory combustion is the ability of briquettes to retain their shape during burning, since premature disintegration leads to incomplete combustion. It is for this reason that some low-rank coals are briquetted in carbonized form. Smoke control is also important and choice of binder is a relevant factor. Briquettes are a useful auxiliary fuel in situations where low-rank coal is used as p.f. The raw coal undergoes only limited drying during milling, so to start up a furnace or to stabilize combustion, milled briquettes can, with advantage, be substituted for milled raw coal.

6.8.3 Coal–water and coal–oil fuels

The impetus for the development of coal–water slurries was pipeline delivery of coal to the place of use. Slurry can be pipelined long distances, and some electricity utilities obtain coal in this form. In typical use [14] finely ground coal is made into a slurry of 40–50% coal by weight, and at its destination the slurry is dewatered with a centrifuge to give *coal cake* of $\approx 20\%$ moisture, which can be milled and used as p.f. In a suitable climate, solar drying is a feasible alternative to centrifuging; the slurry is simply pumped into inclined pans to dry in the sun. However, this is as yet a reality only on a pilot scale. Distinct from centrifugally dewatered slurries are coal–water mixtures (CWM) which contain about 70% coal by weight and are burnt as liquid fuels in place of fuel oil.

Superior, therefore, in terms of heat value, are coal–oil mixtures. Their use is by no means new and plant designed primarily for heavy fuel oils can be adapted straightforwardly to use coal–oil mixtures. Bituminous coal or lignite can be used. Experience in steam raising with this class of fuels is favorable; good conversion to CO_2 (that is, complete combustion), good smoke control, and low ash are features of coal–oil fuels properly burnt. As regards safety, they have the following advantages:

(a) They do not self-heat in storage, which coals, especially low-rank ones, often do.

(b) Since they are denser than fuel oil, an accidental fire is easier to extinguish than one originating in fuel oil alone.

6.9 Conclusion

To the contribution to world energy requirements from direct combustion of coal by the various techniques discussed in this chapter should be added the contribution that coal makes in gasified form (Chapter 5). Coal consumption in world terms includes that used in making coke for metallurgical use. Less so now than before the advent of petrochemicals, coal is also a means of producing chemicals, either by gasification to CO/H_2 mixtures and subsequent conversion to hydrocarbons or alcohols, or from tars.

References

[1] Howard J. B., Essenhigh R. H., 'Mechanism of solid-particle combustion with simultaneous gas-phase volatiles combustion', *Eleventh Symposium (International) on Combustion*, 399, Pittsburgh: The Combustion Institute (1967)
[2] Loison and Chauvin (1964), cited in: Field M. A., Gill D. W., Morgan B. B., Hawksley P. G. W., *Combustion of Pulverised Coal*, Leatherhead: British Coal Utilisation Research Association (1967)
[3] Smith I. W., 'The kinetics of combustion of pulverised semi-anthracite in the temperature range 1400–2200° C', *Combustion and Flame* **17** 421 (1971)
[4] Hamor R. J., Smith I. W., Tyler R. J., 'Kinetics of combustion of a pulverised brown coal char between 630 and 2200° C', *Combustion and Flame* **21** 153 (1973)
[5] Baum M. M., Street P. J., 'Predicting the combustion behaviour of coal particles', *Combustion Science and Technology* **3** 231 (1971)
[6] Badzioch S., Hawskley P. G. W., 'Kinetics of thermal decomposition of pulverised coal particles', *Industrial Engineering and Chemistry Process Design and Development*, **9** 521 (1970)
[7] Suuberg E. M., Peters W. A., Howard J. B., 'Product compositions and formation kinetics in rapid pyrolysis of pulverized coal—implications for combustion', *Seventeenth Symposium (International) on Combustion*, 117, Pittsburgh: The Combustion Institute (1979)
[8] Essenhigh R. H., Csaba J., 'The thermal radiation theory for plane flame propagation in coal dust clouds', *Ninth Symposium (International) on Combustion*, 111, New York: Academic Press (1963)
[9] Nusselt W., V.D.I. **68** 124 (1924). Cited by Essenhigh and Csaba [8]
[10] Papers printed in *Journal of the Institute of Fuel*, **11** (1937)
[11] Eapen T., Blackadar R., Essenhigh R. H., 'Kinetics of gasification in a combustion pot: comparison of theory and experiment', *Sixteenth Symposium (International) on Combustion*, 515, Pittsburgh: The Combustion Institute (1977)
[12] Avedesian M. M., Davidson J. F., 'Combustion of carbon particles in a fluidized bed', *Transactions of the Institution of Chemical Engineers* **51** 121 (1973)
[13] Prins, W., cited in *Fluidized Bed Combustion* (Radovanovic M. Ed.), Washington: Hemisphere Publishing Corporation (1986)
[14] Elliott M. A. (Ed.), *Chemistry of Coal Utilization*, Second Supplementary Volume, New York: Wiley (1981)

CHAPTER 7

Combustion of Less Common Fuels

Abstract

A number of fuels other than those derived from coal, oil or natural gas are discussed, including wood, bagasse, municipal solid waste and peat. Some mechanistic information for wood combustion is given and the relative merits of wood and more orthodox solid fuels for particular applications are discussed. Accounts of peat combustion and municipal solid waste combustion are largely in the context of steam raising. Methanol, ethanol and vegetable oils as substitute transport fuels are discussed. Shale products also receive attention, emphasis being given to the way in which the carbon remaining in the shale after retorting can itself be used as a fuel to heat the retort. Tar sands are discussed briefly.

7.1 Wood

7.1.1 Introduction

Until about one hundred years ago, wood was the predominant fuel in the USA. Its unit of quantity for sale was the cord; one cord is a 128 ft^3 (3.6 m^3) bulk volume of wood and in 1875, for every ton of anthracite mined, three cords of wood was used for combustion [1]. Woods for combustion can be classified according to how many cords are thermally equivalent to one ton of anthracite. One to two cords will release the amount of heat that a ton of anthracite can release and the differences between woods in these terms are due primarily to different moisture contents. Also, a cord is specified only in terms of bulk volume and wood fuels will settle to different bulk densities. Current usage of fuel wood is considerable, and wood ignition is of obvious importance in forest fire control and the fire safety of timber products used in building.

7.1.2 Characteristics of wood

Wood is composed mainly of cellulose, hemicellulose, and lignin, and elemental analyses of different woods are somewhat variable; carbon in the range 48–55 percent (for dry wood), hydrogen from 5–7 percent. These differences are reflected in small differences in heat values which, for dry woods, are usually in the range 19.5 to 23 MJ kg^{-1}. Woods with heat values at the higher end of the range contain an appreciable quantity of resin in addition to the primary constituents. It is possible, though rare, for the resin content of a wood to cause its heating value to be as high as 26 MJ kg^{-1}. Pitch pine is an example of a high-resin wood. Seasoned wood, which has been allowed to equilibrate its moisture with that of the atmosphere, will have a heat value of about 17 MJ kg^{-1}. Global kinetic studies of wood combustion leading to expressions of the type:

$$\text{Rate} = A\sigma \exp[-E/RT], \qquad (7.1)$$

where A = pre-exponential factor (s^{-1})
E = activation energy (J mol^{-1})
R = Universal Gas Constant
= 8.314 J K^{-1} mol^{-1}
T = temperature (K)
σ = density (kg m^{-3}),

reveal that while E is always of the order of 10^2 kJ mol^{-1}, it can vary between wood samples by a factor of about two, and that A can differ between wood samples by orders of magnitude. Methods outlined in Chapter 1 for obtaining kinetic information from thermal ignition experiments are in fact very suitable for woody materials.

7.1.3 Wood pyrolysis

In proximate analysis of woods, *volatile matter* accounts for about 80 percent of the dry weight. Pyrolysis precedes or accompanies combustion and most of the pyrolysis products are capable of subsequent combustion. Hemicellulose pyrolysis occurs to a significant degree at temperatures above about 230° C, cellulose pyrolysis above about 330° C and lignin pyrolysis above about 230° C. Cellulose and lignin, one a polysaccharide which breaks down to glucose when treated with acid, and the other a macromolecule which on delignification breaks down to a complex mixture of phenols, differ considerably in their pyrolysis behavior. In particular, lignin pyrolysis results in larger proportions of char (>50%) than cellulose pyrolysis. In addition to small amounts of char, and more abundant amounts of tar, cellulose pyrolysis produces a miscellany of simple oxygenated organic compounds in small amounts, for example, acetaldehyde, methanol and furan.

7.1.4 Kinetics of wood combustion

A number of investigations of wood combustion have been undertaken, including that by Tinney [2] which compares experimental findings and results developed from a model. This work deals with small cylinders of fir (dowels) heated in a furnace. The cylinders were much longer than wide, meaning that heat transfer could be regarded as purely radial. Experimentally, dowels had a thermocouple at the center for determining the internal temperature, and weight loss was also followed.

When the heat balance equation for a dowel was solved numerically for various furnace temperatures, it was found that to simulate experimental temperature–time behavior it was necessary to change the kinetic parameters at a time corresponding to fractional weight loss of 0.33 to 0.5. This point, at which the kinetic parameters need to be changed, is called the breakpoint and is believed to correspond physically to the time when evolution of volatiles ceases. The change required was about a threefold increase in A and about a 35 percent increase in E.

The above treatment considers the kinetics only as a single Arrhenius function with parameters requiring modification at the breakpoint, whereas it is possible to simulate the temperature and weight loss behavior by use of a detailed mechanism with several processes (cellulose pyrolysis, char burnout, etc.) each with its own Arrhenius parameters. In view of the different temperature sensitivities of the various processes, and the relative prevalence of the respective processes depending on temperature, such a scheme has the potential to represent the behavior without the need for a step change in the kinetic parameters. This was the approach taken by Vovelle, Mellottee and Delbourgo [3].

These workers investigated the weight loss of various wood samples under flowing air in a thermobalance and obtained weight loss/temperature curves. For analysis of their results they viewed the wood as comprising two components: the cellulose (50%) and lignin and hemicellulose together (50%). They assembled kinetic equations and integrated them numerically with parameters (A,E) adjusted to fit the experimental results, those for cellulose having been obtained from experiments on pure cellulose. We will consider their data for a wood product termed particle board, for which a match of the experimental and calculated weight-loss curves was achieved by reason of the kinetic scheme given in Table 7.1 and the associated differential equations. Figure 7.1 shows the experimental and calculated weight-loss curves.

The features of the kinetics include competing zero-th and first-order decomposition of the cellulose component and significant decomposition of the lignin/hemicellulose component at lower temperatures. The hemicellulose is believed to account for this.

7.1.5 Gas-phase reactions

Weight-loss experiments do not enable combustion of the devolatilization products of wood to be described, yet this is of major importance as the pyrolysis products account for a very substantial proportion of the heat value of the wood. The current view of the combustion of wood volatiles is that radicals are generated initially during pyrolysis:

$$R - R' \rightarrow R + R'.$$

Table 7.1 Kinetic scheme for particle board (wood) combustion (Vovelle, Mellottee and Delbourgo [3])

Process	*A*	*E* (kJ mol^{-1})	*Order*
Cellulose → decomposition products	$6 \times 10^{18}\ s^{-1}$	230	1
Cellulose → decomposition products	$4 \times 10^{17}\ kg\ s^{-1}$	230	0
Residual solid → combustion products	$4 \times 10^{2}\ s^{-1}$	63	1
Lignin/hemicellulose → decomposition products	$5 \times 10^{5}\ s^{-1}$	84	1
Residual solid → combustion products	$8 \times 10^{1}\ kg^{0.5}\ s^{-1}$	63	0.5

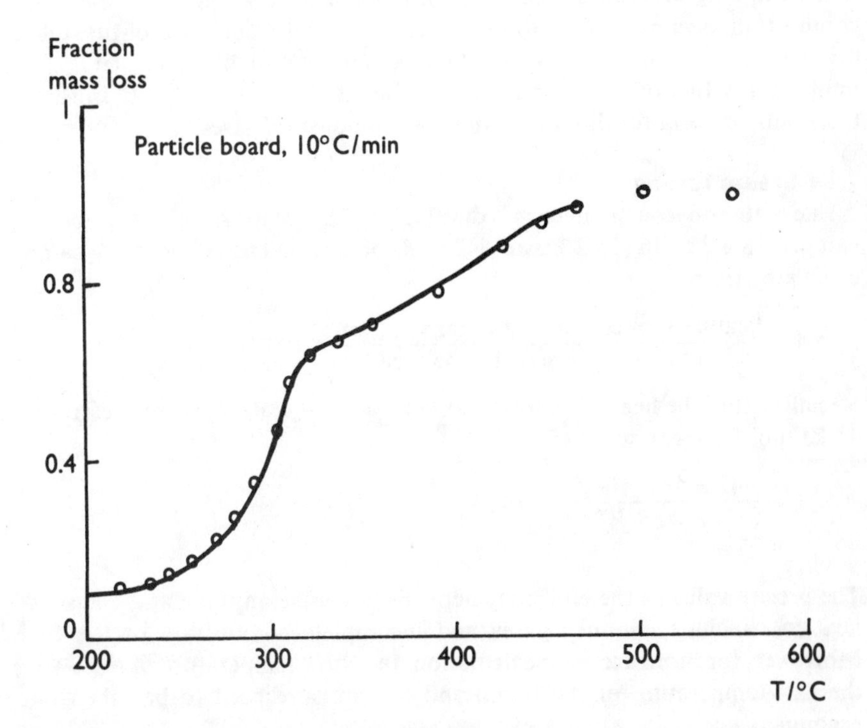

Figure 7.1 Experimental and calculated weight-loss curves for particle board (wood) in a thermobalance (Vovelle, Mellottee and Delbourgo [3])
(Reproduced with the permission of the Combustion Institute.)

For $R = CH_3$, the following process is an example of a reaction sequence in combustion initiated by a radical generated by pyrolysis:

$$CH_3 + O_2 + M \rightarrow CH_3O_2 + M,$$

where M is a third body, followed by:

$$CH_3O_2 \rightarrow CH_2O + OH.$$

The OH radical so generated is involved in further attack and CH_2O (formaldehyde) goes on to form HCO and CO. The relatively low temperature of wood flames favors participation of CH_2O as a key intermediate.

7.1.6 Relative merits of wood fuels

Wood for industrial fuel use can be divided into two categories:

- wood with moisture $< 15\%$ by weight and a heat value of about 20 MJ kg^{-1}, and
- wood with moisture $> 15\%$ with a proportionately lower heat value.

Low-moisture wood fuel includes waste from the furniture and other wood-utilizing industries, as well as sawdust and other fines commonly termed wood flour. Wood fuel of higher moisture may originate from sawmills, or from wood-chipping processes. One tonne of low-moisture wood fuel releases, on combustion, heat equivalent to that released by about 500 litres of fuel oil or 0.5–0.6 tonnes of bituminous coal. One positive feature of wood fuel is that, unlike many fuel oils and bituminous coals, it contains negligible sulfur so there will be no sulfur dioxide in the post-combustion gases.

7.1.7 Steam raising

A rule of thumb from plant tests is that 5 lb (2.27 kg) of low-moisture wood fuel can generate 33.4 lb (15.2 kg, or 842 mol) of steam. The efficiency Φ can be calculated from:

$$\Phi = \frac{\text{heat used in converting water to steam}}{\text{total heat released by the fuel}} \times 100\%. \tag{7.2}$$

Recalling that the heat of vaporization of water (at 1 atmosphere pressure) is 43 kJ mol^{-1}, we can write:

$$\Phi = \frac{842 \times 43 \times 10^3}{2.27 \times 20 \times 10^6} \times 100$$
$$= 80\%.$$

The precise value of the efficiency depends of course on the plant. Causes of heat loss include chimney gases and incomplete combustion of the fuel. Moreover, for more accurate calculation, the entry temperature of the water, the exit temperature of the steam and the pressure need to be taken into account.

It is relevant to outline a representative case study [4] in which the finishing plant at a major textile factory in Alabama, USA, was supplied with steam generated solely by wood combustion. The wood used was purchased entirely as waste from sawmills; no trees were destroyed to provide the wood fuel, which was used in quantities of the order of 10^5 tonnes per year. The saving from the use of wood rather than fuel oil at prices prevailing at the time of installation of the wood-burning plant was such that the plant paid for itself in the first four to five years of its operation. The ash was used for landfill, though there is potential for the use of wood ash in soil treatment.

7.1.8 Domestic use

Wood is a good fuel for domestic heating and certain designs of stove [5] are suitable for wood combustion. In some situations wood obtained from thinning and removal of dead wood in timber areas can be put to such use. This material would otherwise have been wasted. A cord (see Section 7.1.1.) of wood produces only one-fifth to one-third of the weight of ash that a ton of anthracite produces, and as mentioned above, wood ash can be used in soil treatment.

When burnt correctly, wood produces little smoke or soot. Whereas a fireplace delivers only up to 25 percent of the fuel heat into the room, a properly operated wood stove can deliver up to 70 percent. Part of the reason for this is that a wood stove draws much less room air than a fireplace. A fireplace will draw typically 6 m^3 of air per minute, whereas a small wood stove may draw less than 1 m^3 per minute.

7.2 Bagasse

7.2.1 Source, composition and properties

Bagasse is sugar cane refuse and is a cellulosic material. It is widely used in sugar mills and often meets their entire fuel needs. Some sucrose remains in the bagasse and the moisture content following processing is 45–55 percent. When dry, its heat value is about 17 MJ kg^{-1}. Green (high-moisture) bagasse has about 10 MJ kg^{-1} heat value. Ash content is usually low, 1–3 percent. When used in steam raising, bagasse is characterized by high combustion rates and efficiencies of up to 60 per cent.

7.2.2 Thermal ignition

Spontaneous ignition is a problem with stockpiled bagasse and extrapolated Frank–Kamenetskii plots from basket experiments to predict maximum safe stockpiling dimensions (see Chapter 1) grossly overestimate these dimensions for bagasse. While this is partly due to the participation of water in exothermic processes, the fact that the material is well above ambient temperature when it is assembled as a pile also needs to be examined. It is interesting to note that this difficulty with bagasse stimulated work in thermal ignition theory which enabled initial temperature excesses above ambient to be accounted for in the critical conditions [6], as outlined below.

A stockpile of bagasse is treated as an approximate infinite slab with one dimension (height) much smaller than the other two. The classical ignition condition is therefore (notation as in Chapter 1):

$$\delta_{crit} = 0.88 \tag{7.3}$$
$$\theta_{max} = 1.19 \tag{7.4}$$
$$\theta(0) = 0, \tag{7.5}$$

where $\theta(0)$ refers to the time of assembly of the pile. In the modified treatment [6]:

$$\theta(0) = \theta_i > 0, \tag{7.6}$$

and therefore ignitions are possible for lower values of δ than the classical critical one. For given δ, there will be a critical value of θ_i, $\theta_{i(crit)}$, signifying the threshold between a thermal steady state and ignition. The definition of θ_i is:

$$\theta_i = \frac{E}{R(T_o)^2}[(T_o + T_i) - T_o], \tag{7.7}$$

where T_o = ambient temperature (K), and
$T_i + T_o$ = initial pile temperature, assumed uniform. (7.8)

Now a pile of bagasse with the shortest dimension ≈ 4 m and stockpiled at 298K will have $\delta \approx 0.002$ and the corresponding calculated value [6] of $\theta_{i(crit)}$ is 11.6. Assigning a value of 125 kJ mol^{-1} to E and substituting in equation 7.7 gives:

$$(T_i + T_o) = 366\text{K} = 93°\text{C}.$$

Recalling that bagasse is the residue from a process occurring in water very close to its boiling point, this result indicates the possibility of ignition due to a large initial temperature excess.

7.3 Municipal solid waste (MSW)

7.3.1 Nature of MSW

MSW incineration is necessary to reduce its bulk volume and destroy micro-organisms. It is possible for the heat released to be put to use in steam raising. In this respect it should be noted that the heat value of MSW is of the order of 10 MJ kg^{-1} and that is somewhat higher than the heat value of some low-rank coals in their bed-moist state. However, the fact that MSW fluctuates in composition and contains some noncombustibles adds to the difficulty of utilizing MSW as a fuel. The potential of MSW as a fuel to make steam and hence electricity is best understood by reference to a recent case study [7].

7.3.2 Energy from MSW

In the study chosen for illustration, MSW from the local area was blended with citrus peel to give high-moisture fuel of ≈ 6 MJ kg^{-1} heat value. MSW cannot be subjected to laboratory-scale calorimetric determinations of heat value, which use gram quantities, because it is non-uniform. Hence the boiler itself must be used as a calorimeter by measurement of certain temperatures and the feed rate, etc. A procedure for *boiler-as-calorimeter* has been proposed by the American Society of Mechanical Engineers and was applied to the MSW/ citrus peel to yield a value of 6.1 MJ kg^{-1}. When receiving 263 tonnes per day of this fuel, 400 tonnes (2.21×10^7 mol) per day of steam was generated (1.66×10^4 kg h^{-1}). The water phase change occurred at 573 psig (39×10^5 N m^{-2}). The temperature of saturated steam at this pressure is approximately 250° C and the heat of vaporization is 1.73×10^6 J kg^{-1}. This steam was then superheated to 393° C and a heat capacity of ≈ 2000 J kg^{-1} K^{-1} is appropriate to the superheating. From these and supplementary data, the efficiency of the steam-raising process was calculated as 56 percent. We will examine the method of such calculations in more detail in Section 7.4 which discusses steam raising with peat, which, like MSW, is a low-heat-value fuel.

7.4 Peat

7.4.1 Combustion properties

Peat was mentioned in Chapter 6 as the precursor of coal. When raw it has a moisture content of about 90 percent, which can be reduced to 30–35 percent by air drying. In world terms its use as fuel is significant and includes electricity generation. Combustion properties of certain peats relative to those of selected coals were examined [8] to establish the degree of suitability of peat as p.f. in steam raising or as a gas-turbine fuel. One way to assess the potential of a solid fuel for use as p.f. is for 1 g quantities of the milled fuel to be swept by a current of pure oxygen into a small chamber at 700° C, resulting in ignition. The test is then repeated with mixtures of the fuel plus an inert mineral, and the minimum percentage by weight of the inert mineral necessary for suppression of ignition is termed the Godbert number. Coals have Godbert numbers generally in the range 3–15, correlating quite strongly with volatile matter. Values in excess of ≈ 7 are necessary for their use as p.f. in traditional boiler designs.

This value is believed to apply approximately to peats as well and peats require pretreatment to give them a satisfactory degree of combustibility. Pretreatment consists of drying and milling, each of which has a strong influence on the Godbert number. For a particular UK peat, Figure 7.2 shows the proportion of the peat sample that has passed through a 200 mesh (76 μm) screen, against the Godbert number at two moisture contents reflecting different degrees of drying. Benefits of both milling and drying are evident. However, some peats from different deposits dried to the same extent and, with the same range of values on the vertical axis, gave curves well to the left of those in Figure 7.2, with Godbert numbers down to 3. These findings indicate that peats from different sources may require significantly different pretreatment for interchangeability. Factors responsible for these differences include [8]:

(a) Presence of intact root material, which tends to enhance combustibility.
(b) Degree of decomposition, manifest in terms of the physical nature of the peat and describable only in rather imprecise or subjective terms. For example, peats with a small degree of decomposition will release colorless

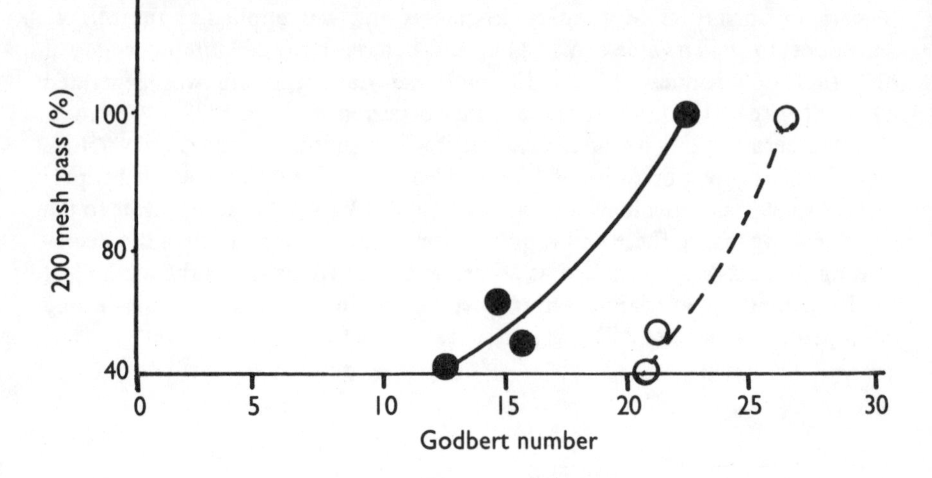

Figure 7.2 Particle size against Godbert number for UK peat. Filled circles: 30% moisture, Open circles: 15% moisture (Lambie [8])
(Reproduced with the permission of the Institute of Energy.)

water when squeezed, whereas those advanced in decomposition will release darkly colored water. A high degree of decomposition appears to favor combustibility.

(c) Elemental analysis. Carbon content varies for different peats, an average value being 58% dry basis. A high carbon content favors combustibility, as does a high hydrogen content (average 5–6% dry basis).

The heat value of a typical dry peat is about 22 MJ kg^{-1}, and for an air-dried peat, about 15 MJ kg^{-1}. We will consider its large-scale use in beds and as a p.f., two of the combustion techniques for coals discussed in Chapter 6.

7.4.2 Peat combustion for power

Electricity generation with peat in a bed [9] used lumps of the fuel of sizes from less than 2 inches to in excess of 12 inches. As in our discussion of steam raising with wood or MSW, we will discuss performance in the context of efficiency.

The plant received 12 469 kg h^{-1} of peat with a moisture content of 27.6% and a heat value of 16 MJ kg^{-1}. The heat released per hour by the fuel (H_{tot}) is therefore 2.00×10^{11} J (= 0.200 TJ). Water at 200°C was evaporated at 58 964 kg h^{-1} and at this temperature the latent heat is 1.95×10^6 J kg^{-1}. Assigning to the heat per hour going into the steam in this way the symbol H_{s1}:

$$H_{s1} = 1.95 \times 10^6 \times 58\,964 \times 10^{-12} \text{ TJ} = 0.115 \text{ TJ}. \qquad (7.9)$$

The steam is then superheated to 425°C and a heat capacity of 2210 J kg^{-1} K^{-1} applies. Calling this component of the heat H_{s2}:

$$H_{s2} = 58\,964 \times (425 - 200) \times 2210 \times 10^{-12} \text{ TJ} = 0.029 \text{ TJ}. \qquad (7.10)$$

Efficiency was enhanced by using an economizer, utilizing sensible heat from the post-combustion gases which had raised the temperature of the water before evaporation by about 60° C. A heat capacity of ≈ 4460 J kg^{-1} K^{-1} is appropriate for water at the economizer temperatures. Calling this component H_{s3}:

$$H_{s3} = 58\ 964 \times 60 \times 4460 \times 10^{-12}\ \text{TJ} = 0.016\ \text{TJ}. \tag{7.11}$$

The efficiency ϕ is calculated from:

$$\phi = \frac{H_{s1} + H_{s2} + H_{s3}}{H_{tot}} \times 100\% = 80\%. \tag{7.12}$$

The data used in these calculations relate to acceptance trials [9] of the plant purchased for use with peat. The efficiency is very satisfactory although it tended to be somewhat lower during longer term use. Combustion of peat lumps for steam raising to make electricity is therefore practicable. Problems include possible difficulties in maintaining ignition, especially if the peat admitted to the plant has a moisture content greater than its air-dried value, and 40% moisture is an upper limit. On the other hand, with peat below its air-dried moisture value there may be a spontaneous ignition hazard before admittance. Lack of uniformity of the lumps is also a problem, the smaller lumps burning out much more rapidly than the larger ones, creating nonuniformity of combustion. Substitution of p.f. firing for a combustion bed, using p.f. plant designed for brown coal, also yielded acceptable performance, with efficiencies of 67% [9]. The use of p.f. size particles instead of large pieces has the advantage that the moisture content of the fuel as admitted can be up to 55% without ignition instability.

7.5 Process gas

7.5.1 Background

Many industrial processes produce effluent gases of considerable heat value, sometimes the same as that of producer gas (≈ 4 MJ m^{-3}). Such a gas can be burnt without the need for auxiliary fuel. A process gas of lower heat value can be incinerated if mixed with small amounts of natural gas. Benefits accruing from such afterburning of process gas are:

(a) Elimination of hydrocarbons. Effluent gases released into the atmosphere must comply with local emission standards for hydrocarbons.

(b) Odor control. Odorous contaminants will be broken down in the incineration flame.

7.5.2 The refinery flare

While many afterburning processes will use industrial burners such as might be used with producer gas, refineries and petrochemical plants use a flare. Waste gases from petroleum plants have high heat values. Oxygen for combustion is supplied by the surrounding air. Combustion is rarely complete in such a system, so care has to be taken that gases subject to emission regulation, for re reduced in the flare to a sufficiently low level. Other features

ame to prevent extinction through flame instability.
ration of steam or air jets when hydrocarbons of high C:H ratio
nt, to control smoke emission.
(c) Velocity control of the influx fuel gas to prevent flashback of the flame.

A typical refinery flare can dispose of 50 tonnes of waste gas per hour.

7.6 Alcohol utilization in spark ignition (SI) engines

Methanol is attractive as an alternative to gasoline in the powering of motor vehicles since it can be made from synthesis gas using feedstocks which are abundant and cheap, for example low-rank coal and biomass. Methanol has received attention as a motor fuel both neat and blended with gasoline. The performance of methanol-powered vehicles is best understood by reference to an example of carefully evaluated exploratory use [10].

In California in the nineteen-eighties, several hundred vehicles in government use were powered by neat or almost neat methanol and performance was closely monitored. Since methanol has a heat value of 22.7 MJ kg^{-1} and a density of 790 kg m^{-3}, and gasoline has a heat value of approximately 45 MJ kg^{-1} and a density $\approx$ 700 kg m^{-3}, 1 gallon of gasoline is equivalent in energy terms to about 1.76 gallons of methanol.

Results with 506 small saloon cars of the same make and model, each having clocked a mileage of the order of 50 000, were that an average of 30.2 mpg was achieved with gasoline and 18.0 mpg with methanol, equivalent to 31.6 mpg with methanol using an energy equivalence factor of 1.76. While the comparison is therefore favorable in energy terms, the range of a methanol-powered vehicle is correspondingly shorter and this is a practical limitation. Performance (in terms of acceleration) and engine durability over 50 000-mile usage with methanol were found to be at least as good as with gasoline. In terms of emissions, inhalable particulates (diameter $\approx$ 10 microns) were reduced by an estimated 20 percent by comparison with gasoline usage, and NO_x by 10 percent. There is a potential to reduce ambient levels of ozone by 17 percent by the substitution of methanol for gasoline. These figures represent very significant environmental advantages for methanol utilization over that of gasoline. On the debit side, formaldehyde in the exhaust emissions from methanol-powered cars has to be considered.

It is to be expected that countries with a large sugar industry will be interested in the production of ethanol by fermentation for fuel use. Pre-eminent in this respect is Brazil where about 3 million motor vehicles are powered by ethanol and over 5 million by gasoline/ethanol blends.

As fuels for spark ignition engines, both methanol and ethanol have good octane ratings, that is, a low propensity to knock. Now that lead is proscribed as an octane enhancer in many parts of the world for environmental reasons, methanol and ethanol can fulfill the same function. Currently some 3 percent of US gasoline usage in weight terms is accounted for by oxygenated octane enhancers including methanol and ethanol.

7.7 Vegetable oils as diesel substitutes

Whereas alcohols can be substituted for gasoline in spark ignition engines, they are not suitable as diesel substitutes for use in compression ignition engines. Stationary diesel engines can be operated on alcohol fuels if *spark assisted* [11] but a compression ignition engine can run *as such* on certain vegetable-oil fuels. Such a material, called Honne oil, was evaluated as a substitute for diesel in a compression ignition engine [12]. Honne oil is denser and more viscous than the diesel for which it is proposed as a substitute and has about 70 percent of the heat value.

In discussing the performance of the engine with diesel and with Honne oil it is necessary to introduce the terms *indicated power* and *brake power*. The indicated power of an engine is that developed in the cylinders and is measured by a pressure indicator. The brake power is the power available at the output shaft of the engine and is measured by application of a brake. It is always lower than the indicated power because of the engine's own friction. The fuel consumption can be expressed in terms of either of these following quantities:

$$\text{Brake specific fuel consumption} = \frac{\text{weight of fuel used per hour}}{\text{brake power}} \qquad (7.13)$$

$$\text{Indicated specific fuel consumption} = \frac{\text{weight of fuel used per hour}}{\text{indicated power}}. \qquad 7.14$$

In Figure 7.3(a) each of these quantities is shown as a function of power for the test engine running on diesel and on Honne oil. The fact that in each case the Honne oil curve lies above the diesel curve simply reflects the lower heat value as noted and when the brake-specific fuel consumption is re-expressed on an energy basis, as in Figure 7.3(b), the curves are very close together, with Honne oil having a slightly lower consumption at higher loads. Exhaust gas temperatures were somewhat higher with Honne oil. Ease of engine starting was satisfactory with the Honne oil and, in contrast to some vegetable materials previously examined as diesel substitutes, it did not cause problems due to solid residue formation.

7.8 Fuels from shale

7.8.1 Background

In the early eighteen-seventies [13] workmen constructing a railway in western USA lit a cooking fire and placed some rocks excavated during the construction round the fire to shelter it. The rocks themselves ignited and the scene of the incident was named Burning Stone Cut.

The workmen had discovered a shale deposit and this was the first such discovery in North America although shale products were by that time being used in Europe. Samples of the rock from the railway construction were

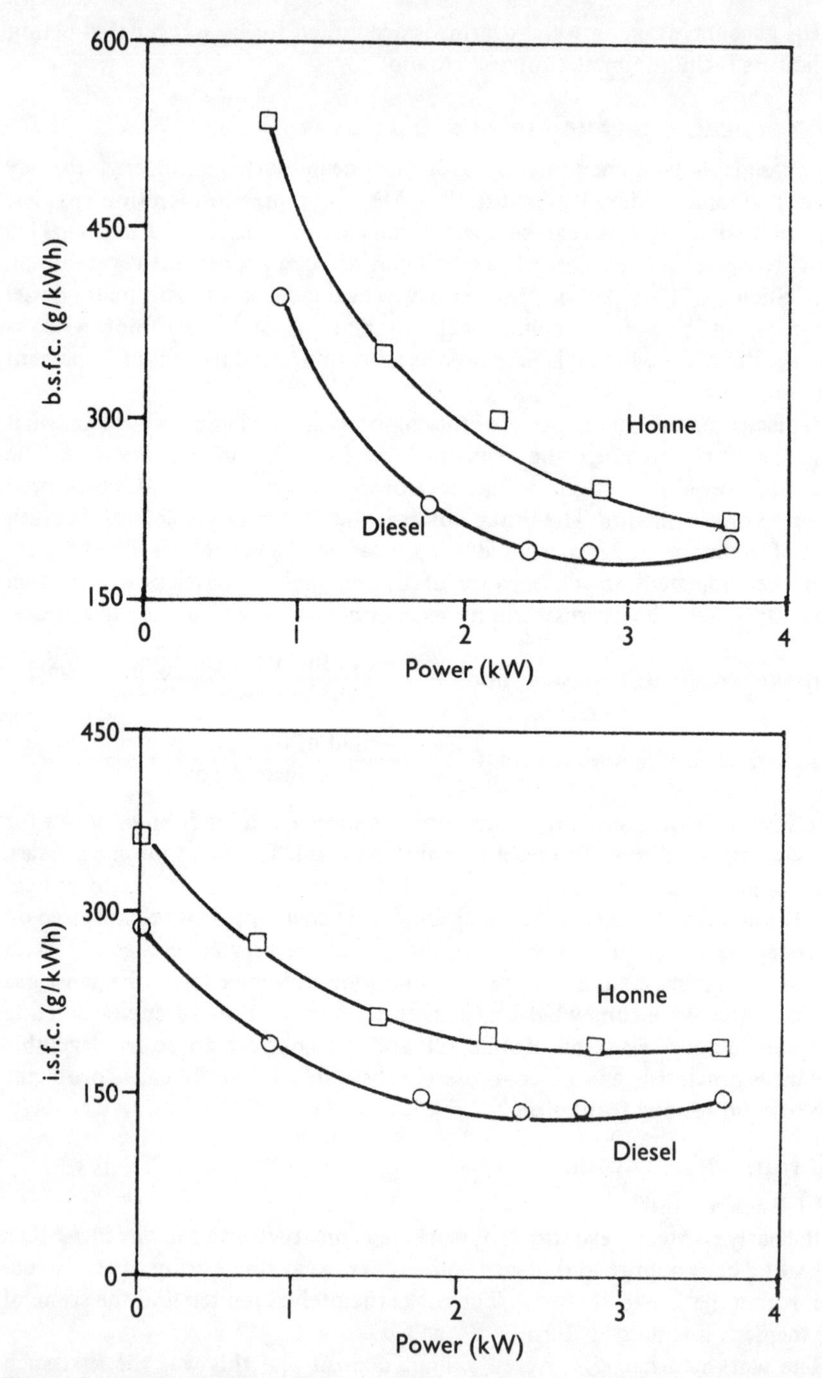

Figure 7.3 (a) Brake-specific and indicated-specific fuel consumption for a compression ignition engine operating on diesel and on Honne oil (Samaga [13])
(Reproduced with the permission of the Indian Institute of Petroleum.)

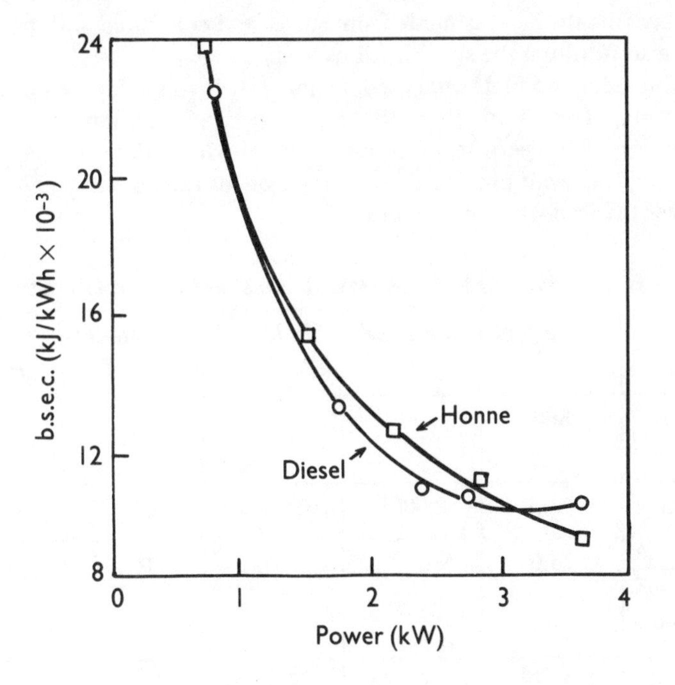

Figure 7.3 (b) Brake-specific energy consumption for a compression ignition engine operating on diesel and on Honne oil. (Samaga [13])
(Reproduced with the permission of the Indian Institute of Petroleum.)

subjected to scientific tests and it was found that a ton of rock yielded, on processing, thirty-five gallons of oil suitable for fuel or lubrication use. In world terms, 20 to over 100 gallons of oil may be yielded per ton of shale, meaning that vast amounts of rock have to be handled, producing large amounts of spent shale waste.

The organic ingredient of shale is kerogen, which yields oil on thermal decomposition in a retort. Hence the formation of oil from shale involves a chemical process, unlike fractionation of crude petroleum which is basically a physical separation of materials with different boiling ranges. Shale oil produced in a retort requires refining in the same way as crude petroleum and as the hydrogen:carbon ratio of kerogen is lower than that of crude petroleum, hydrogenation to upgrade the shale oil may be required. Shale oil so upgraded is called *syncrude*. Shale products tends to be higher in heteroatoms (S,N) than petroleum.

7.8.2 Substitution for petroleum products

Refining, upgrading and sulfur/nitrogen control enable shale oil to be engineered into equivalents of petroleum products. Equivalence can be understood by reference to a US Navy investigation in which jet fuel, conforming to very stringent specifications, was made from shale, and compared with petroleum products also fulfilling the specifications [14].

Three shale-derived fuels were used, denoted A, B and C for the purposes of this summary. They were made by retorting, fractionation and catalytic hydrogenation. The petroleum products with which they are compared, denoted 1 and 2, were obtained from crude oil in the usual way. Table 7.2 summarizes the important properties.

Table 7.2 Comparison of jet fuels made from shale oil and crude petroleum

	Shale fuel A	*Shale fuel B*	*Shale fuel C*	*Petroleum fuel 1*	*Petroleum fuel 2*
Nitrogen (ppm)	860	79	5	—	—
Sulfur %	0.05	0.0001	0.0005	0.11	<0.4
Aromatic fraction (% by volume)	26.0	23.7	21.3	16.2	18.2
Boiling range (°C)	171–282	177–288	177–288	176–258	176–263
Heat value (MJ kg^{-1})	43.10	—	—	43.07	43.07

Fuels A, B and C are counterparts of the kerosene components of crude petroleum. Other product boiling ranges—light naphtha, heavy naphtha, diesel—are also obtainable from shale.

7.8.3 Combustion of spent oil shale

Not all of the organic component of shale is removed on retorting. Organic loss on retorting varies widely between different deposits, being, for example, 75% for Colorado shale and 47% for Moroccan shale. This means that spent shale, having retained some organics, can be burnt to provide heat for the retorting process. Combustion kinetics for examples of spent Australian shale in a fluidized bed have been deduced experimentally [15]. In the underlying analysis, the time taken for the carbon content to burn to a fractional extent X is the sum of two components, related to oxygen diffusion and chemical kinetics.

The kinetic contribution is:

$$t_k = \frac{-\ln(1-X)}{bk'C}, \qquad (7.15)$$

where k' = rate constant
= A exp $[-E/RT]$ (order 1 in oxygen assumed)
b = stoichiometric coefficient
= 1 for conversion to CO_2
C = oxygen concentration (mol m^{-3}).

The meaning of equation 7.15 is that under conditions such that the burning of the spent shale is governed purely by kinetics without any effects due to pore diffusion, t_k is the time required for conversion of a fraction X of the initial quantity of material to combustion products.

The time t_d for conversion of the same fraction X if internal diffusion controls the rate is given by:

$$t_d = \frac{P(X)d^2\sigma}{24bCD'} + \frac{d^2\sigma X}{6bC\mathrm{Sh}D'}, \tag{7.16}$$

where $P(X) = 1 - 3(1 - X)^{0.66} + 2(1 - X)$
d = particle diameter (m)
σ = molar concentration of carbon in the particle (mol m^{-3})
D = diffusion coefficient (m^2 s^{-1})
D' = effective diffusivity (m^2 s^{-1})
Sh = Sherwood number.

D' has a different value for each spent shale material examined and depends on the porosity of the particles and the tortuosity, that is, the factor by which the route taken in the transport of oxygen into the particles is longer than the most direct route.
The Sherwood number is given by:

$$\mathrm{Sh} = \frac{k_c d}{D},$$

where k_c = convection mass transfer coefficient (m s^{-1}).

The total reaction time is given by:

$$t_{tot} = t_k + t_d. \tag{7.17}$$

Experimental application of the above approach involves fluidized-bed combustion of spent shale samples and the monitoring of carbon combustion by infrared analysis of oxides of carbon to give measured values of X as a function of time. In Figure 7.4, the time taken for 90 percent conversion of the carbon content of the spent shales to oxides is plotted as a function of d for three temperatures: 600°C, 650°C and 700°C. The larger particles require longer for conversion, d appearing in the numerator of each term in the diffusion component of the conversion time (equation 7.16).

For a particular particle size (for example, 1 mm) at each of the three experimental temperatures, values of T can be inserted into equation 7.17 (via k') as well as D (temperature dependent), C, d, etc., with $X = 0.9$ and t_{tot}, the experimentally deduced times to 90 percent conversion. Three equations in

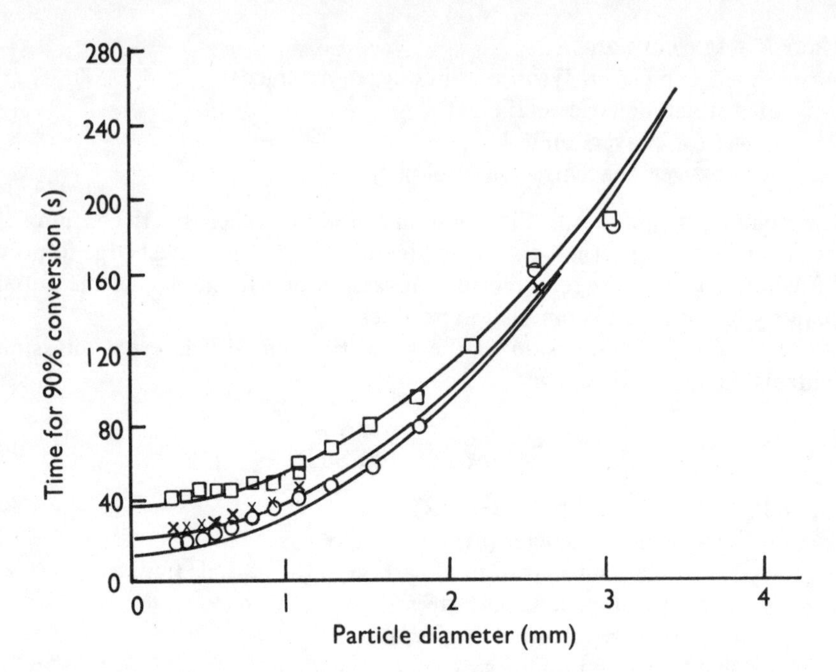

Figure 7.4 Times of conversion of the carbon content of spent Australian shale to oxides of carbon. Squares: 600° C. Crosses: 650° C. Circles: 700° C. (Charlton [15])
(Reproduced with the permission of Butterworth Heinemann.)

three unknowns (A, E and D') will result, giving a means of obtaining values for these quantities. For the spent shale, (see the results in Figure 7.4) the following were thus calculated:

$$D' = 1 \times 10^{-5}\ \mathrm{m^2\ s^{-1}}$$
$$E = 78\ \mathrm{kJ\ mol^{-1}}$$
$$A = 0.13\ \mathrm{Pa\ s^{-1}}.$$

The spent shale to which these kinetics apply had 14.9% carbon.

7.9 Fuels from tar sands

7.9.1 Introduction

Tar sands are sand or sandstone whose pores are filled with bitumen. Mild heating (for example, with hot water) is required to separate the bitumen which can then be processed to make liquid fuels. The most abundant deposits of tar sands are in North America, particularly Alberta, Canada. The bitumen content of Canadian tar sands is typically 12 percent.

7.9.2 In-situ heating

Heat can be generated in a tar-sand reservoir by igniting some of the deposit in situ to provide heat for oil recovery. Laboratory-scale tests [16] were performed on a Canadian tar sand in which a charge of the tar sand was subjected to a temperature program of 5° C min^{-1} up to 1000K with flowing oxidizing gas.

Oxides of carbon and unused oxygen were measured in the efflux gas. The distinctive feature of the combustion is that it displays two regions of reactivity separated by a minimum in oxygen consumption. Figure 7.5 shows oxygen consumption as a function of time (lower horizontal scale) and hence temperature (upper horizontal scale) for the tar sand under the heating conditions described, with 15% oxygen as the influx. The minimum forms a boundary between low-temperature reaction, with a maximum in terms of oxygen consumption at ≈ 740K, and high-temperature reaction with a maximum at ≈ 850K. Similar experiments were performed under conditions of greater oxygen abundance (up to 40%), and in these experiments, oxygen balance indicates that the hydrogen content of the bitumen is oxidized almost entirely in the low-temperature regime.

7.10 Historical perspective

While there has been relatively recent impetus for research into these less-common fuels owing to growing awareness of the fact that crude oil quantities are finite, in fact use of some of the fuels discussed in this chapter predates use

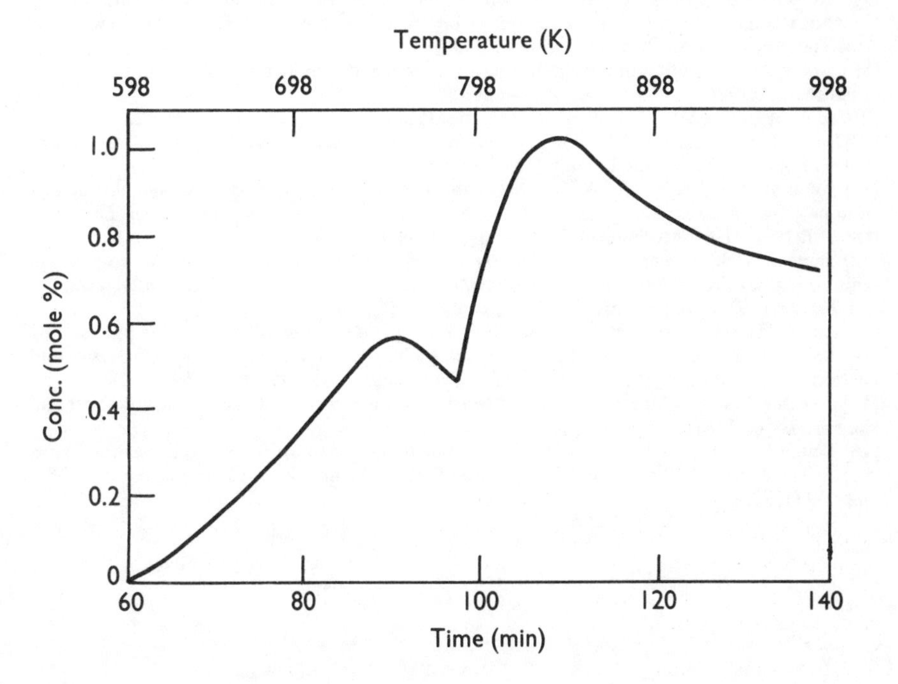

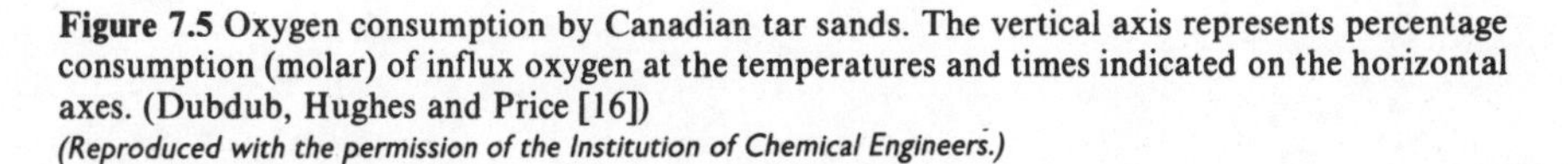

Figure 7.5 Oxygen consumption by Canadian tar sands. The vertical axis represents percentage consumption (molar) of influx oxygen at the temperatures and times indicated on the horizontal axes. (Dubdub, Hughes and Price [16])

(Reproduced with the permission of the Institution of Chemical Engineers.)

of petroleum products. For example, events regarded as marking the creation of the petroleum industry were the discovery of crude oil in Pennsylvania by E. Drake in 1859 and the founding of the Standard Oil Company by J. D. Rockefeller in 1870. However, shale was being retorted and refined in Europe by 1850. Perhaps the correct way to view the pressing into service of these fuels in place of petroleum products is as revival or rediscovery rather than novelty in fuel technology.

References

[1] Johnson A. J., Auth G. H. (Eds), *Fuels and Combustion Handbook*, New York: McGraw-Hill (1951)

[2] Tinney E. R., 'The combustion of wooden dowels in heated air', *Tenth Symposium (International) on Combustion*, 925, Pittsburgh: The Combustion Institute (1965)

[3] Vovelle C., Mellottee H., Delbourgo R., 'Kinetics of thermal degradation of cellulose and wood in inert and oxidative atmospheres', *Nineteenth Symposium (International) on Combustion*, 797, Pittsburgh: The Combustion Institute (1983)

[4] Russell B., 'Wood-fired steam plant at Russell Corporation', *Wood Energy: Proceedings of Governor William G. Milliken's Conference*, 113, Ann Arbor Science Publishers, Michigan (1977)

[5] Cheremisinoff N. P., *Wood for Energy Production*, Ann Arbor Science Publishers, Michigan (1980)

[6] Gray B. F., Scott S. K., 'The influence of initial temperature excess on critical conditions for thermal explosions', *Combustion and Flame* **61** 227–236 (1985)

[7] Ettehadieh B., Lee S. Y., 'Incineration of low-quality MSW in the rotary water-cooled O'Connor combustor', *Fourteenth Biennial Conference on Waste Disposal*, 193, New York: American Society of Mechanical Engineers (1990)

[8] Lambie R., 'Estimation of the combustibility of coal and peat by the Godbert inflammability apparatus', *Journal of the Institute of Fuel* **26** 250 (1953)

[9] Cronin W., Lang J. F., 'Peat-fired power stations', *Journal of the Institute of Fuel* **27** 545 (1954)

[10] Smith K. D., 'California's methanol prográm', *Seventh International Symposium on Alcohol Fuels*, 36–48, Paris: Editions Technip (1986)

[11] Johns R. A., Henham A. W. E., Newnham S. K. C., 'The combustion of alcohol fuels in stationary spark-assisted engines', *Proceedings of the Institution of Mechanical Engineers* (IMechE 1991-1), 61, Mechanical Engineering Publications Ltd (1991)

[12] Samaga B. S., 'Use of Honne oil (Kino seed oil) as an alternative fuel for the compression ignition engine', *Proceedings of the IX National Conference on Internal Combustion Engines*, **Vol 1 A-20**, Dehra Dun: Indian Institute of Petroleum (1985)

[13] Anon., 'The oil deposits of the Great West', *Scientific American*, March 14, 1874 pp. 160–161

[14] Solash J., Hazlett R. N., Hall J. M., Nowack C. J., 'Relation between fuel properties and chemical composition. I: Jet fuels from coal, oil shale and tar sands', *Fuel* **57** 521 (1978)

[15] Charlton B. G., 'Comparative fluidized bed combustion kinetics of some Australian spent oil shales', *Fuel* **66** 384 (1987)

[16] Dubdub I., Hughes R., Price D., 'Kinetics of in situ combustion of Athabasca tar sands studied in a differential flow reactor', *Transactions of the Institution of Chemical Engineers* **68A** 342–349 (1990)

CHAPTER 8

The Behavior of Sulfur in Combustion Processes

Abstract

Sulfur reactions in flames are discussed from the viewpoints of the H_2S/air flame and sulfur added to other flames as H_2S or as SO_2. The formation of metal sulfates in combustion plant due to reaction with fuel sulfur is also discussed. The use of metal compounds to fix fuel sulfur and thereby prevent its release with the post-combustion gases as SO_2 is discussed, giving examples.

8.1 Introduction

Products of combustion of solid, liquid and gaseous fuels in industrial plant are of great environmental importance and are very properly the subject of a great deal of research and discussion. Our concern in this chapter is to outline the combustion behavior of sulfur which can lead to emission of sulfur oxides. Sulfur dioxide has arguably received more scientific attention than any other pollutant owing to its role in acid rain and visibility reduction, as well as its effects on health. If a particular fuel has a sulfur content sufficiently high for the sulfur oxide content of the post-combustion gases to be a potential problem (and many do), control is possible under three broad strategies:

(a) Reduce the sulfur content of the fuel (pre-combustion control), that is, desulfurize it.

(b) Incorporate into the combustion system a material such as limestone, capable of trapping sulfur oxides as sulfate (combustion control).

(c) Remove sulfur oxides from the effluent gases by scrubbing (post-combustion control).

We will examine sulfur combustion chemistry mainly in terms of the organic sulfur in fuels. In coals, some sulfur is usually present as pyrite (FeS_2) in addition to organic sulfur and this also contributes to the sulfur dioxide release at sufficiently high temperatures.

8.2 Sulfur reaction in flames

8.2.1 The hydrogen sulfide/air flame

The oxidation of organic sulfur in solid and liquid fuels can be understood by reference to a laboratory study of the oxidation of hydrogen sulfide as a model compound [1]. In this study two hydrogen sulfide flat flames (see Chapter 4) were examined, termed by the investigators flame 5 (gas influx 0.061 mole fraction H_2S, 0.2 mole fraction oxygen, balance nitrogen) and flame 6 (influx 0.05 mole fraction H_2S, 0.139 mole fraction oxygen, balance argon). Temperature profiles in the flames were measured by thermocouple techniques and species in the flame were examined by mass spectrometry. Sulfur oxides (SO, SO_2 and SO_3) were also determined by passing subsamples of the analysis stream to the mass spectrometer through a cold trap and chemically analyzing the condensate.

Discrepancies between amounts of sulfur dioxide measured by mass-spectrometric and chemical methods led to the belief that in the flame, some of the sulfur dioxide existed as S-O-O, a superoxide structure, which in time became the normal structure O-S-O. In flame 6 in its steady state, about 13 percent of the sulfur dioxide existed as S-O-O, whereas in flame 5 the percentage was 28. Flame 6 had the greater proportion of inert gas and independent work [2] indicates that inert gas suppresses S-O-O formation. The flame results therefore show the correct trend.

Figure 8.1 shows concentration (for all species other than SO_3, which will be discussed separately) and temperature profiles for flame 5. Conversion of hydrogen sulfide to the various oxidized forms of sulfur is complete. Rate information is available from these profiles by means of equation 4.20 given in the discussion of flames in Chapter 4:

$$\frac{\sigma_o v_o}{M_i}\frac{dG_i}{dz} = K_i.$$

The flux G_i can be calculated from the experimental plots of mole fraction x_i against z in the following way [3]. The *mass* fraction f_i is available from the mole fraction from:

$$f_i = x_i \frac{M_i}{M_m}, \tag{8.1}$$

where M_m = mean molecular weight (kg mol^{-1}).

Now, recalling equation 4.18 and incorporating the continuity condition (equation 4.3):

$$G_i = \frac{N_i M_i (v + V_i)}{\sigma v}$$

$$= \frac{N_i M_i (v + V_i)}{\sigma_o v_o}.$$

Since N_i = mole number per unit volume:

$$\frac{N_i M_i}{\sigma} = f_i, \tag{8.2}$$

therefore:

$$G_i = f_i \frac{v + V_i}{v}. \tag{8.3}$$

The diffusional component of the flux V_i is simply:

$$V_i = -\frac{D}{X_i}\frac{dX_i}{dz}, \tag{8.4}$$

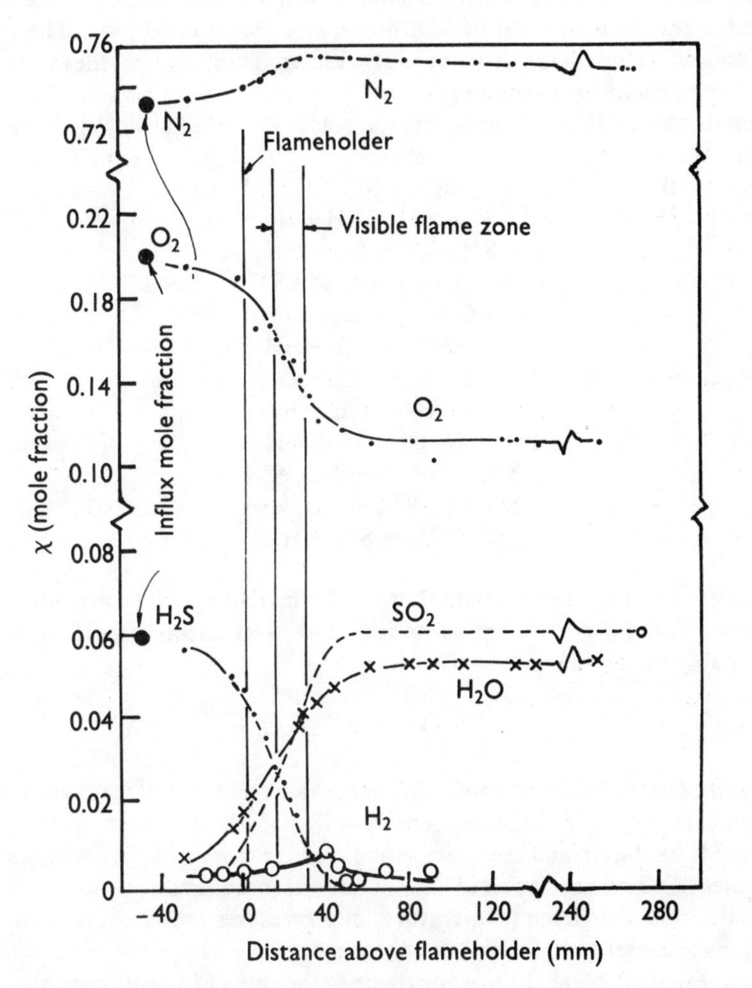

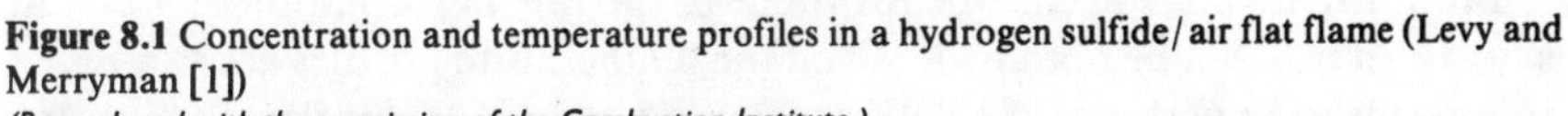

Figure 8.1 Concentration and temperature profiles in a hydrogen sulfide/air flat flame (Levy and Merryman [1])
(Reproduced with the permission of the Combustion Institute.)

and the diffusion coefficient D is straightforward to estimate from knowledge of the gas composition and temperature at any particular z-value. Values of x_i and hence f_i (equation 8.1) as a function of z are obtainable from experimental measurements such as those in Figure 8.1, together with the velocity profile. The quantity $G_i(z)$ can be calculated and its differential obtained from the slopes. From these, $K_i(z)$ can be calculated by substitution of the differentials into equation 4.20, and K_i can be plotted against z for the species of interest, that is, the *rate* of formation of the different species at different points along the flame.

In Figure 8.2 the rates of formation or disappearance of the various species, including sulfur trioxide, thus calculated for flame 5, and the rates of formation of SO and SO_2, are seen to have steep maxima in the hotter region of the flame, while the rate of formation of SO_3 has only a gentle maximum. The corresponding curves for flame 6 are broadly similar. These rate profiles provide a basis for mechanistic discussion.

Steps in the oxidation (numbering them as in the original work [1]) include:

$$H_2S \rightarrow SH + H \qquad 5$$
$$H_2S + H \rightarrow SH + H_2 \qquad 6$$
$$SH + O_2 \rightarrow SO + OH \qquad 7$$
$$OH + H_2S \rightarrow H_2O + SH \qquad 8$$
$$SO + O_2 \rightarrow SO_2 + O \qquad 9$$
$$O + H_2S \rightarrow OH + SH \qquad 10$$
$$H + O_2 \rightarrow OH + O \qquad 18$$
$$O + H_2 \rightarrow OH + H \qquad 19$$
$$OH + H_2 \rightarrow H_2O + H \qquad 20$$
$$SO_2 + O + M \rightarrow SO_3 + M \qquad 21$$
$$SO + O_2 + M \rightarrow SO_3 + M \qquad 22$$
$$SO_3 + H_2 \rightarrow SO_2 + H_2O \qquad 23$$

At much lower temperatures than those in the flat flame and at subatmospheric pressures, S_2O [4] is believed to be involved in branching (see Chapter 2), presumably by the step:

$$S_2O + O \rightarrow 2SO,$$

followed by steps 9 and 10 above. SO is a biradical and very reactive. Two of the branching steps in the above scheme—18 and 19—are common to hydrogen sulfide oxidation and the hydrogen–oxygen reaction. The branching step 7 generates SO. From the point of view of understanding sulfur oxidation in fuel utilization, the flat flame temperatures and pressures are of course superior, being much closer to industrial conditions.

From Figure 8.2 and the one for flame 6, the rate of formation of water at, say, $z = 20$ mm can be obtained. When the temperature profiles are examined, this means a temperature of 1250K for flame 5 and 1350K for flame 6. Product

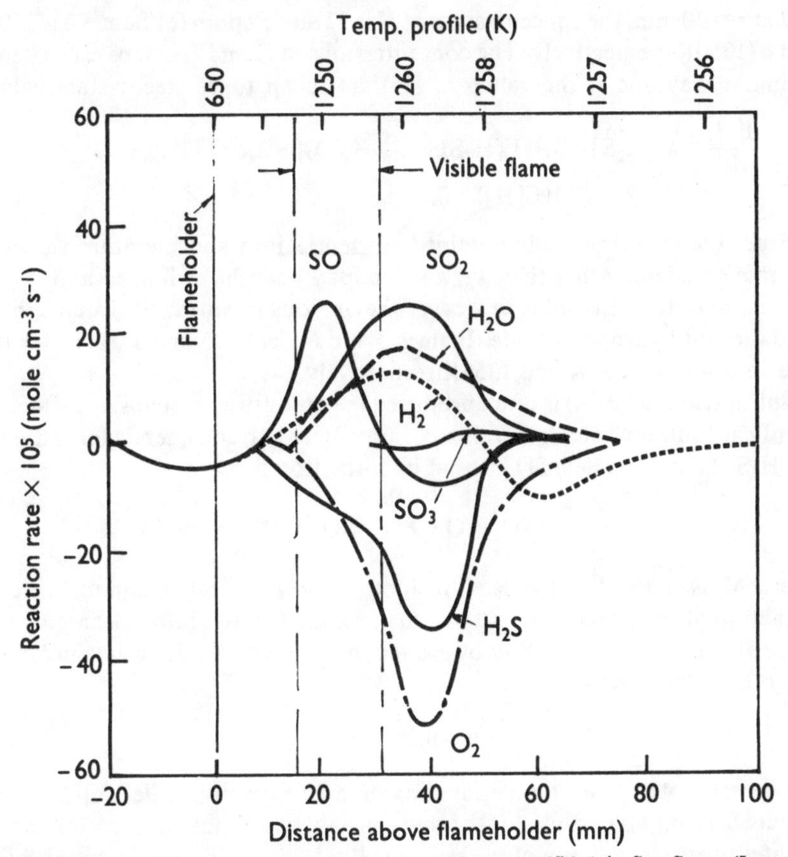

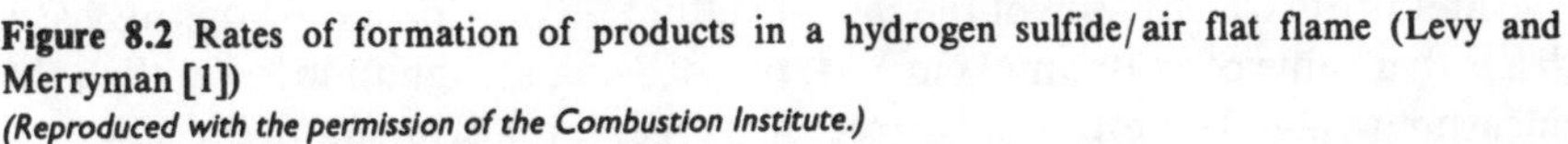

Figure 8.2 Rates of formation of products in a hydrogen sulfide/air flat flame (Levy and Merryman [1])
(Reproduced with the permission of the Combustion Institute.)

concentrations can be measured at these z-values and substituted into the expression for water formation rate from the above kinetic scheme, that is:

$$\frac{d[H_2O]}{dt} = k_8[OH][H_2S] + k_{20}[OH][H_2], \qquad (8.5)$$

where the subscripts in the rate constants refer to steps in the above scheme. From the measured rate (Figure 8.2) and literature values of the rate constants, together with knowledge of $[H_2S]$ and $[H_2]$ from the mass spectrometric results (Figure 8.1) a value for [OH] at any value of the coordinate z can be deduced and at $z \approx 20$ mm the value of OH so calculated is 4 ppm for flame 5 (1250K) and 8 ppm for flame 6 (1350K). From this and the rate of formation of hydrogen at the same z-value (Figure 8.2) the value of [H] can be calculated from:

$$\frac{d[H_2]}{dt} = k_6[H][H_2S] - k_{20}[OH][H_2], \qquad (8.6)$$

and at $z \approx 20$ mm, the concentration of H is 11 and 19 ppm for flames 5 (1250K) and 6 (1350K) respectively. The concentrations of O and H can be estimated in a similar way and if the values of [H] are taken to be steady-state values:

$$\frac{d[H]}{dt} = k_5[H_2S] - k_6[H][H_2S] - k_{18}[H][O_2] + k_{19}[O][H_2] + k_{20}[OH][H_2] = 0. \quad (8.7)$$

From the experimentally obtained concentrations and literature values of the rate constants other than k_5, a value for k_5 can be estimated and it is of interest to obtain this value in view of the obvious importance of step 5 in the oxidation of hydrogen sulfide. Flames 5 and 6 yield 595 and 1180 s^{-1} for this rate constant at 1250K and 1350K respectively.

Sulfur trioxide (SO_3) is often produced in combustion systems using fuels of significant sulfur content, therefore it is instructive to consider its formation in the H_2S flame. The belief [1] is that its formation is by:

$$SO_2 + O + M \rightarrow SO_3 + M,$$

where M is a third body, leading to the experimental maximum rate of formation of $\approx 10^{-5}$ mol cm^{-3} s^{-1} shown in Figure 8.2. It is fairly clear from this value of the rate that reaction of atomic, not molecular, oxygen is involved. The rate of the reaction:

$$SO_2 + 0.5O_2 \rightarrow SO_3,$$

can be estimated from the sulfur dioxide and oxygen profiles in the flame (Figure 8.1); reliable values for the rate constant at various temperatures are in the literature. Calculation of the rate of sulfur trioxide formation on this basis leads to a value of $\approx 10^{-7}$ mol cm^{-3} s^{-1}, two orders of magnitude lower than the measured rate. The established role of oxygen atoms in sulfur trioxide formation, as well as in chain-branching, necessitates information on the oxygen atom profile in the steady-state flame. This can be done in essentially the way outlined for OH and H: the differential rate expression for [O] is set up and the steady-state approximation ($d[O]/dt = 0$) applied at different z-values and hence temperatures. At each z-value chosen there will be an algebraic equation with different coefficients of the concentrations because of the temperature dependence of the rate constants. Their solution yields the O atom profile and this was done [1] for two mechanisms as follows:

(a) That for which the formation of O is due to step 9 and direct dissociation of O_2, that is:

$$O_2 + M \rightarrow 2O + M \qquad (M = \text{second body}),$$

the net formation being expressible by:

$$SO + 2O_2 \rightarrow SO_2 + 3O.$$

(b) That for which formation of O is due to steps 18 and 20, together with oxygen dissociation, the net formation being:

$$H_2 + 2O_2 \rightarrow H_2O + 3O.$$

Mechanism (b) involves no sulfur species. Figure 8.3 shows calculated O atom profiles in flame 5 for (a) and (b) respectively and Figure 8.4 shows the experimental sulfur trioxide profile for the same flame (previously omitted from Figure 8.1) and it is clear that on the basis of mechanism (a) the O atom concentration drops to zero at almost the precise *z*-value where the sulfur trioxide concentration forms a plateau. This is very convincing evidence of the participation of step 9 in O atom and hence sulfur trioxide formation. On the

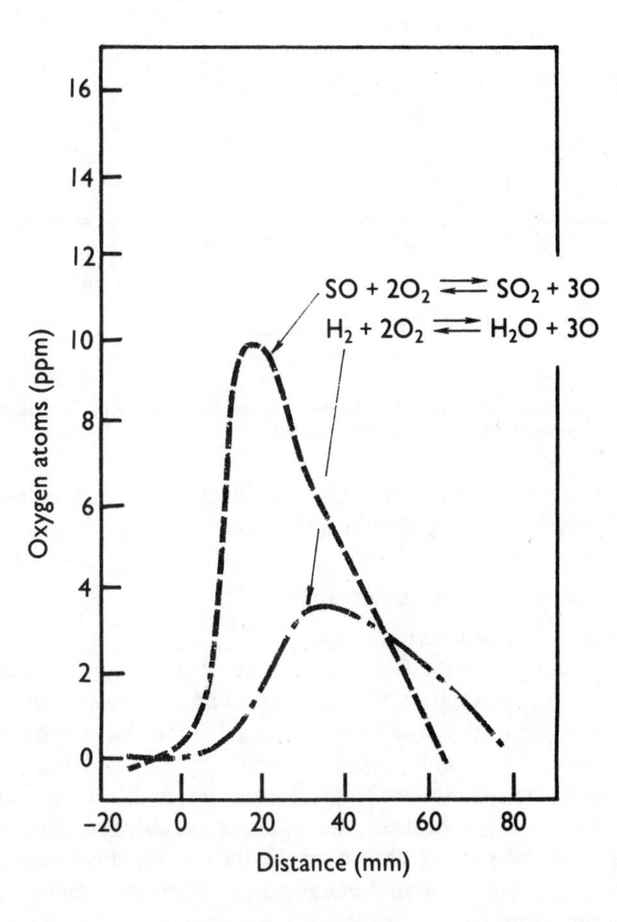

Figure 8.3 O atom profiles in a hydrogen sulfide/air flat flame, calculated according to two proposed mechanisms (Levy and Merryman [1])
(Reproduced with the permission of the Combustion Institute.)

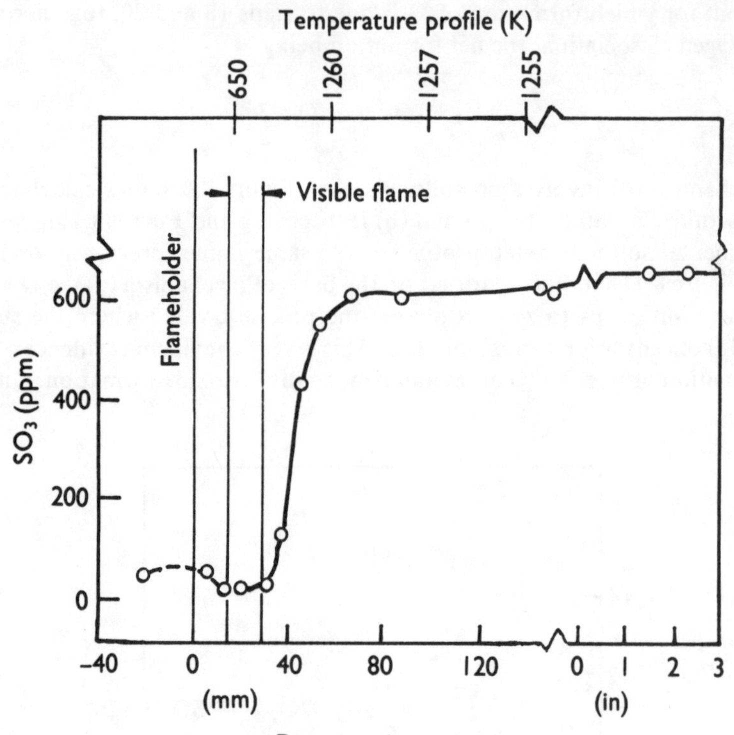

Figure 8.4 Sulfur trioxide profile in a hydrogen sulfide/air flat flame (Levy and Merryman [1]) *(Reproduced with the permission of the Combustion Institute.)*

basis of mechanism (b) the sulfur trioxide level would not be expected to plateau until considerably further into the flame.

8.2.2 Sulfur introduced into other flames

- **Sulfur added as sulfur dioxide**

We have seen that an understanding of sulfur behavior in flames requires consideration of the interaction with hydrogen and oxygen species, so it is not surprising that mechanistic studies of sulfur in flames have sometimes used sulfur—either as sulfur dioxide or as hydrogen sulfide—artificially introduced into a hydrogen/oxygen flame. We will consider such an investigation by Fenimore and Jones [5]. Whereas in the hydrogen sulfide flat flame studies [1] steady states of certain species were *assumed*, the work by Fenimore and Jones examines the possibility of actual chemical equilibrium in some of the processes involving sulfur.

For a flat flame of composition in mole fraction terms: argon 0.42, hydrogen 0.47, oxygen 0.1 and sulfur dioxide 0.01 (that is, a fuel-rich flame) removal of

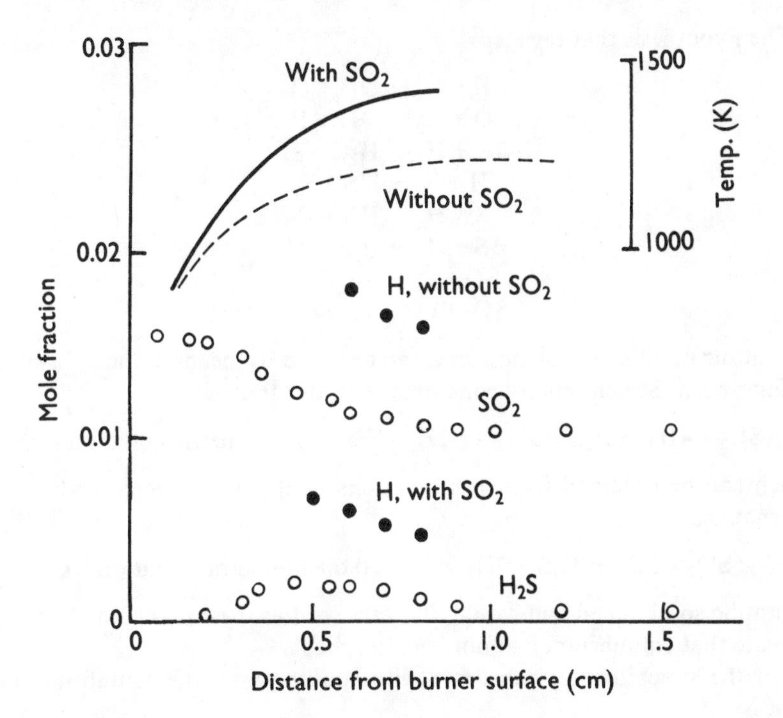

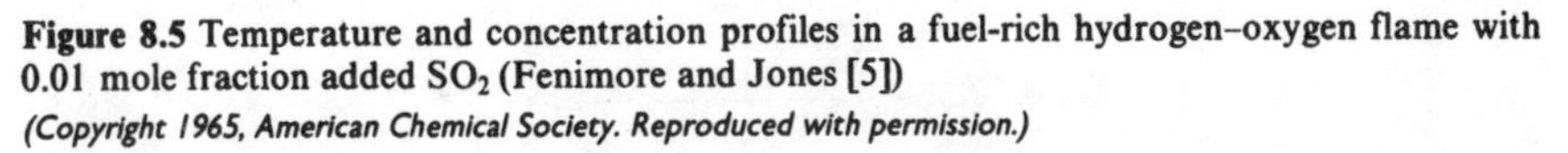
Figure 8.5 Temperature and concentration profiles in a fuel-rich hydrogen–oxygen flame with 0.01 mole fraction added SO_2 (Fenimore and Jones [5])

gas samples and analysis led to the profiles shown in Figure 8.5. The H profiles, like those of the stable products, were experimentally determined. From the temperature profile, by means of the theory discussed in Chapter 4, it was possible to calculate heat release rates at various points in the flame, and this is shown in Figure 8.6.

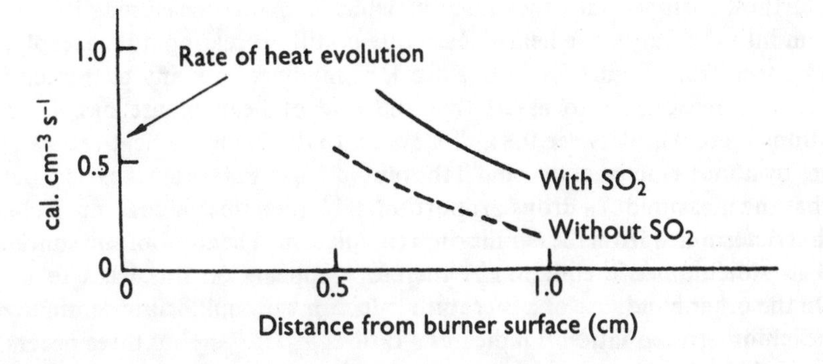

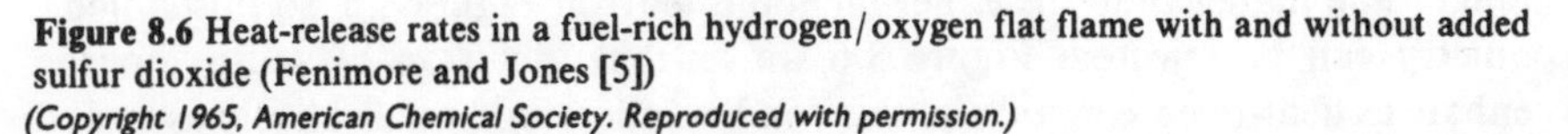
Figure 8.6 Heat-release rates in a fuel-rich hydrogen/oxygen flat flame with and without added sulfur dioxide (Fenimore and Jones [5])

The hypothesis that the steps:

$$\begin{aligned} H + O_2 &\rightarrow OH + O \\ O + H_2 &\rightarrow OH + H \\ OH + H_2 &\rightarrow H_2O + H \\ H + S_2 &\rightarrow HS + S \\ S + H_2 &\rightarrow HS + H \\ HS + H_2 &\rightarrow H_2S + H \\ SO + O &\rightarrow S + O_2 \\ SO_2 + O &\rightarrow SO + O_2, \end{aligned}$$

each attain equilibrium in the flame, can be tested by means of the information in Figure 8.5. Stoichiometry consistent with the above is:

$$SO_2 + 4H_2 \rightarrow H_2S + 2H_2O + 2H \qquad \text{(stoichiometric equation A)},$$

which can be obtained from combinations of the above steps, whereas the alternative:

$$SO_2 + 3H_2 \rightarrow H_2S + 2H_2O \qquad \text{(stoichiometric equation B)},$$

cannot be so obtained and would, if established as the correct stoichiometry, indicate that equilibrium had not been reached.

For the stoichiometry consistent with equilibration, the equilibrium constant is:

$$K_{eq} = \frac{X_H^2 \, X_{H_2O}^2 \, X_{H_2S}}{X_{SO_2} \, X_{H_2}^4}, \tag{8.8}$$

where X denotes mole fraction, and this can be rearranged to give a calculated value of X_H (using a literature value of K_{eq}) which can be compared with the experimental one at any chosen position in the flame.

The value of X_H so calculated 1.2 cm into the flame, where X_{H_2S} has levelled (Figure 8.5), is about a factor of 3.7 lower than the measured value at 0.8 cm, the farthest distance into the flame at which X_H was measured. However, 0.8 cm into the flame, the heat-release rate is still decreasing quite steeply, as can be seen from Figure 8.6 and, since H is involved in many heat-releasing steps, it is reasonable to assert that the rate of heat release has a direct relationship to X_H. Between 0.8 and 1.2 cm into the flame the heat-release rate drops by about a factor of two and if the plot of X_H in Figure 8.5 is extrapolated so that the measured X_H drops proportionately, then this value is much closer to that determined from the equilibrium calculation. The equilibrium summarized in stoichiometric equation A therefore appears on this basis to hold.

On the other hand, use of a literature value for the equilibrium summarized in stoichiometric equation B indicates a ratio X_{H_2S}/X_{SO_2} about three orders of magnitude higher than the experimental one from Figure 8.5, so this stoichiometry can be rejected. Figure 8.6 shows that inclusion of sulfur dioxide enhances heat-release rates, and this can be understood qualitatively as due to the effect of sulfur dioxide on the strongly exothermic radical recombination step:

$$H + H + M \rightarrow H_2 + M.$$

Any mechanism by which the rate of this step is increased will increase the heat-release rate, and sulfur dioxide is believed to effect such an increase by the reaction:

$$H + SO_2 + M \rightarrow HSO_2 + M,$$

followed by:

$$HSO_2 + H \rightarrow H_2 + SO_2,$$

the net process being simply H recombination, but at a faster rate than without sulfur dioxide, manifest in the heat-release behavior shown in Figure 8.6.

The work of Fenimore and Jones as outlined so far relates to fuel-rich conditions. Similar experiments under fuel-lean conditions (for example, 0.2 mole fraction of hydrogen in the influx) result in conversion of a small proportion of the sulfur dioxide to the trioxide. By contrast there is no hydrogen sulfide, as there is in parts of the flame under fuel-rich conditions (Figure 8.5). However, even under fuel-rich conditions the proportion of hydrogen sulfide is small and most of the sulfur is in an oxidized state. The point was therefore addressed as to whether the sulfur trioxide participates in any equilibrium in the flame, for example:

$$SO_2 + 0.5O_2 = SO_3$$

or

$$SO_2 + O_2 = SO_3 + O$$

or

$$SO_2 + O + M = SO_3 + M,$$

where = denotes chemical equilibrium. This could be verified from measured estimates of the O radical amounts and the equilibrium constants: the first two of the three equilibria above predict a $[SO_3]/[SO_2]$ ratio much smaller than the measured one, and the third predicts that the ratio will be much larger than the measured one. This indicates that the SO_3 formation observed under fuel-lean conditions does not take its origin from any of these equilibria. The explanation furnished is that the forward process:

$$SO_2 + O + M \rightarrow SO_3 + M,$$

is the correct one, but that once SO_3 is formed it reacts by a more rapid process than the reverse of the formation step, probably one or both of:

$$SO_3 + O \rightarrow SO_2 + O_2,$$

and

$$SO_3 + H \rightarrow SO_2 + OH,$$

and this would result in a lower experimental ratio of $[SO_3]$ to $[SO_2]$ than that predicted by equilibrium formation, which would of course, require that forward and reverse rates were equal. Sulfur trioxide formation in a hydrogen/oxygen flame containing added sulfur is therefore by the same step as sulfur trioxide formation in the hydrogen sulfide flame itself (Section 8.2.1).

- **Sulfur added as hydrogen sulfide**

Laser fluorescence techniques enabled many intermediate species including S, SO and SH, to be monitored in a hydrogen/oxygen–nitrogen flat flame into which small amounts of hydrogen sulfide were artificially introduced [6]. Like the work of Fenimore and Jones [5], this was concerned with checking for chemical equilibria in sulfur processes. Fenimore and Jones combined certain steps into a stoichiometric relationship whose verification from analyses in the

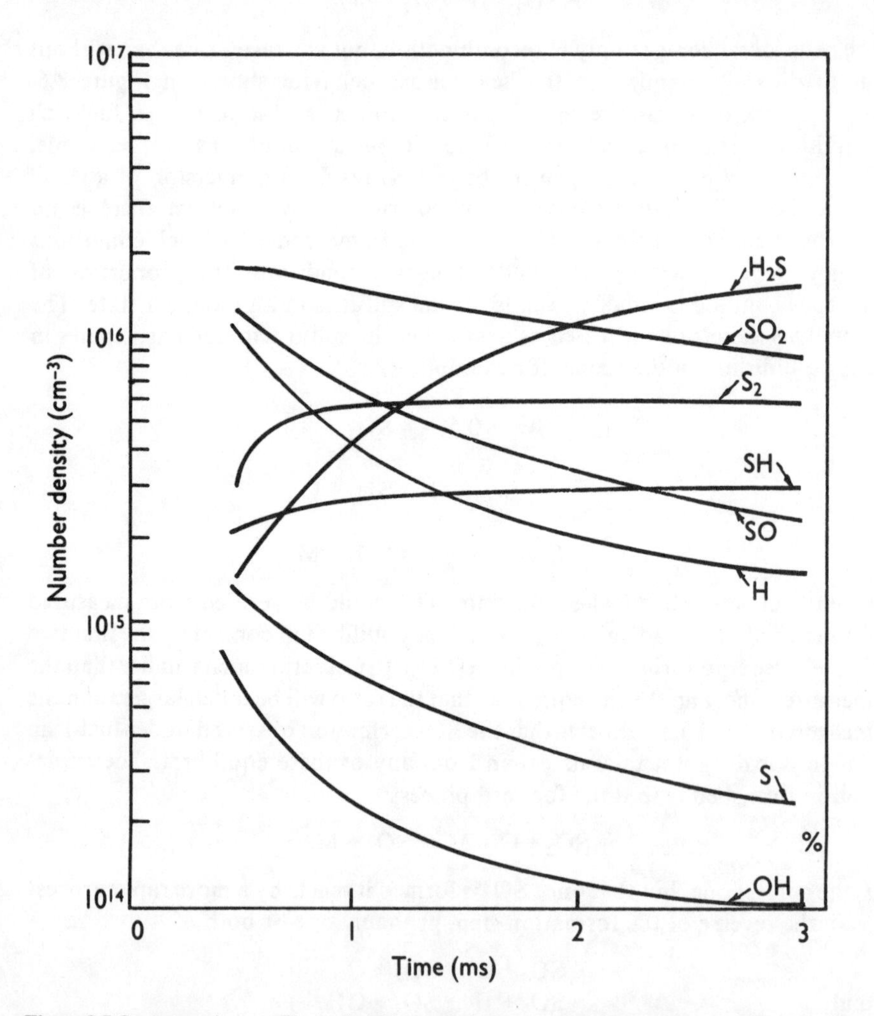

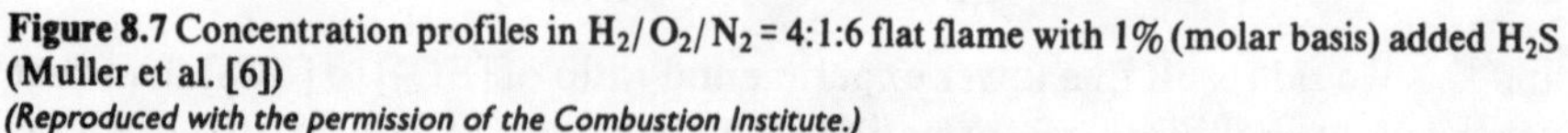
Figure 8.7 Concentration profiles in $H_2/O_2/N_2$ = 4:1:6 flat flame with 1% (molar basis) added H_2S (Muller et al. [6])

(Reproduced with the permission of the Combustion Institute.)

flame supported the view that the steps comprising the stoichiometry were in equilibrium. This was also done in the laser fluorescence work and Figure 8.7 shows number densities (number per unit volume) of the species measured in this way as a function of time from the luminous zone. The time is of course easily convertible into distance in the direction of propagation from the velocity profile, hence Figure 8.7 is actually equivalent to space profiles of the respective species. Temperature profiles were also determined and the sharp decline of OH and H in this region of the flame should be noted. Figure 8.7 is for $H_2/O_2/N_2 = 4:1:6$, that is, a fuel-rich flame, and different proportions were used in other experiments giving different temperature profiles. Hence from multiple experiments, there is the possibility from the measured concentrations and temperature profiles of examining possible equilibria at a wide range of temperatures. Clearly, even in such a fuel-rich flame, with relatively large amounts of molecular hydrogen present, sulfur dioxide is the dominant sulfur species immediately after the luminous zone. Also, all the concentration profiles are significantly modified after exit from the luminous zone.

Consider the stoichiometry:

$$H_2 + SO_2 = SO + H_2O.$$

This would be consistent with equilibration of the following steps:

$$H + SO_2 \rightarrow SO + OH$$
$$OH + H_2 \rightarrow H_2O + H.$$

The reaction quotient K given by:

$$K = \frac{[H_2O][SO]}{[SO_2][H_2]}, \tag{8.9}$$

would, if the proposed equilibria in the flame were established, be an equilibrium constant and would therefore conform to the expression:

$$\ln K \propto -\frac{1}{T}.$$

In Figure 8.8, the reaction quotient in the post-luminous part of the flame is plotted against $1/T$ for five clusters of points representing different influx gas compositions (H_2:O_2:N_2) as indicated: there is some temperature spread within each cluster because of the temperature profiles. The aggregate (solid line) from the experimental points and that of the calculated equilibrium constants (from thermodynamic data) against $1/T$ (dashed line) are in quite close agreement, indicating that the steps examined are indeed at equilibrium. Other stoichiometric relationships were examined and when plotted similarly to Figure 8.8 were all found to lie close to the curve of the calculated equilibrium constant against $1/T$. These are summarized in Table 8.1.

It is not possible to be precise about which equilibria are occurring, in view of the fact that each stoichiometric relationship examined fits more than one

Table 8.1 Equilibria involving sulfur species in flames [6], 1 percent (molar basis) of H_2S added to $H_2/O_2/N_2$ flames

Stoichiometry examined	*Contributing equilibria*	*Expression* for K*
$SO_2 + H_2 \rightarrow SO + H_2O$	$H + SO_2 \rightarrow SO + OH$ $OH + H_2 \rightarrow H_2O + H$	$K = \frac{[SO][H_2O]}{[H_2][SO_2]}$ (See Fig. 8.8)
$S_2 + H_2 \rightarrow 2SH$	$H + S_2 \rightarrow SH + S$ $S + H_2 \rightarrow SH + H$ **or** $H + S_2 \rightarrow SH + S$ $SH + H_2 \rightarrow H_2S + H$ $S + H_2S \rightarrow 2SH$ **or** $H + S_2 \rightarrow SH + S$ $S + H_2S \rightarrow 2SH$ $H_2O + SH \rightarrow OH + H_2S$ $OH + H_2 \rightarrow H_2O + H$	$K = \frac{[SH]^2}{[H_2][S_2]}$
$SH + OH \rightarrow SO + H_2$	$S + OH \rightarrow SO + H$ $H + SH \rightarrow S + H_2$ **or** $S + OH \rightarrow SO + H$ $H + H_2S \rightarrow SH + H_2$ $2SH \rightarrow S + H_2S$ **or** $S + OH \rightarrow SO + H$ $2SH \rightarrow S + H_2S$ $OH + H_2S \rightarrow H_2O + SH$ $H_2O + H \rightarrow OH + H_2$ **or** $SH + O \rightarrow SO + H$ $OH + H \rightarrow O + H_2$ **or** $SH + O \rightarrow SO + H$ $S + H_2 \rightarrow HS + H$ $H_2S + H \rightarrow SH + H_2$ $2SH \rightarrow S + H_2S$ $OH + H \rightarrow O + H_2$	$K = \frac{[H_2][SO]}{[OH][SH]}$
$SO_2 + 2H_2 \rightarrow SH + OH + H_2O$	$H + SO_2 \rightarrow SO + OH$ $H + SO \rightarrow SH + O$ $O + H_2 \rightarrow OH + H$ $OH + H_2 \rightarrow H_2O + H$ **or** $H + SO_2 \rightarrow SO + OH$ $H + SO \rightarrow S + OH$ $S + H_2 \rightarrow SH + H$ $OH + H_2 \rightarrow H_2O + H$	$K = \frac{[SH][OH][H_2O]}{[SO_2][H_2]^2}$

*Giving a linear plot against $1/T$ similar to Figure 8.8

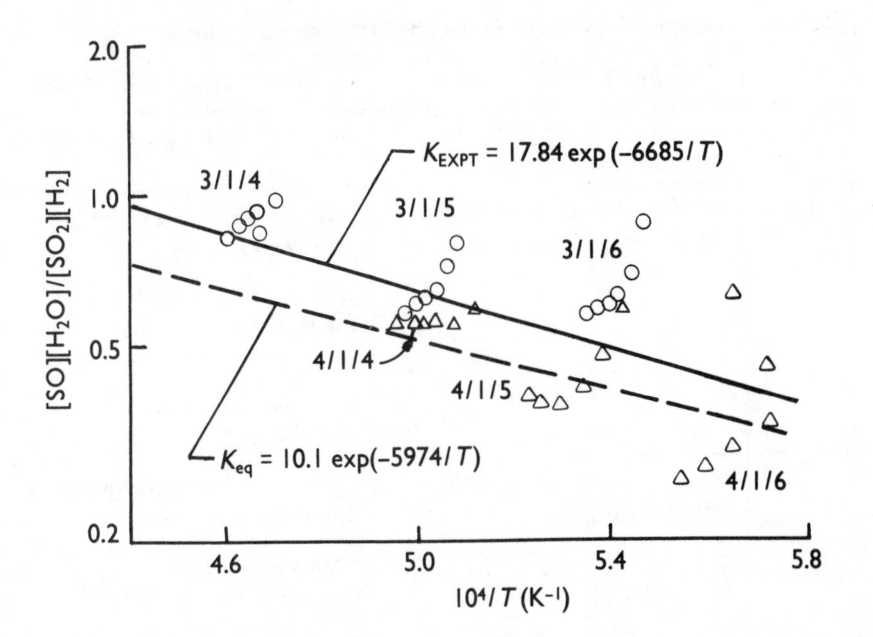

Figure 8.8. Measured and calculated equilibrium constant against reciprocal temperature for $SO_2 + H_2 \rightarrow SO + H_2O$ in $H_2/O_2/N_2$ flames with 1% (molar basis) added H_2S (Muller et al. [6]) *(Reproduced with the permission of the Combustion Institute.)*

scheme, but the following bimolecular steps occur in several of the equilibria, shown to be consistent with the stoichiometry:

$$H + S_2 \rightarrow SH + S$$
$$S + H_2 \rightarrow SH + H$$
$$SH + H_2 \rightarrow H_2S + H$$
$$S + H_2S \rightarrow 2SH$$
$$OH + H_2S \rightarrow H_2O + SH$$
$$H + SO_2 \rightarrow SO + OH$$
$$S + OH \rightarrow SO + H$$
$$SH + O \rightarrow SO + H,$$

and these steps are therefore believed to account for the behavior of sulfur in the region of the flame examined. Apart from its role in catalyzing H recombination noted by Fenimore and Jones, HSO_2 is considered not to be a significant contributor to the mechanism.

- **Summary**

Table 8.2 summarizes the important mechanistic conclusions drawn from the experimental work discussed [1, 4–6]. It is worth reiterating the discovery that oxidized forms of sulfur predominate even under fuel-rich and therefore reducing conditions, a fact likely to have practical consequences in the combustion of solid and liquid fuels containing sulfur. A widely held view is

Table 8.2 Summary of sulfur flame chemistry investigations

Ref.	*Experimental conditions*	*Salient mechanistic features*
[4]	H_2S/O_2, low temperature, subatmospheric pressure	S_2O, a branching intermediate
[1]	H_2S/O_2–N_2 and H_2S/O_2–Ar, flat flame	[H] and [O] taken to be in a steady state Unimolecular rate constant for H_2S dissociation calculated O formation via O_2 dissociation
[5]	H_2/O_2–Ar flat flame with introduced SO_2	Existence of several equilibria in the flame established Catalysis of H + H recombination by SO_2 Predominance of SO_2 even under fuel-rich conditions
[6]	H_2/O_2–N_2 flat flame with introduced H_2S	Several equilibria established in the post-luminous region Several bimolecular steps identified as determining sulfur kinetics Predominance of SO_2 even under fuel-rich conditions

that hydrogen sulfide combustion simulates realistically the fate of fuel sulfur in combustion, for example in coal volatiles.

8.2.3 The reaction of sulfur oxides with metal compounds in coal combustion

Once sulfur oxides have been produced by the types of process outlined, a further aspect of the combustion chemistry is the possibility of reaction between the oxides and any metal compounds present to form sulfates. The consequences in utilization practice are twofold:

(a) Oxides of sulfur formed by oxidation of a coal's sulfur content may, in the combustion reaction system, go on to react with the coal's own mineral and inorganic content (that is, its ash-forming constituents) to form a sulfate deposit, particularly evident on surfaces which experience flame impingement.

(b) Where a coal has a sufficiently high sulfur content for potential sulfur oxide emission to be a problem, one way of reducing the emissions is to include in the reaction system a metal compound capable of converting the sulfur oxides to sulfate, thereby preventing their release with the post-combustion gases.

- **Sulfate deposition on combustion plant surfaces**

Ash from p.f. (see Chapter 6) is very fine and is termed fly ash. A particular fly ash was examined [7] for reaction with sulfur-bearing post-combustion gases under conditions designed to represent those in an industrial furnace by assembling a layer of the fly ash of ≈ 6 mm thickness with its lower surface in contact with metal at 370° C and its upper surface exposed to radiation, and attaining temperatures in excess of 1000° C. The radiation simulates a near-by flame. Simulated post-combustion gas (up to 4% SO_2 molar basis, 2–10% O_2, 15% CO_2, balance N_2) was passed over the fly ash. Clearly the layer of fly ash will display a steep internal temperature gradient and, as the representative results in Table 8.3 show, a sulfur gradient is created by the temperature gradient but with an opposite sign. The results relate to a 0.3–1.1% sulfur dioxide simulated post-combustion gas passed for various times over a layer of fly ash whose primary constituents were silica, aluminum oxide, calcium oxide and iron III oxide, together with smaller amounts of alkali metal compounds.

Table 8.3 Sulfur uptake by fly ash in a simulated post-combustion gas [7]

% SO_2 in gas	*Exposure time (hours)*	*% sulfur after exposure**		
		Bottom layer	*Middle layer*	*Top layer*
0.3	3	6.2	4.5	2.2
1.0	56	7.9	5.4	2.6
1.1	163	9.8	6.5	2.3

*Initial (zero time of exposure) S content of the fly ash was 3.2%. Some of this was lost by decomposition above 800° C.

This indicates that uptake is relatively rapid at first, then much slower; the difference in uptake between 56 and 163 hours exposure is small but measurable. A number of isothermal tests were conducted using very thin layers of the fly ash to eliminate temperature gradients and Figure 8.9 shows uptake against temperature (each temperature point representing a single isothermal experiment) for 2.5% sulfur dioxide with 10% oxygen (upper curve) and 2.2% oxygen (lower curve) in the simulated post-combustion gas. The uptake shows a maximum at about 600° C in each case and depends on the amount of oxygen present as well as on the amount of sulfur dioxide. This can be readily understood if fixation of the sulfur is viewed as conversion of SO_2 (oxidation

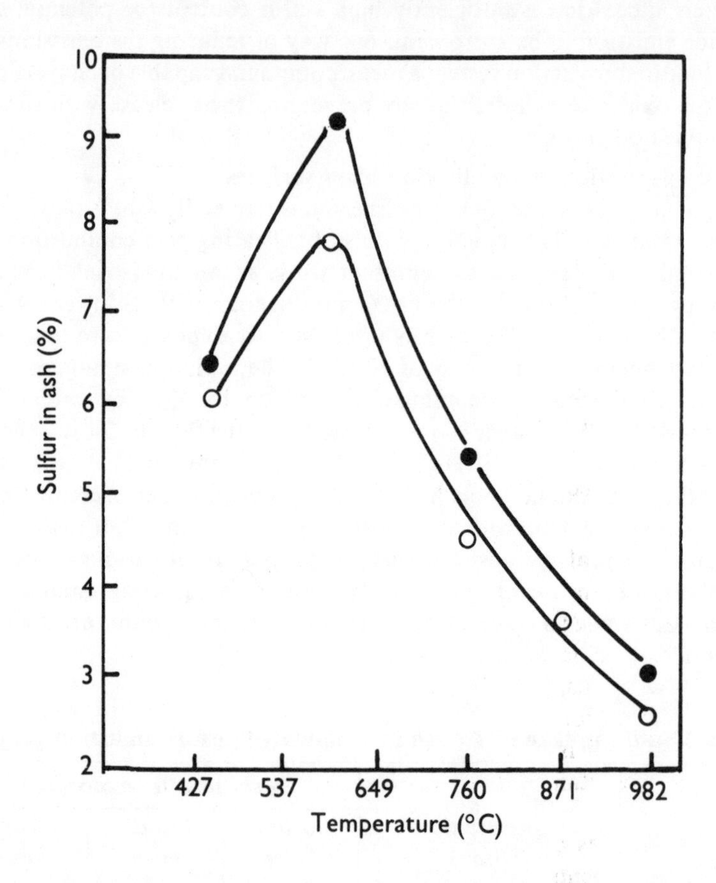

Figure 8.9 Sulfur fixation by fly ash: Upper curve, atmosphere of 2.5% SO_2, 10% O_2, 15% CO_2, balance N_2. Lower curve, 2.5% SO_2, 2.2% O_2, 15% CO_2, balance N_2 (Weintraub et al. [7]) *(Reproduced with the permission of the American Society of Mechanical Engineers.)*

state of sulfur + 4) to SO_4^{2-} (oxidation state of sulfur + 6), thus requiring oxygen as an oxidizing agent. The maximum is due to there being some decomposition of metal sulfates at the higher temperatures. However, this decomposition does not necessarily result in elimination of sulfur in + 6 oxidation state from the system since SO_3 occurs in this high temperature region and is believed to take part in multiple equilibria involving both sulfate decomposition and the gas-phase SO_2/SO_3 interconversion.

There is apparent inconsistency between the results in Figure 8.9 and those in Table 8.3, in that in the former there is maximum sulfur fixation at about 600° C, whereas in the latter, the maximum is close to the metal surface at 370° C. The origin of the difference is twofold:

(a) The fact that the isothermal tests (Figure 8.9) use a thin layer of material with a consequently high degree of gas–particle contact. This leads to a catalytic effect of the particles on the oxidation of sulfur, most marked at the higher temperatures. It is probable that the isothermal tests below 600° C did not come to equilibrium, while those at the higher temperatures did, as previously discussed.

(b) Migration of sulfate in the non-isothermal tests (Table 8.3). Migration requires melting and alkali metal or aluminum sulfates (not forgetting the existence of several complex sulfates as well as the simple ones under these conditions) would be molten at the temperature of the outer surface close to the radiation source and could therefore undergo net movement into the deposit towards the cooler surface, eventually hardening and consolidating the deposit.

- **Summary**

Having discussed earlier in the chapter how fuel sulfur reacts in the gas phase to form sulfur dioxide, the above section details how some of the sulfur dioxide is fixed as sulfates in the ash and points out that the trioxide is a participant in sulfur chemistry equilibria. Sulfur trioxide is of course a factor in plant corrosion.

8.2.4 Use of metal compounds to fix sulfur

A recent study of the use of added metallic compounds in sulfur fixation during coal combustion [8] can be used to illustrate the underlying scientific principles. The metallic compounds fall into two groups: sorbents*, usually calcium compounds, which react to form calcium sulfate with sulfur oxides in the post-combustion gas, and promoters, which contain metals other than calcium. Their role is to modify the metal compound lattice in such a way as to make it more resistant to thermal decomposition or to facilitate sulfur dioxide penetration.

Small briquettes made from pulverized bituminous coal were burnt, with and without sorbents and promoters, in an electrically heated tube furnace, and also in combustion furnaces, and sulfur dioxide release was measured. A number of coals were examined in these ways and they ranged in sulfur content from 1% to 4.8%. They were all high in ash and it is to be expected that the ash-forming constituents and the introduced sorbents and promoters will all have participated in sulfur fixation.

Results for one coal in the tube furnace are shown in Figure 8.10. The added sorbents and promoters are shown to have a strong effect below 800° C. Sulfur fixed as sulfates is released again above 1200° C. Partial analysis of the residue by X-ray methods indicated that fixation had resulted in products other than simple sulfates, including $Na_6Ca_2Al_6Si_6O_{24}(SO_4)_2$, the sodium, aluminum and silicon originating from the coal's own mineral and inorganic content.

In the combustion furnace trials, striking retention results were obtained, for

*Although the recognized term for a sulfur-fixing agent in coal combustion technology, the word *sorbent* has the potential to mislead. *Sorption* generally means a physical process—absorption or adsorption—whereas the action of so-called sorbents in sulfur fixation is sulfate formation and is unquestionably chemical.

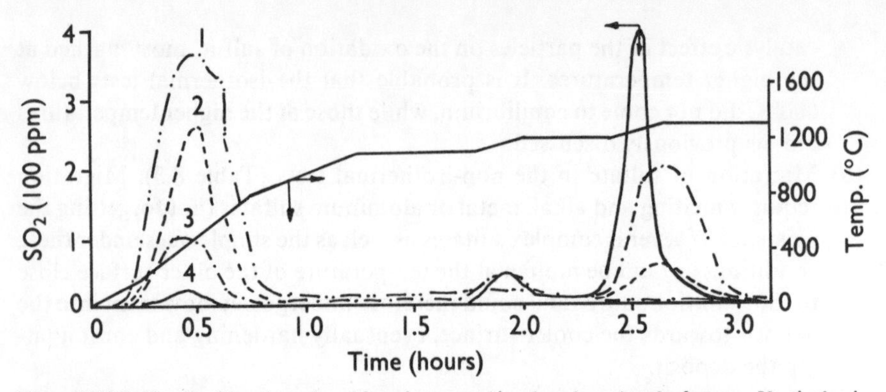

Figure 8.10 Sulfur dioxide release by a bituminous coal in a bench-scale tube furnace. Vertical axis left, ppm SO_2 in the post-combustion gas. Vertical axis right, temperature. Curve 1, coal only. Curve 2, coal + limestone. Curve 3, coal + CaO + MgO + V_2O_5. Curve 4, coal + limestone + $MgCO_3$ (Zhaung et al. [8])

(Reproduced with the permission of Elsevier Science Publishers.)

example, a high-sulfur coal displayed 15–20 percent sulfur retention in the ash without a sorbent, and 52–68 percent retention when burnt with calcium hydroxide and sodium carbonate.

In relatively large-scale furnace trials it was found to be necessary to add calcium in amounts equivalent to 2–2.5 × the sulfur amount in the coal (molar basis) to achieve high retentions, whereas stoichiometry requires only 1 mol calcium per mol sulfur. Calcium compounds as sulfur-fixation agents have found application to fluidized-bed combustion and there, also, it is found that about 2 mol calcium per mol sulfur is required [9]. One reason is that, as well as fixing sulfur, calcium is involved in reaction with minerals in the coal, including silica and alumina.

References

[1] Levy A., Merryman E. L., 'The microstructure of hydrogen sulphide flames', *Combustion and Flame* **9** 229 (1965)

[2] Norrish R. G. W., Oldershaw, G. A., cited by Levy and Merryman [1]

[3] Levy A., Merryman E. L., 'A progress report—fundamentals of thermochemical corrosion reactions', *Engineering for Power* **87** 116 (1965)

[4] Marsden D. G. H., 'A mass-spectrometric study of the oxidation of hydrogen sulphide. I, Observation and identification of some intermediates', *Canadian Journal of Chemistry* **41** 2607 (1963)

[5] Fenimore C. P., Jones G. W., 'Sulphur in the burnt gas of hydrogen–oxygen flames', *Journal of Physical Chemistry* **10** 3593 (1965)

[6] Muller C. H., Schofield K., Steinberg M., Broida H. P., 'Sulphur chemistry in flames', *Seventeenth Symposium (International) on Combustion*, 867, Pittsburgh: The Combustion Institute (1979)

[7] Weintraub M., Goldberg S., Orning A. A., 'A study of sulphur reactions in furnace deposits', *Engineering for Power* **83** 444 (1961)

[8] Zhuang Y. H., Shen D. X., Xiao P. L., 'Sulphur fixation during coal briquette combustion', *Processing and Utilisation of High-Sulphur Coals*, III, (R. Markuszewski and T. D. Wheelock, Eds), 653, Amsterdam: Elsevier (1990)

[9] Anders R., Sauer H., 'Removal of SO_2 from coal combustion systems', *Symposium Series No. 106, 'Desulphurisation in Coal Combustion Systems*, Rugby: Institution of Chemical Engineers (1989)

CHAPTER 9

The Behavior of Nitrogen in Combustion Processes

Abstract

The formation of nitrogen oxides in combustion reactions is discussed, giving examples including methane/air flames with added nitro-compounds to simulate fuel nitrogen. The formation of NO_x from fuel nitrogen in coal combustion is also discussed. Thermal NO_x, which originates not from fuel nitrogen but from atmospheric nitrogen, is also dealt with and two routes to its formation—the Zeldovich mechanism and the prompt *mechanism—are discussed, giving examples.*

9.1 Introduction

When elemental sulfur is burnt, the product is sulfur dioxide and even under fuel-rich conditions hydrogen sulfide goes to sulfur dioxide on combustion. We have seen that hydrogen sulfide is a good model material for sulfur in fuels, for example in petroleum products or coal volatiles, and the behavior of the nitrogen content can similarly be understood by reference to ammonia.

When ammonia is burnt, the dominant reaction (see Chapter 5) is:

$$2NH_3 + \frac{3}{2}O_2 \rightarrow N_2 + 3H_2O.$$

The alternative process:

$$2NH_3 + \frac{5}{2}O_2 \rightarrow 2NO + 3H_2O,$$

requires a catalyst if it is to achieve preponderance. Accordingly, some fuel nitrogen forms elemental nitrogen N_2 on combustion and thus conversion to

NO is less than stoichiometric; depending on combustion conditions, NO formation may account for only a small proportion of the nitrogen present. Even so, amounts are large enough to require control since nitric oxide (NO) is readily converted by atmospheric oxygen to nitrogen dioxide (NO_2) which is toxic in ppm quantities and is also involved in smog formation. NO and NO_2 together are referred to as NO_x. Nitrous oxide, N_2O, is also produced in combustion (notably in forest fires) but has an atmospheric chemistry cycle independent of NO and NO_2.

Apart from NO_x from fuel nitrogen, nitrogen from air in a combustion system can, at high temperatures, form NO_x. This is known as thermal NO_x to distinguish it from fuel NO_x.

9.2 Simulation of fuel nitrogen by nitrogen-containing additives

In experimental investigations [1] yields of NO_x in excess of 50 percent of stoichiometric were obtained when methane was burnt in air and controlled amounts of methylamine (CH_3NH_2) were blended into the system to simulate fuel nitrogen. In some experiments air and methane were premixed before entry into the burner and in others they were admitted separately and required to diffuse together, that is, there were premixed and diffusion flames respectively. The respective NO_x results were significantly different. Figure 9.1 shows conversions to nitric oxide for a range of amounts of added methylamine and the premixed flame displays greater conversion. The air supply was greater than the stoichiometric requirement in each case.

Viewing the mechanism of NO formation as involving, first of all, the stripping of nitrogen from the amine functional group in a similar fashion to the oxidation of ammonia:

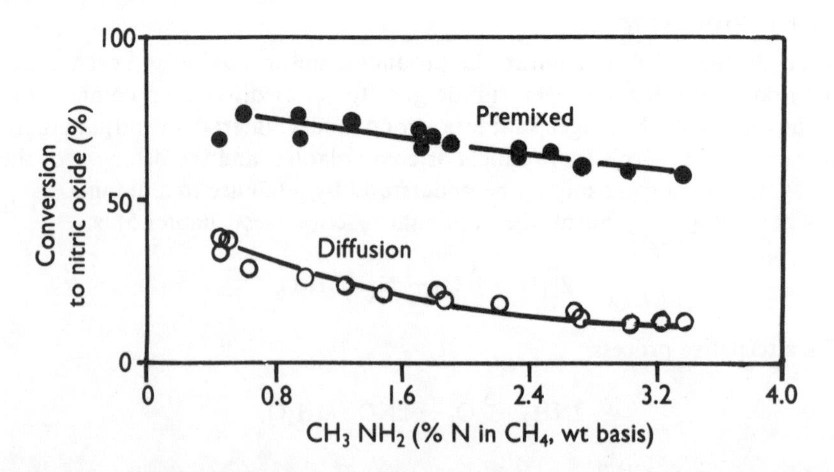

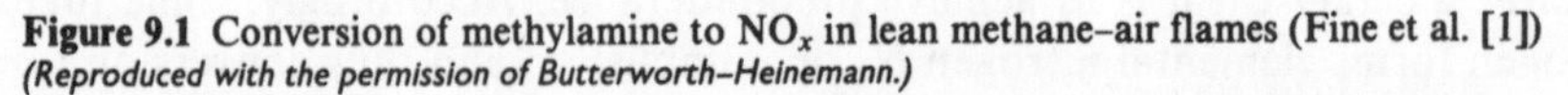

Figure 9.1 Conversion of methylamine to NO_x in lean methane–air flames (Fine et al. [1]) *(Reproduced with the permission of Butterworth–Heinemann.)*

$$NH_2 \rightarrow NH + H$$
$$\downarrow$$
$$N + H$$

followed by:

$$N + N \rightarrow N_2$$

or

$$N + O \rightarrow NO,$$

it is not difficult to accept that premixedness will favor the second of the two reactions of the nitrogen atoms to stable products, and this is consistent with the pattern in Figure 9.1.

Other schemes for NO_x formation from fuel nitrogen have been investigated, as in work on the effects of a number of nitrogen compounds as additives on laminar premixed methane–air flames diluted with nitrogen or argon [2]. Additives were pyridine, ammonia and acetonitrile, representing different nitrogen bonding in fuel, and, for mechanistic elucidation, higher proportions of nitrogen were blended with the methane than would occur naturally in fuels. The mechanism examined was somewhat different from simple stripping of NH_2 to nitrogen atoms.

Discussion of the results requires introduction of the equivalence ratio ϕ:

$$\phi = \frac{\text{fuel : oxidant ratio}}{\text{stoichiometric fuel : oxidant ratio}} . \tag{9.1}$$

Clearly $\phi > 1$ denotes fuel-richness and $\phi <$ denotes fuel-leanness. Measurements of nitrogen conversion to oxides were by analysis of post-combustion gas by a 'NO_x meter'. Such devices convert any NO present to NO_2 before analysis, thus all NO_x is measured as NO_2 irrespective of relative amounts of NO and NO_2 initially present..

For acetonitrile (CH_3-C≡N) as additive, the findings at various values of ϕ are given in Figure 9.2(a). The numerical results are expressed as actual NO_x measured against amount of NO_x at 100 percent acetonitrile conversion. The yields are seen to be close to stoichiometric at low values of the abscissa (that is, at lower amounts of added acetonitrile) and to depart significantly from stoichiometric at higher values. Points for the lean mixtures lie on almost the same curve, whereas those for the one rich mixture are on a distinctly lower curve. As ϕ changes, so does the amount of heat released per unit amount of total gas (in molar or volume terms) and so, therefore, does the flame temperature. The view is expressed [2] that the very close results for all the fuel-lean values of ϕ arise from cancellation of effects due to varying ϕ and to consequently varying flame temperature. Calculated adiabatic flame temperatures for ϕ = 0.7 and ϕ = 0.95 are respectively 2350K and 2540K. These are rather higher than those discussed in Chapter 5 because although the experiments use O_2:Ar in the proportion which simulates air, the heat capacity of argon (a monatomic gas) is significantly lower than that of nitrogen (a diatomic gas) and this difference is reflected in the flame temperatures.

The adiabatic flame temperature can therefore be controlled and made equal between experiments with different ϕ-values by using other O_2:Ar ratios, and Figure 9.2(b) shows the effects of ϕ on NO_x formation from acetonitrile at a single value of the peak temperature. NO_x yields are seen to decrease with increasing ϕ and in Figure 9.2(c) it is shown that at a fixed value of ϕ of 0.95, NO_x conversion increases with increased flame temperature. Ammonia and pyridine displayed NO_x conversions very close to those displayed by acetonitrile under the same conditions.

The kinetic scheme used in analyzing these results was previously proposed by Fenimore [3]. It postulates that the fuel nitrogen (denoted RN in the scheme below) reacts with a radical intermediate (denoted X) to form a nitrogen-containing intermediate (denoted N(i)) which can then react either with an oxidizing species (denoted Ox) to form NO or with NO itself to form N_2. In generalized terms, therefore:

$$\mathrm{RN} + X \xrightarrow{k_1} \mathrm{N(i)}$$

$$\mathrm{N(i)} + \mathrm{Ox} \xrightarrow{k_2} \mathrm{NO}$$

$$\mathrm{N(i)} + \mathrm{NO} \xrightarrow{k_3} \mathrm{N_2}.$$

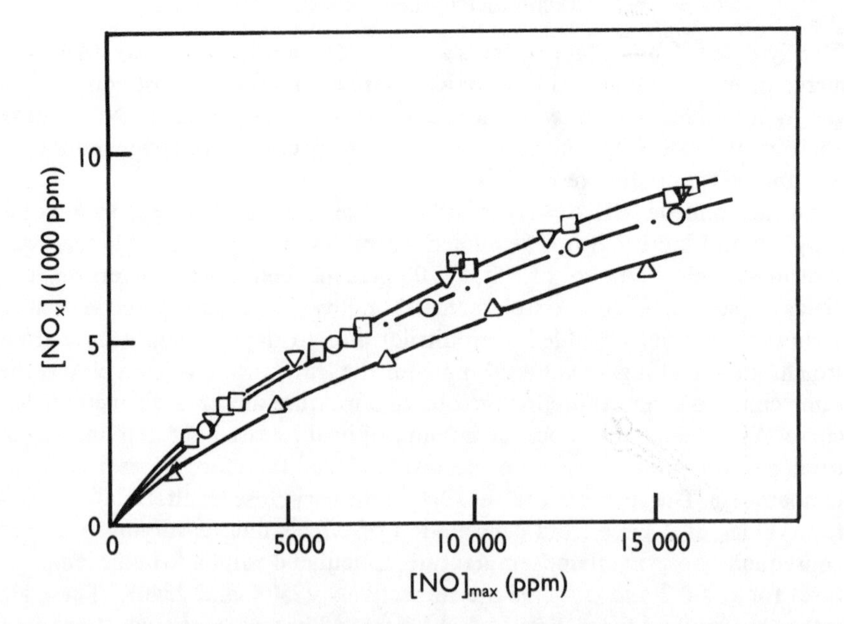

Figure 9.2 (a) Conversion of acetonitrile to NO_x in methane–oxygen–argon flames. O_2:Ar = 21:79. Circles ϕ = 0.70. Inverted triangles (∇) ϕ = 0.82. Squares ϕ = 0.95. Triangles (Δ) ϕ = 1.16 (Crowhurst and Simmons [2])

(Reproduced with the permission of the Combustion Institute.)

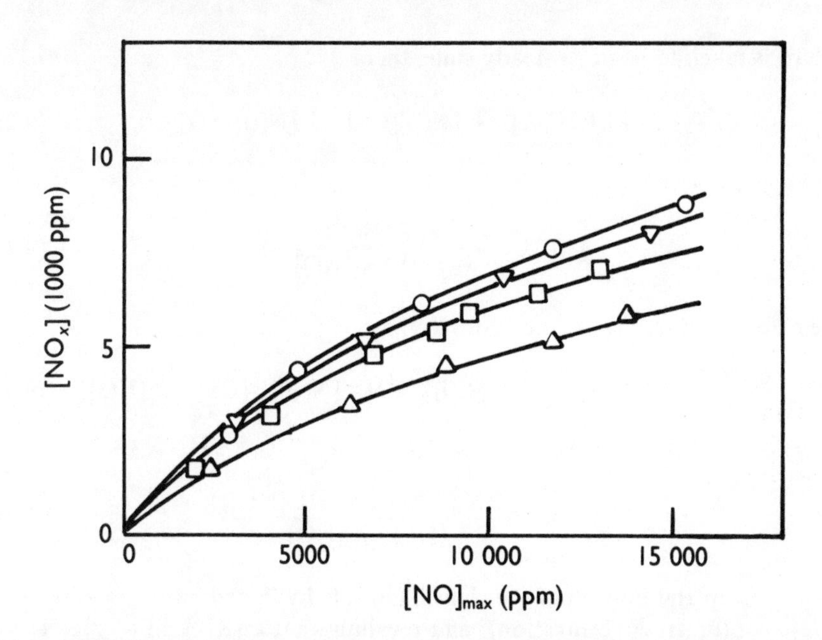

Figure 9.2 (b) Conversion of acetonitrile in methane–oxygen–argon flames. O_2:Ar varied between the experiments to keep the calculated adiabatic flame temperature the same for all the experiments at (2215 ± 35)K. Circles, triangles etc. denote ϕ values as in Figure 9.2(a) (Crowhurst and Simmons [2])

(Reproduced with the permission of the Combustion Institute.)

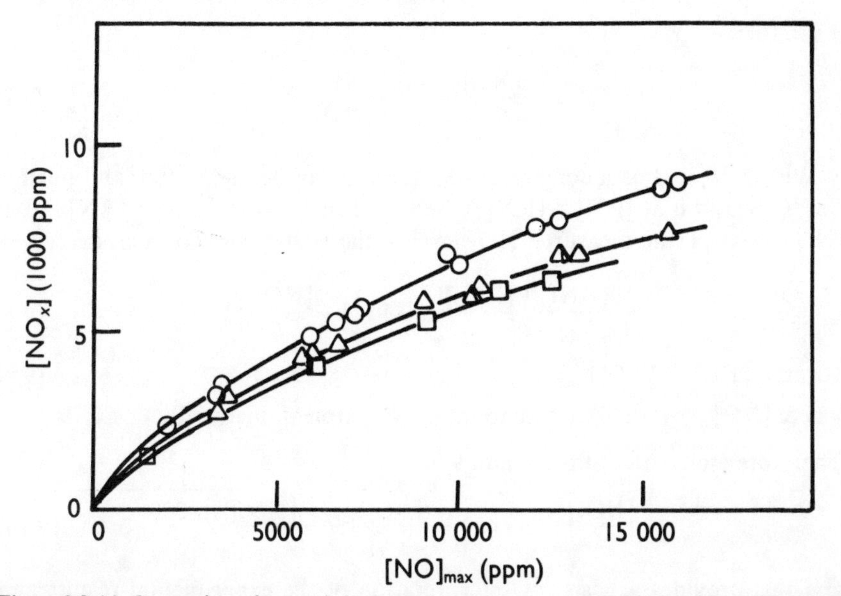

Figure 9.2 (c) Conversion of acetonitrile in methane–oxygen–argon flames with ϕ = 0.95 and O_2:Ar adjusted to give different adiabatic flame temperatures: Circles 2540K. Triangles 2415K. Squares 2365K (Crowhurst and Simmons [2])

(Reproduced with the permission of the Combustion Institute.)

If N(i) is taken to be in its steady state, then:

$$\frac{\mathrm{dN(i)}}{\mathrm{d}t} = k_1[\mathrm{RN}][X] - k_2[\mathrm{N(i)}][\mathrm{Ox}] - k_3[\mathrm{N(i)}][\mathrm{NO}] = 0, \tag{9.2}$$

which gives:

$$[\mathrm{N(i)}] = \frac{k_1[\mathrm{RN}][X]}{k_2[\mathrm{Ox}] + k_3[\mathrm{NO}]}. \tag{9.3}$$

We can write a rate expression for [NO] as:

$$\frac{\mathrm{d[NO]}}{\mathrm{d}t} = k_2[\mathrm{N(i)}][\mathrm{Ox}] - k_3[\mathrm{N(i)}][\mathrm{NO}] = [\mathrm{N(i)}]\{k_2[\mathrm{Ox}] - k_3[\mathrm{NO}]\}. \tag{9.4}$$

Combining equations 9.3 and 9.4 gives:

$$\frac{\mathrm{d[NO]}}{\mathrm{d}t} = \frac{k_1[\mathrm{RN}][X](k_2[\mathrm{Ox}] - k_3[\mathrm{NO}])}{k_2[\mathrm{Ox}] + k_3[\mathrm{NO}]}. \tag{9.5}$$

Dividing top and bottom of the right-hand side by k_3 and letting $(k_2[\mathrm{Ox}]/k_3) = L$ (units, concentration), and recalling that $k_1[\mathrm{RN}][X] = -\mathrm{d[RN]}/\mathrm{d}t$ gives:

$$\frac{\mathrm{d[NO]}}{\mathrm{d}t} = -\frac{\mathrm{d[RN]}}{\mathrm{d}t} \times \frac{L - [\mathrm{NO}]}{L + [\mathrm{NO}]}. \tag{9.6}$$

Therefore:

$$\frac{\mathrm{d[NO]}}{\mathrm{d[RN]}} = -\frac{L - [\mathrm{NO}]}{L + [\mathrm{NO}]}. \tag{9.7}$$

With L treated as a constant, the equation can be integrated, and putting limits [NO] = 0 at [RN] = $[\mathrm{RN}]_o$ (the initial fuel nitrogen), and [RN] = 0 at [NO] = $[\mathrm{NO}_x]$ (the measured NO_x level in the post-combustion gases), gives:

$$\frac{\{[\mathrm{RN}]_o + [\mathrm{NO}_x]\}}{2L} = -\ln\left\{1 - \frac{[\mathrm{NO}_x]}{L}\right\}. \tag{9.8}$$

Clearly, $[\mathrm{RN}]_o = [\mathrm{NO}_x]_{\mathrm{max}}$, (9.9)

where $[\mathrm{NO}]_{\mathrm{max}}$ = maximum amount of NO_x stoichiometrically possible.

Therefore from equations 9.8 and 9.9:

$$\frac{[\mathrm{NO}_x]}{L} = 1 - \exp\frac{-([\mathrm{NO}_x]_{\mathrm{max}} + [\mathrm{NO}_x])}{2L}, \tag{9.10}$$

and this provides a basis for interpretation of the experimental results and trends when two limiting cases are identified:

(a) $[\mathrm{NO}_x]_{\mathrm{max}}$ small, $[\mathrm{NO}_x] \rightarrow [\mathrm{NO}_x]_{\mathrm{max}}$. This is the initial part of the experimental yields where there is full conversion to NO_x.

(b) $[NO_x]_{max}$ very large, $[NO_x] \rightarrow L$. This is the plateau region in the experimental yields and L is the ceiling value of $[NO_x]$.

It is pointed out [2] that the experimental yields fit an equation of form:

$$[NO_x] = L\frac{[NO_x]_{max}}{1 + [NO_x]_{max}}, \qquad (9.11)$$

and it should perhaps be reiterated for clarity that $[NO_x]_{max}$ is the maximum stoichiometric amount of NO_x possible at a particular concentration of fuel nitrogen and L is the limiting value of $[NO_x]$ beyond which no increase in $[NO_x]$ is found to occur with increasing amounts of fuel nitrogen.

Rearranging equation 9.11:

$$\frac{1}{[NO_x]} = \frac{1}{L}\left\{\frac{1}{[NO_x]_{max}} + 1\right\}. \qquad (9.12)$$

Thus sets of yield data such as those in Figures 9.2(a)–(c) replotted as $1/[NO_x]$ against $1/[NO_x]_{max}$ will yield a straight line of intercept $1/L$ and this is a good means of obtaining experimental values of L. They can be checked against values on the basis of equation 9.10 in the following way. Equation 9.10 can be rewritten:

$$[NO_x] = L\left(1 - \exp\left\{-([NO_x]_{max} + [NO_x])/2L\right\}\right), \qquad (9.13)$$

and so $[NO_x]$ plotted against $1 - \exp\{-([NO_x]_{max} + [NO_x])/2L\}$ yields L from the slope; L in the exponential is taken as the experimental value obtained as indicated above. Thus, there are routes to L from experiment (equation 9.11) and from the proposed Fenimore mechanism (equation 9.10) and agreement will clearly provide support for the Fenimore mechanism.

Agreement was actually extremely close for all the experimental tests with acetonitrile and Table 9.1 gives representative results. The Fenimore mechanism, with its step whereby NO can disappear by reaction with a nitrogen-containing intermediate to form N_2, being thus vindicated by the results, it remains to identify species involved in the Fenimore mechanism, discussed so far only in generalized terms.

With acetonitrile as the additive in a lean flame, N(i) is believed to be OCN, formed by:

$$CH_3CN + OH \rightarrow CH_2CN + H_2O$$
$$CH_2CN + O_2 \rightarrow OCN + CH_2O.$$

Ox in the generalized scheme is the O atom and formation of NO from OCN is by:

$$OCN + O \rightarrow NO + CO,$$

and the nitric oxide removal step is therefore:

$$NO + OCN \rightarrow CO_2 + N_2.$$

Table 9.1 Values of the limiting concentrations of $[NO_x]$ from methane flames with nitrogen additives (From Crowhurst and Simmons [2])

	L(ppm)	
Combustion conditions	*Experimental* (Equation 9.11)	*From Fenimore mechanism* (Equation 9.10)
$\phi = 0.70$ O_2:Ar = 21:79 CH_3CN additive Adiabatic flame temperature 2350K	15 250 ± 450	15 150
$\phi = 0.95$ O_2:Ar = 21:79 CH_3CN additive Adiabatic flame temperature 2540K	16 900 ± 900	17 200
$\phi = 0.82$ O_2:Ar = 21:79 NH_3 additive Adiabatic flame temperature 2460K	17 500 ± 800	17 400

This is believed to be the mechanism for nitrogen in cyano-bonding in a fuel-lean flame, but there are different mechanisms for fuel-rich flames or for other nitrogen bonding in the additive which may or may not fit the general Fenimore scheme. Methylamine as an additive in a diffusion flame with a methane/hydrogen blend as fuel is believed to proceed to NO_x by a route of the Fenimore type but with different intermediates [4]. It is believed that in the very early stages of the flame the methylamine forms HCN, which further breaks down to a variety of intermediates including N, NH and NH_2 which are subsequently N(i) in the Fenimore scheme. Experiments were carried out at various pressures and a sharp decline in NO_x in the post-combustion gas with pressure was observed, from 106 ppm at 1 atmosphere to 37 ppm at 16 atmospheres.

To understand this effect we return to Figure 4.4, a schematic diagram of a diffusion flame, and we superimpose on it a fuel profile which peaks in the center and has a negative slope at either side. HCN is formed very close to the center and diffuses outward towards the flame surface and as it does so it breaks down to N, NH, etc. as noted. A point is reached at which the rate of diffusion becomes negligible in comparison with the rate of breakdown. This is the region of the flame at which the Fenimore mechanism operates and it is argued [4] that the higher the pressure, the closer to the center will be this region, and because of the fuel profile, the less oxidizing will be the conditions. This is consistent with the observed NO_x trends: lower yields at higher pressures.

9.3 Fuel NO_x from coal combustion

Determination of fuel NO_x yields from coals and other fuels can be evaluated if formation of thermal NO_x is excluded by using oxygen diluted with gases other than nitrogen. In this way the NO_x measured can be assumed to have originated entirely from the fuel nitrogen. This was the approach taken by Chen at al. [5] in their study of NO_x emission from 48 coals from various parts of the world: Ar/CO_2–O_2 mixtures were used instead of air. Coal samples were burnt as p.f. at O_2:Ar/CO_2 equivalent to 5% excess air and NO_x in the post-combustion gases was determined. As in the work of Heberling and Boyd [4] it was measured as $NO + NO_2$ and it is assumed that NO is the primary product, some of which is converted to NO_2. NO and NO_x are thus synonymous in this discussion.

We first focus on results for 19 of the coals, which had previously been subjected to high-temperature pyrolysis tests to determine the nitrogen distribution between volatiles and residual solid, nitrogen in the latter being termed *nonvolatile* N. The amount of HCN in the nitrogen in volatiles was measured and expressed as ppm under reference conditions, and the nonvolatile N as ppm N_2 equivalent. The total volatile N was clearly the nitrogen content of the coal, minus that of the char. Yields of NO were found to fit the correlation:

$$\mathrm{NO\ (ppm)} = 318 + 702\ \%\mathrm{N} + 0.188\ [\mathrm{HCN}] - 0.347[\mathrm{nonvolatile\ N}], \tag{9.14}$$

where %N is the percentage fuel N content and the nitrogen concentrations are in ppm equivalents, as discussed. This correlation was obeyed very closely by the nineteen coals, which had N values from 0.68 to 2.5 percent and ranged in rank from lignite to anthracite. Measurements of NO_x were in the range 400–1400 ppm. The indications are twofold:

(a) That fuel nitrogen devolatilization and conversion to HCN promotes NO_x.
(b) Any tendency for nitrogen to be retained in the solid phase instead of being devolatilized works against NO_x formation.

This can be understood by detailed examination of results for two of the nineteen coals: a Pennsylvania anthracite and an Australian lignite. Table 9.2 sets out the data. The coal with the higher nitrogen produces much less NO_x because of its higher rank and consequently much smaller extent of devolatilization. This indicates that in prediction and control of fuel NO_x coal volatile content is likely to be an important factor. The fact that [HCN], but not [volatile nitrogen], appears in the correlation illustrates the potential of this and similar nitrogen species, for example CH_3CN, to react to form fuel NO_x.

Whereas trials with O_2:Ar/CO_2 ratios equivalent to 5% excess air produced

Table 9.2 NO production by two coals (Chen et al. [5])

	% N	[HCN] (ppm)	[nonvolatile N] (ppm equivalent)	[NO] (ppm)* (Equation 9.14)
Anthracite, (Pa, USA)	0.84	43	1267	476
Lignite, (Victoria, Australia)	0.68	752	556	744

*[NO] values thus calculated in good agreement with measured values [5].

up to 1400 ppm NO_x as discussed, the range was 150 to < 400 ppm when the technique of *staged combustion* was used and this has obvious value in control. In staged combustion, instead of admitting the p.f. and air together, the p.f. is admitted with only a proportion of the air needed, which means that devolatilization and combustion of volatiles occurs under fuel-rich conditions. As we have seen, devolatilized nitrogen is powerful in forming NO_x, so burning under fuel-rich conditions favors NO_x reduction. Some distance down from the point of admission of p.f., the remainder of the air requirement is added and this enables combustion of the devolatilized fuel to burn out and is sometimes referred to as the burnout stage.

The efficacy of this procedure in reducing NO_x depends on the proportion of the air admitted at the first stage. Chen at al. [5], in examining different coals under these conditions, continued to use 5% excess air and varied the proportion of that added at stage 1 from 0.45 upward. Calling this proportion S_{R1}, clearly the proportion added at the second stage is $(1 - S_{R1})$. For a Canadian lignite with about 1000 ppm NO_x when $S_{R1} = 1$ the results were as shown in Figure 9.3.

The NO_x exiting the combustion system is seen to decrease with decreasing S_{R1} down to a minimum, then to start to increase. The minimum is 300 ppm, a very substantial reduction on $S_{R1} = 1$. Gas exiting the first stage zone of combustion showed a monotonic decrease in NO with decrease in S_{R1}. Twenty-six of the coals were subjected to staged combustion trials, and NO at this point had a positive correlation with rank. Coal nitrogen at this point is considered to consist of five groups: that converted to NO in stage 1; that converted to N_2 in stage 1; that devolatilized in stage 1 as HCN; that devolatilized in stage 1 as NH_3; and that retained in the solid residue. Distribution of nitrogen between these categories is important in determining the fate of nitrogen in stage 2 and for all the coals, nitrogen speciation at the exit from stage 1 was determined by analytical techniques.

Model materials (char/ammonia, propane/ammonia or propane/NO of various proportions) were also used to investigate the second stage. They were

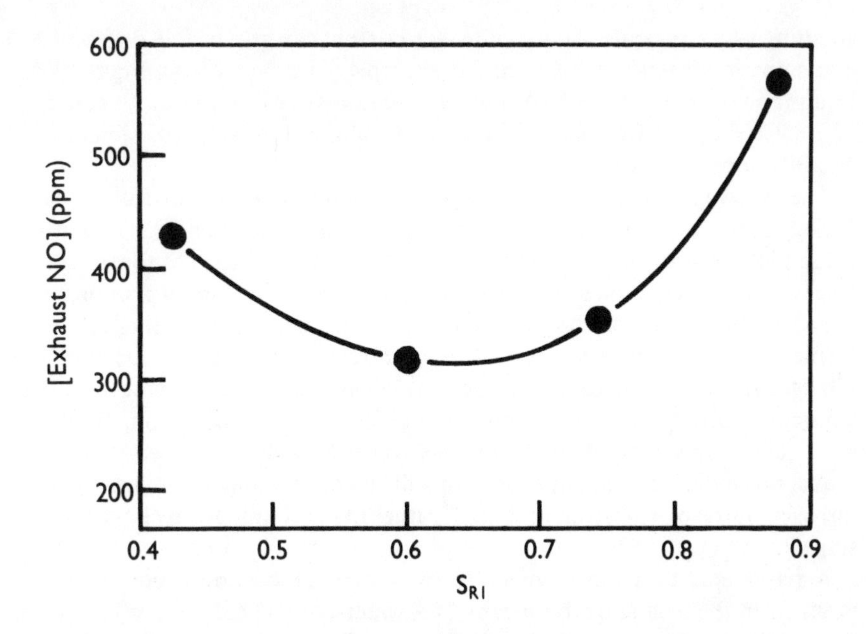

Figure 9.3 NO_x release by a Canadian lignite as a function of proportion of oxidant admitted at the first stage of a two-stage combustion process (Chen et al. [5])
(Reproduced with the permission of the Combustion Institute.)

burnt initially with various S_{R1} and, for $S_{R1} > 0.75$, NO was the dominant nitrogen species exiting stage 1. For all of the model materials under these conditions the NO was retained in stage 2; there was no significant reduction in the second stage of NO formed in the first stage. For smaller values of S_{R1} (richer mixtures in the first stage) HCN and NH_3 were the dominant gaseous nitrogen species exiting the first stage and there was less NO exiting the second stage. With large S_{R1} the fuel in stage 2 is largely CO, whereas with small S_{R1} there is considerable propane, or char as the case may be. Char nitrogen conversion to NO in the second stage also decreased with increasing S_{R1}. These findings all indicate the potential for decline in NO from the total process by staged combustion.

The 26 coals themselves gave the following correlation of exhaust NO with nitrogen speciation at the end of phase 1, with all concentrations given on a ppm or ppm equivalent basis:

$$\mathrm{NO} = 0.87S_{\mathrm{R1}}[\mathrm{NO}] + \frac{0.67\{[\mathrm{HCN}] + [\mathrm{NH_3}]\}}{1 + 0.004\{[\mathrm{HCN}] + [\mathrm{NH_3}]\}} + 0.27S_{\mathrm{R2}}^{0.5}[\mathrm{char\ N}], \qquad (9.15)$$

where $S_{R2} = (1 - S_{R1})$.

This equation is the counterpart of equation 9.14 for single-stage combustion and the first term on the right will be very small at low S_{R1} and the third term very small at high S_{R1}. The correlation also provides a basis for understanding why two coals of equal nitrogen content can give widely different NO emissions in staged combustion; for example [5] a Norwegian high-volatile bituminous coal and a USA (Alabama) medium-volatile bituminous coal each of 1.85% (dry, ash-free basis) nitrogen gave minimum NO_x emissions of 260 and 380 ppm respectively.

The Norwegian coal, because of its high volatile matter, lost much of its nitrogen by devolatilization during stage 1 where the fuel-rich conditions caused less of it to be converted to NO than in single-stage combustion. The large extent of devolatilization in stage 1 means that only a small proportion of the coal nitrogen entered stage 2 as char nitrogen and, correspondingly, the NO formation in this stage was small. By contrast, the Alabama coal lost less nitrogen in stage 1 and carried more through as char to stage 2 and the higher emission is attributable to the resulting greater char contribution. While for both coals a lowering of NO emissions was achieved by the two-stage technique, this indicates that improvement will be greatest for coals releasing more nitrogen by devolatilization in stage 1, rather than retaining it in chars entering stage 2.

A refinement that can be made to two-stage combustion is cooling of the gases from the first stage by means of a water-cooling column. When this is done at the *exit* from stage 1, so that it has no effect on the speciation, the result is nevertheless a significant reduction in the NO release by the total combustion process. This is due to the effect on the temperature of stage 2 of the lower sensible heat of the influx gases to it, and the effect of this in turn on the rate of conversion of NH_3 and HCN to NO. For the Canadian lignite for which results are shown in Figure 9.3, the minimum NO emission at about $S_{R1} = 0.7$ was reduced by some 50 ppm by cooling the exit gases from stage 1 by 200K.

9.4 Thermal NO_x

9.4.1 Background

Thermal NO_x is formed by reaction of atmospheric nitrogen with oxygen, and occurs only at higher temperatures. Natural gas flames are of quite sufficiently high temperature for thermal NO_x to occur and its control is therefore important in natural gas utilization [6]. Two mechanisms are believed to contribute to thermal NO_x: the Zeldovich mechanism and the Fenimore mechanism (prompt NO) and we will consider them in turn.

9.4.2 The Zeldovich mechanism

This is simply:

$$N_2 + O \rightarrow NO + N \quad (1)$$
$$N + O_2 \rightarrow NO + O, \quad (2)$$

and it was proposed that O atoms be made available by the *equilibrium*:

$$O_2 = 2O. \quad (3)$$

The rate of formation of NO is:

$$\frac{d[NO]}{dt} = 2k_1[N_2][O]. \qquad (9.16)$$

However there is now evidence that when the Zeldovich mechanism operates, [O] values substantially exceed their equilibrium value and this is an important point in view of the dependence of the mechanism on there being O atoms. It was examined in experiments on laminar hydrogen–oxygen–nitrogen flames [7] by taking NO analyses at various points from the burner. A number of mixture strengths were used, from 0.9 to 1.4 mol oxygen per 2 mol of hydrogen, with nitrogen varied from 4.4 to 7.6 mol per 2 mol hydrogen. Flame temperatures were accordingly in the range 1685K to 2150K and flame heights about 5 mm above the burner port. From the initial compositions, flame temperatures and thermodynamic data, equilibrium mole fractions of O could be calculated and they varied between the flames in the range 7.5×10^{-6} to 3.8×10^{-4}.

The differential dT/dz (z = distance from burner) is expected to be positive along the flame and the view taken in NO formation is that the Zeldovich reactions (1) and (2) above occur only at large z, that is, beyond the primary, luminous part of the flame. The question with regard to O availability is then whether in this region [O] is equal to or greater than its equilibrium value. The value of [O] at various points along the direction of propagation can be obtained from calculated values of [O] at various times, since t is easily converted to z from knowledge of the velocity. Calculations of the values of [O] at different times require coupling of the rate expressions for (1) and (2) to those for the following steps, several of which we met in Chapter 2 when discussing the hydrogen–oxygen reaction *per se* (though in that discussion we were not concerned chiefly with the post-luminous region, as we are here).

$$H + O_2 \rightarrow OH + O \qquad (4)$$
$$O + H_2 \rightarrow OH + H \qquad (5)$$
$$O + H_2O \rightarrow 2OH \qquad (6)$$
$$H + H_2O \rightarrow H_2 + OH \qquad (7)$$
$$O + O + M \rightarrow O_2 \qquad (8)$$
$$H + H + M \rightarrow H_2 + M \qquad (9)$$
$$H + OH + N_2 \rightarrow H_2O + N_2 \qquad (10)$$
$$H + OH + H_2O \rightarrow 2H_2O. \qquad (11)$$

An expression for $d[O]/dt$ and corresponding expressions for the other intermediates can be set up on the basis of the above mechanism and numerical integration of the system of interdependent equations gives *inter alia* $[O](t)$; literature values of the rate constants are available. $[NO](t)$, and therefore $[NO](z)$, is then obtainable from substitution of [O] values into the integrated form of equation 9.16. $[NO](z)$ is, of course, the calculated [NO] profile along the flame, which can be compared with the experimental profile. The numerical integration requires a value of [O] at z = limit of the primary part of the flame; in other words an initial value of the atom concentration in the post-luminous region where NO formation is believed to occur. This affects [O](t) and hence the profile [NO](z). For one particular flame the equilibrium initial value of [O] in percentage terms was 0.033 and the [NO] profiles were calculated with this value and also with the higher values of 0.1% and 0.3%. The results are shown in Figure 9.4(a) where the experimental NO profile is seen to lie between those calculated for initial [O] values 0.1 and 0.3, well above the profile calculated on the basis of equilibrium [O]. This indicates that there are greater than equilibrium values of [O] exiting the primary region, leading to enhancement of NO yields by the Zeldovich mechanism.

The experimental profile becomes parallel to the one for equilibrium $[O]_{initial}$ at large z, but does not match at all well the shape of either of the others. The kinetic scheme was modified to improve this by incorporating the following steps:

$$H + O_2 + N_2 \rightarrow HO_2 + N_2 \qquad (12)$$

$$H + O_2 + H_2O \rightarrow HO_2 + H_2O \qquad (13)$$

$$HO_2 + OH \rightarrow H_2O + O_2. \qquad (14)$$

For the same flame as that for which the profiles are shown in Figure 9.4(a) the [NO](z) profiles, when calculated from equation 9.16 via the [O] values from the extended mechanism, are shown in Figure 9.4(b). The experimental points are close to the profiles for $[O]_{initial}$ = 0.3% and 1%, much higher than the equilibrium value. Moreover, the calculated curves resemble in shape the experimental [NO] profile. The following statements can therefore be made:

(a) The extended mechanism, like the simpler one, predicts that [O] concentrations at the exit from the primary zone are well above equilibrium ones at the flame temperature.

(b) H atom recombination via HO_2 occurs to a significant degree in the post-luminous region of the flame where the Zeldovich route to NO formation is taken.

9.4.3 Prompt NO

Prompt NO requires a hydrocarbon fragment to participate in the mechanism and was examined by introducing small, controlled amounts of hydrocarbons into $H_2/O_2/N_2$ flames [8]. Figure 9.5(a) shows NO measurements along a flame for various amounts of added acetylene. The lowest curve, from experiments with nil additive, must therefore display Zeldovich NO, and the curve is seen to go to zero at the reaction zone, in accordance with what has already been said about NO formed by this route. The other curves, showing increasing NO with additive amount for any axial distance from the reaction zone, show that significant amounts of NO can be expected at the primary zone. Hence Zeldovich NO is formed outside this zone and prompt (Fenimore) NO inside and outside it. This point of contrast is important.

Using methane as the additive and subtracting Zeldovich NO determined

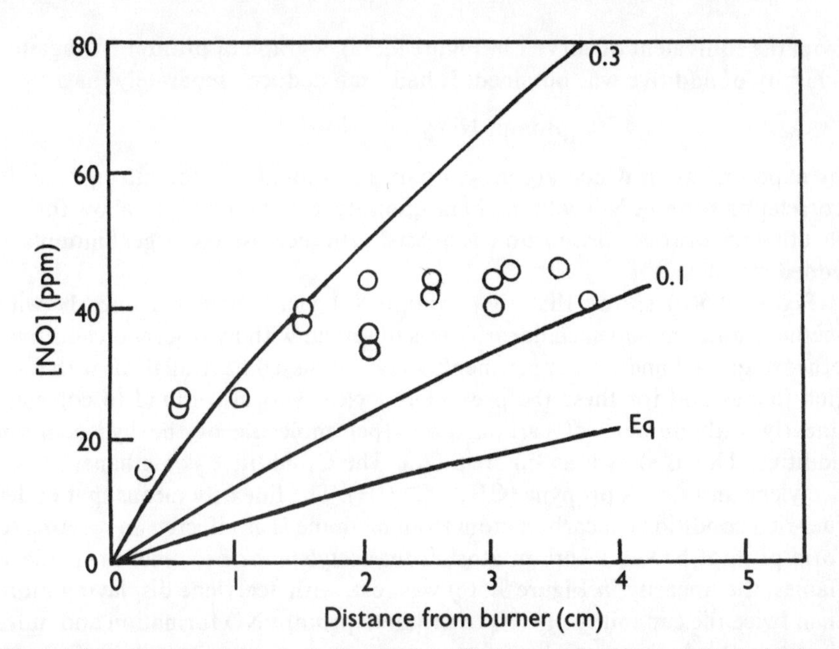

Figure 9.4 (a) Experimental and calculated NO profiles in a premixed laminar H_2–O_2–N_2 flame. Circles, experimental points. Lines, calculated profiles at the values of %$[O]_{initial}$ given (Homer and Sutton [7])

(Reproduced with the permission of the Combustion Institute.)

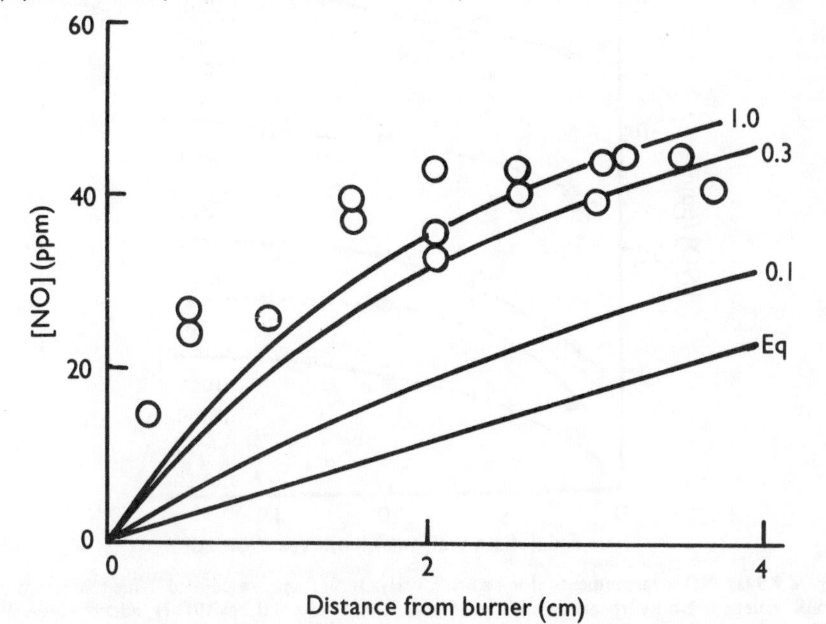

Figure 9.4 (b) As for (a), but with $[O]_{initial}$ calculated from a kinetic scheme including H atom recombination via HO_2 (Homer and Sutton [7])

(Reproduced with the permission of the Combustion Institute.)

from the equivalent of curve 1 in Figure 9.5(a), a graph of prompt NO against quantity of additive was obtained. It had been deduced separately that:

$$\text{prompt NO yield} \propto [N_2]^{1.0},$$

by experiments in which argon was partially substituted for nitrogen, so in correlating prompt NO with methane quantity it was possible to allow for the fact that the nitrogen proportion decreased with successively larger amounts of added methane.

Figure 9.5(b) shows that the prompt NO yield increases linearly with methane amount and such linearity was observed with hydrocarbon additives other than methane. The experiments so far discussed have all dealt with fuel-rich flames and for these the prompt NO yield was also found to correlate linearly with number of carbon atoms per molecule of the hydrocarbon additive. This is shown as Figure 9.5(c). The C_1 additive is methane, the C_2 acetylene and the C_3 propyne ($CH_3—C{\equiv}CH$). The linearity means that under fuel-rich conditions, a carbon atom from methane is as effective in reacting to form prompt NO as a carbon atom from acetylene or propyne. In fuel-lean flames, the linearity in Figure 9.5(c) was lost, with acetylene displaying more than twice the capability of CH_4 to promote prompt NO formation and more than two-thirds the capability of propyne.

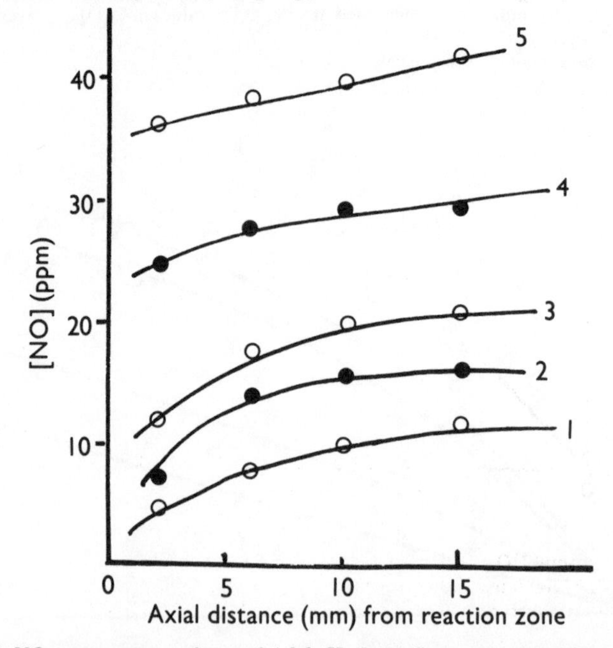

Figure 9.5 (a) NO measurements along a 3:1:3.3 H_2:O_2:N_2 flame, calculated flame temperature 2350K. Curve 1: No hydrocarbon additive. Curve 2: 0.98% (molar basis) C_2H_2 added. Curve 3: 3.02% C_2H_2 added. Curve 4: 4.43% C_2H_2 added. Curve 5: 6.08% C_2H_2 added. Reaction zone referred to on the horizontal axis is a luminous zone about 1 mm from the burner mouth (Hayhurst and Vince [8])

(Reproduced with the permission of Macmillan Magazines Ltd and Dr. A.N. Hayhurst.)

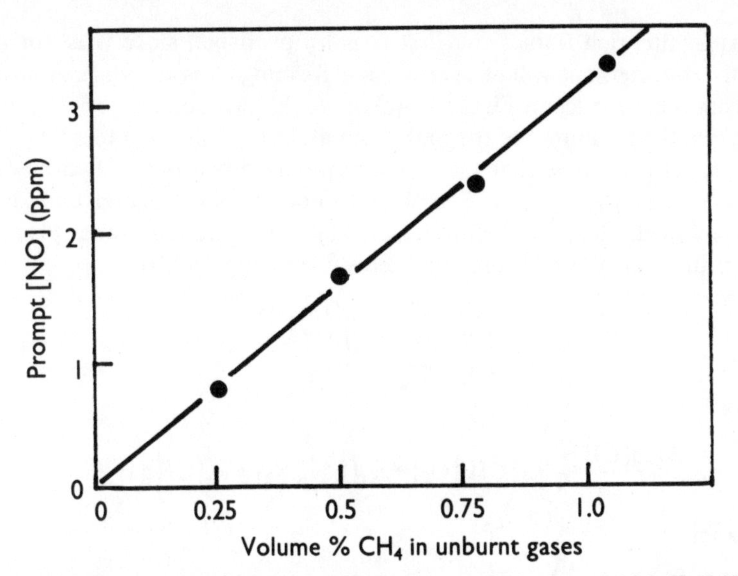

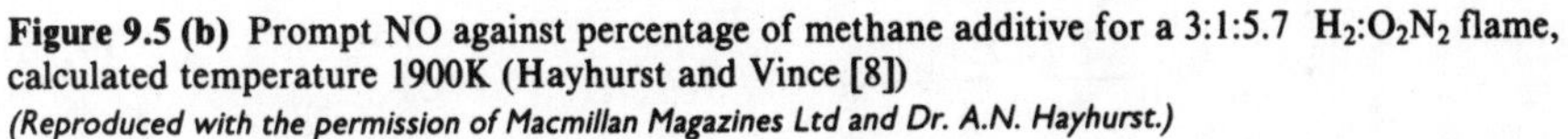

Figure 9.5 (b) Prompt NO against percentage of methane additive for a 3:1:5.7 H_2:$O_2$$N_2$ flame, calculated temperature 1900K (Hayhurst and Vince [8])
(Reproduced with the permission of Macmillan Magazines Ltd and Dr. A.N. Hayhurst.)

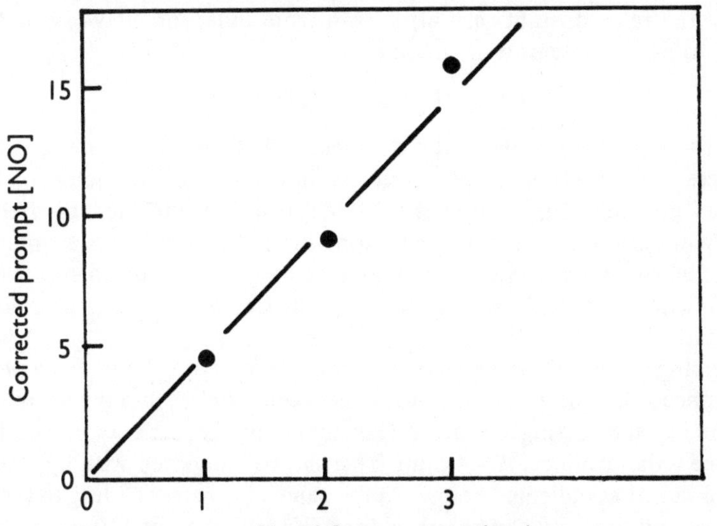

Figure 9.5 (c) Prompt NO yield with C_1, C_2, and C_3 additives in a H_2:O_2:N_2 flame (proportions as for Figure 9.5(a)) (Hayhurst and Vince [8])
(Reproduced with the permission of Macmillan Magazines Ltd and Dr. A.N. Hayhurst.)

Of the two primary steps suggested [8] in the formation of prompt NO:

$$CH + N_2 \rightarrow HCN \rightarrow N$$

or

$$C_2 + N_2 \rightarrow 2CN,$$

at least in fuel-rich flames the first is more probable, since it is not to be expected that methane will be as efficient at forming C_2 as acetylene or propyne but, from the evidence in Figure 9.5(c), it would have to be if the second step was in fact the beginning of the path from molecular nitrogen to prompt NO.

The belief [8], [9] is that there is a stripping down of the hydrocarbons present to CH_i, where $i = 1, 2, 3$, or 4, and a pooling of these, under conditions approaching equilibrium, before significant CO production takes place. CO production is then one of three processes competing for CH:

$$CH + O_2 \rightarrow CO + OH \quad (X)$$

$$CH + N_2 \rightarrow HCN + N \quad (Y)$$

$$CH + H_2 \rightarrow CH_3. \quad (Z)$$

Clearly:

$$-\frac{d[CH]}{dt} = k_X[CH][O_2] + k_Y[CH][N_2] + k_Z[CH][H_2]. \quad (9.17)$$

Rearranging:

$$-\frac{d\ \ln[CH]}{dt} = k_X[O_2] + k_Y[N_2] + k_Z[H_2]. \quad (9.18)$$

Now the rate constants are all known from independent work and their substitution, and integration, gives:

$$[CH] = [CH]_o \exp(-1.95 \times 10^8 t), \quad (9.19)$$

where t = time (s) and subscript o denotes $t = 0$. Times can be obtained from distances from the burner plus velocity information. The value of $[CH]_o$ depends not only on the quantity of hydrocarbon admitted to the flame initially but also on the equilibria existing in the CH_i pool. It is estimated [9] that 1 in 200 of the carbon atoms admitted as acetylene in the flames (see Figure 11.5(a)) will lead to CH. Any HCN formed by step Y is taken to be converted quantitatively to NO.

Referring to the 0.98% acetylene experiments in Figure 9.5(a), for a particular distance along the axis of the burner between 5 and 10 mm the prompt NO, measured by subtracting the total NO from that on the curve for no additive at the same axial distance, is 4.4 ppm. This can be compared with a calculated value obtained as indicated below. At the value of t corresponding to the axial distance, equation 9.19 can be solved for [CH] at that point. [NO] can then be obtained from the rate expression for step Y:

$$\frac{d[NO]}{dt} = 2k_Y[CH][N_2], \quad (9.20)$$

since HCN and N are both taken to be quantitatively converted to NO. This equation can be integrated with initial conditions [NO] = 0 at $t = 0$. [CH] is known, $[N_2]$ is easily estimated and, as already noted, k_Y is known. The value of [NO] so calculated at the axial position chosen is 4.3 ppm, almost indistinguishable from the experimental value. This provides further support for the view that in fuel-rich flames, prompt NO formation is via route Y with pooling of CH_i species followed by attack on CH by H_2, O_2 and N_2.

Figures 9.5(a)–(c) all deal with H_2:O_2 3:1, that is, stoichiometric ratio 1.5. With stoichiometric ratios between 1.5 and 1, there is, in experiments of the type described with hydrocarbon additives, a sharp decline in prompt NO [9]. Reaction X, and a similar one involving OH, are accelerated by the increased oxygen in going from fuel-rich to stoichiometric conditions, at the expense of Y and hence of prompt NO formation. Also, from the fact discussed above that under lean conditions one acetylene molecule is more effective in promoting prompt NO than two methane molecules or two-thirds of a propyne molecule, C_2 participation in prompt NO under less fuel-rich conditions is a possibility.

9.5 Conclusion

In this chapter and Chapter 8 we have discussed sulfur and nitrogen oxide formation at some length because of their importance in fuel usage. Annual NO_x emissions in the USA are of the order of 10^{10} kg and limits have to be imposed in the release of NO_x. These vary from place to place, but a good example is the requirement of a maximum of 0.6 lb NO_x (expressed as NO_2) per 10^6 BTU fuel input [10] in steam-raising with coal. Sulfur emission regulations are sometimes similarly expressed, for example, [10], 1.2 lb SO_2 per 10^6 BTU in coal combustion. In SI units these are approximately 0.26 kg NO_2 per GJ, and 0.5 kg SO_2 per GJ, respectively. Clearly an understanding of the principles of S and N combustion discussed is necessary in addressing such practical issues.

References

[1] Fine D. J., Slater S. M., Sarofim A. F., Williams G. C., 'Nitrogen in coal as a source of nitrogen oxide emission from furnaces', *Fuel* **53** 120 (1974)

[2] Crowhurst D., Simmons R. F., 'The formation of NO_x from nitrogen-containing additives in premixed methane flames', *Combustion and Flame* **51** 289 (1983)

[3] Fenimore C. P., (1972) cited by Crowhurst and Simmons [2]

[4] Heberling P. V., Boyd M. G. 'The effect of pressure on NO yield from fuel-N in laminar diffusion flames', *Combustion and Flame* **41** 331 (1981)

[5] Chen S. L., Heap M. P., Pershing D. W., Martin G. B., 'Influence of coal composition on the fate of volatile and char nitrogen during combustion', *Nineteenth Symposium (International) on Combustion*, 1271, Pittsburgh: The Combustion Institute (1983)

[6] Warnatz J., 'NO_x formation in high-temperature processes', *Proceedings, European Applied Research Conference on Natural Gas (Eurogas '90)*, (E. Brendberg, B. F., Magnussen, O. T. Onsager, Eds), 303, Trondheim: Tapir Press (1990)

[7] Homer J., Sutton M. M., 'Nitric oxide formation and radical overshoot in pre-mixed hydrogen flames', *Combustion and Flame* **20** 71 (1973)

[8] Hayhurst A. N., Vince I. M., 'Production of prompt nitric oxide and decomposition of hydrocarbons in flames', *Nature* **266** 524 (1977)

[9] Hayhurst A. N., Vince M., 'The origin and nature of prompt nitric oxide in flames', *Combustion and Flame* **50** 41 (1983)

[10] Wark K., Warner C. F., *Air Pollution: Its Origin and Control*, Second Edition. New York: Harper and Row (1981)

CHAPTER 10

Formation of Combustion-derived Pollutants Other than SO_x and NO_x

Abstract

The formation of carbon particles in combustion processes is discussed, with special reference to hydrocarbon combustion. A treatment of sooting propensity of many hydrocarbons in premixed and diffusion flames based on normalization of data from many different burners is presented and results are given for alkanes, alkenes, alkynes and aromatics. Sooting in methane/air flames, propane/oxygen flames and in a diesel engine is also discussed. Submicron unburnt carbon (SUC) formation from pulverized fuel (p.f.) is also described. Hydrocarbon emissions from combustion of hydrocarbon or alcohol combustion in the presence of motor oils is outlined, as well as polyaromatic hydrocarbon (PAH) emissions from a benzene flame.

10.1 Carbon particles

10.1.1 Propensities of various hydrocarbons to soot formation*

Smoke is defined as *small gasborne particles resulting from combustion* and soot as *an agglomeration of carbon particles* [2]. Factors relevant to sooting include fuel/oxidant ratio, aromatic/aliphatic content of the fuel and extent of mixedness of the fuel and the oxidant. Experimentally determined propensities of different hydrocarbon fuels to sooting are fairly abundant in the literature but comparison is not straightforward since sooting is influenced by heat and mass transfer, and therefore by the burner configuration. Strict comparison is therefore possible only between results obtained with the same apparatus.

However, normalization of data from many laboratories for the sooting characteristics of hydrocarbons was achieved [1] by first recognizing the need

*This section has drawn extensively on a review article by Calcote and Manos [1].

to consider premixed and diffusion flames separately, and defining for each a threshold soot index (TSI). For premixed flames, this definition requires first that of the minimum equivalence ratio for sooting, ϕ_c:

$$\phi_c = \frac{[\text{fuel influx rate/oxidant influx rate}]}{[\text{fuel influx rate/oxidant influx rate}]_{\text{stoichiometric}}}. \tag{10.1}$$

The TSI is defined as follows:

$$\text{TSI} = a - b\phi_c, \tag{10.2}$$

where a and b are constants for data for different hydrocarbons obtained on the same burner. For diffusion flames, the feature with which sooting is correlated is the flame height and the TSI is defined by:

$$\text{TSI} = a\left(\frac{\text{m.w.}}{h}\right) + b, \tag{10.3}$$

where h = minimum flame height for sooting (mm), and
m.w. = molecular weight of the hydrocarbon (g mol^{-1}).

One such set of data [3] for diffusion flames in air of over 70 hydrocarbons has 1-methyl naphthalene as the hydrocarbon with the strongest tendency to soot, and n-hexane as the hydrocarbon with close to the weakest tendency. These can be used to establish a scale of TSI and for 1-methyl naphthalene and n-heptane the arbitrarily assigned TSI values are 100 and 2 respectively. Summarizing:

1-methyl naphthalene, m.w. 142 g mol^{-1}, h = 5 mm, TSI = 100
n-hexane, m.w. = 86 g mol^{-1}, h = 149 mm, TSI = 2.

Substituting these data pairs into equation 10.3 and solving for a and b (denoting these a_1 and b_1) gives:

$$a_1 = 3.52$$
$$b_1 = -0.0331.$$

These are the a and b values for the two hydrocarbons. Two others studied in the same apparatus—n-heptane (m.w. 100 g mol^{-1}) and decalin (m.w. 138 g mol^{-1})—when their experimental h values are substituted into equation 10.3 together with the previously determined a and b values, give:

TSI (n-heptane) = 2.36
TSI (decalin) = 12.8.

Values of a and b obtained from results from the same compounds with a different burner are not the same as the above, nor, therefore, are the TSI values. However, TSI values for the same hydrocarbon with different burners can be normalized by adjusting the constants, and also scaled to give a maximum value of the TSI of 100 and a minimum value of zero. This normalization can be carried out either for premixed flames (TSI = $f(\phi_c$, m.w.))

or for diffusion flames (TSI = $f(h$, m.w.)). The validity of the approach can be understood from the following example.

Adjusted values a_1 and b_1 of the parameters for the data already discussed [3] are:

$$a_1 = 3.10$$
$$b_2 = 1.07.$$

Independent prior work [4] examined 25 hydrocarbon/air diffusion flames in a totally different configuration of burner and adjusted values of a_2 and b_2 for these data are:

$$a_2 = 3.81$$
$$b_2 = -0.76.$$

Consider methyl cyclohexane (m.w. 98) studied in each of the investigations cited. Its h values were 94 mm and 70 mm respectively in the burners of parameters a_1, b_1, and a_2, b_2. Therefore, substituting in equation 10.3 with the adjusted values of the parameters:

$$\mathrm{TSI}_1 = \left(3.1 \times \frac{98}{94}\right) + 1.07 = 4.3$$

$$\mathrm{TSI}_2 = \left(3.81 \times \frac{98}{70}\right) - 0.76 = 4.6.$$

The values are close, illustrating that the adjusted parameters have normalized results with different burners. The values are also low and therefore indicate a small propensity to sooting of this hydrocarbon compound. Consider now an aromatic compound: styrene (m.w. 104). In the burner characterized by a_1, b_1 its smoke point was 4 mm, giving TSI = 82, indicative of a much greater propensity to sooting.

We have considered this approach in some detail in respect of two normalized sets of TSI, h correlations for diffusion flames, whereas the full compilation makes use of 6 sets of TSI, h correlations for diffusion flames and 5 sets of TSI, ϕ_c correlations for premixed flames. Two of the five diffusion-flame correlation sets used not flame height but minimum volumetric flow rate of fuel for sooting V in place of h in equation 10.3. Of course, only rarely will any one hydrocarbon have been studied in all five burners for premixed flames or all six for diffusion flames. Some representative results are given in Table 10.1 from which it is clear that aromaticity promotes sooting. It is possible from the many normalized TSI values found in the way outlined to plot trends of sooting tendency with molecular structure.

Figure 10.1 shows the diffusion flame TSI values as a function of the number of carbon atoms for three homologous series, as well as for a number of alicyclics and aromatics. The following points deserve comment:

1. TSI values tend to differ only slightly, if at all, between isomers, for example, the values for n-butane and isobutane are the same to within

Table 10.1 Examples of normalized data for threshold smoke index (TSI)

(a) Premixed flames, TSI = $f(\phi_c$, m.w.)

Hydrocarbon (MW/g mol^{-1})	C:H *atomic ratio*	*Data set*					*Mean*
		1	2	3	4	5	
ethane (30)	0.333	42			28		35 ± 7
benzene (78)	1	68	93	75	85		80 ± 12
n-hexane (86)	0.429	65		63			64 ± 1
toluene (92)	0.875	78	87	93	72	86	83 ± 11
tetralin (132)	0.833	108	92			95	98 ± 10

(b) Diffusion flames, TSI = $f(h$, m.w.)

	C:H *atomic ratio*	*Data set*						*Mean*
		6	*7**	*8***	*9†*	*10*	*11†*	
acetylene (26)	1				2.7		4.6	3.7 ± 1.0
cyclopentane (70)	0.5		3.0	3.5	3.5			3.3 ± 0.2
methyl-cyclopentane (84)	0.5			5.0	4.8			4.9 ± 0.1
toluene (92)	0.875	52		48	50	48		50 ± 2
ethylbenzene (106)	0.8	61		56				59 ± 3

*Ref. [4]

**Ref. [3]

†Minimum volumetric flow rate for sooting used as a measure of sooting. Minimum flame height for sooting used in the other diffusion flame results.

experimental error. However, consider the isomers of molecular formula C_8H_{18}. This of course could represent n-octane:

$$CH_3(CH_2)_6CH_3,$$

or one of the several isomers of methyl heptane, for example:

$$\begin{array}{l} CH_3\,CH_2\underset{\displaystyle CH_3}{\underset{|}{C}}H(CH_2)_3CH_3 \end{array}$$

or of ethyl hexane, for example:

$$\begin{array}{l} CH_3CH_2\underset{\displaystyle C_2H_5}{\underset{|}{C}}H(CH_2)_2CH_3 \end{array}$$

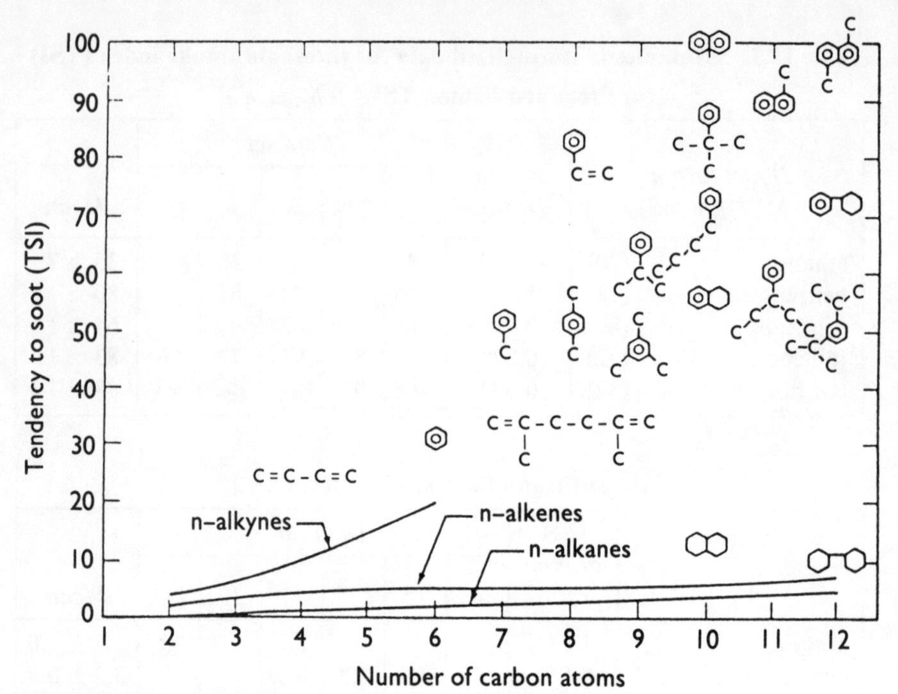

Figure 10.1 TSI against number of carbon atoms for hydrocarbon diffusion flames (Calcote and Manos [1])
(Reproduced with the permission of the Combustion Institute.)

or of dimethyl hexane, for example:

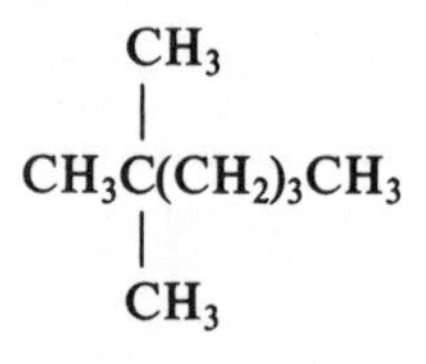

or of trimethyl pentane, for example:

$$\begin{array}{c} CH_3 \\ | \\ CH_3C - CHCH_2CH_3 \\ |\qquad | \\ CH_3\ \ CH_3 \end{array}$$

Within such a variety of isomers there is a tendency for those with the shorter carbon skeleton length to have the higher TSI, that is, molecular compactness promotes sooting. However the values of TSI for the C_8H_{18} isomers are only in the approximate range 3–5 and the differences between them are therefore fairly insignificant when it is recalled that TSI values go up to 100.

2. Apart from the alkanes and alkenes, the trend is for one additional carbon atom to produce a TSI increment of 6–12 units.

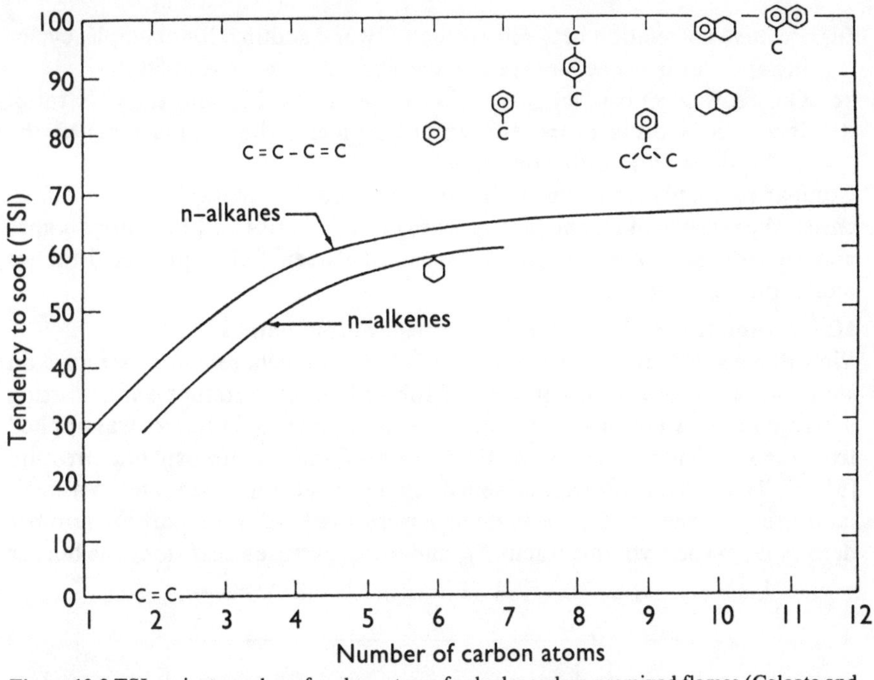

Figure 10.2 TSI against number of carbon atoms for hydrocarbon premixed flames (Calcote and Manos [1])
(Reproduced with the permission of the Combustion Institute.)

3. The non-aromatics have a much lower intrinsic sooting tendency than the aromatics. Butadiene is out of sequence with the other non-aromatics and this is consistent with the belief that this conjugated compound is frequently a precursor of soot formation. The link between conjugation and sooting is also demonstrated by the styrene result in Figure 10.1.

4. There was no systematic correlation between diffusion-flame TSI values and C:H ratio of the hydrocarbon fuel. An interesting instance of this lack of correlation is that four hydrocarbons with C:H ratio unity—acetylene, benzene, styrene and dimethyl naphthalene—have TSI values of respectively 3.7, 31, 81 and 98. For the alkyne homologous series there is a decrease in TSI value with increasing C:H ratio.

The TSI index plotted against number of carbon atoms for premixed flames is shown in Figure 10.2. We should perhaps make the point that the higher TSI values for alkanes and alkenes in premixed flames, than in diffusion flames, has no physical meaning since the definition of TSI is different in the two cases. Features of interest in the results include:

(a) Small differences only between isomers, as with diffusion flames.
(b) A slope of about 7 TSI units per carbon atom. In contrast to the diffusion-flame findings, alkanes and alkenes have this degree of sensitivity to number of carbon atoms, at least up to about C_7.
(c) A high TSI value for butadiene, as in the diffusion flames.

(d) Positive correlation between aromaticity and sooting, for example, cyclohexane and benzene have respective TSI values of 56 and 80.
(e) Only a weak (positive) correlation between the TSI and the C:H ratio.
(f) Increased sooting propensity among isomers with compactness of the molecule, as with diffusion flames.

Comparing trends for diffusion flames and premixed flames the chief difference is the reversal of alkanes and alkenes in Figures 10.1 and 10.2. For alkanes and alkenes there is significant correlation between TSI in premixed flames and in diffusion flames.

10.1.2 Soot formation in laminar methane/air flames

We will consider results reported in the relatively recent research literature on soot formation in the combustion of fuels of practical importance, starting with methane. An optical technique for examining soot in flames was applied to laminar cylindrical methane/air diffusion flames at atmospheric pressure [5], [6]. The fuel and air were supplied separately at controlled rates to a small laboratory burner and measurements were made of soot particle number density N, particle volume fraction f_v and mean particle size d along the burner

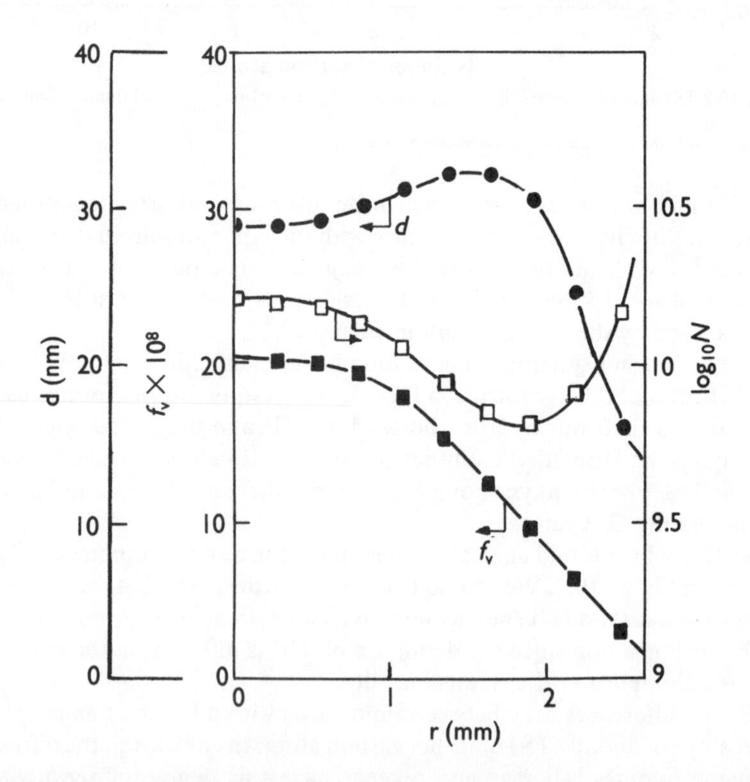

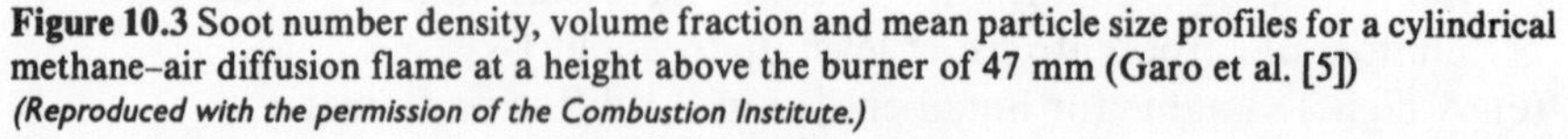
Figure 10.3 Soot number density, volume fraction and mean particle size profiles for a cylindrical methane–air diffusion flame at a height above the burner of 47 mm (Garo et al. [5])
(Reproduced with the permission of the Combustion Institute.)

axis and also radially. In some experiments, polyaromatic hydrocarbon (PAH) was also determined by gas chromatography/mass spectrometry.

Radial profiles of N, f_v and d are shown in Figure 10.3 from which it can be seen that the volume fraction has a maximum at the axis. The subsequent decline is due to destruction of the soot by oxidation. Each of d and N has a turning value 1 to 2 mm from the axis. Comparison of the volume fraction and diameter profiles suggests that growth and destruction processes overlap. Close to the edge of the flame there is a high concentration of small, largely burnt out particles. The axial coordinate to which the radial measurements in Figure 10.3 apply is close to the maxima in the f_v and d profiles in this direction, which are shown in Figure 10.4.

The first stage in sooting is formation of solid carbon from gas-phase

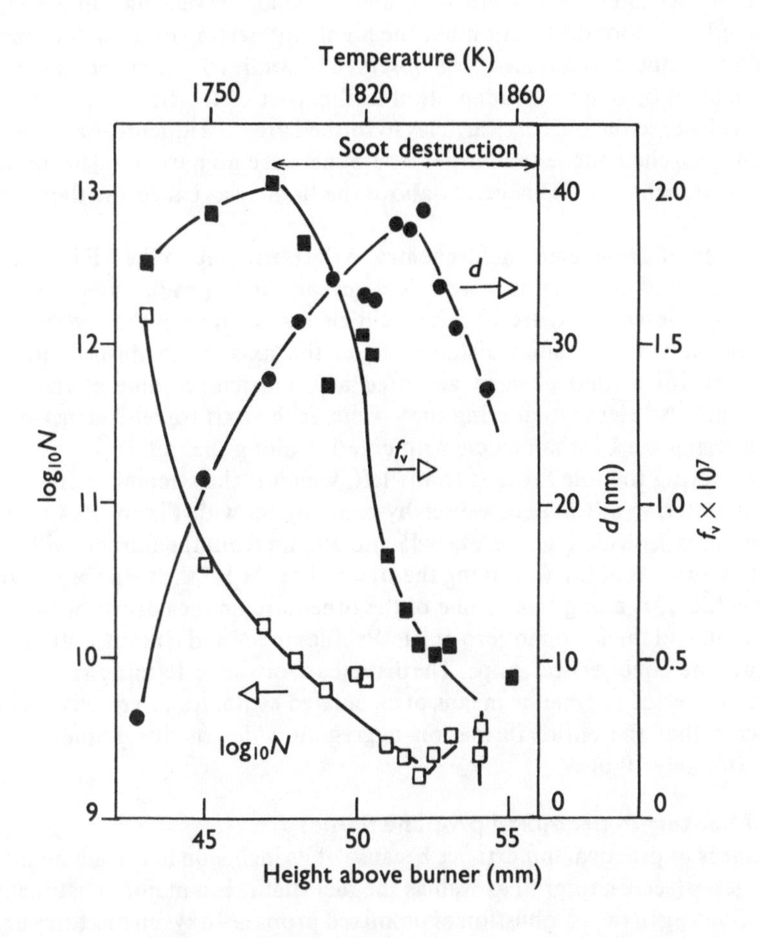

Figure 10.4 Axial profiles of soot number density, volume fraction and mean particle sizes for a cylindrical methane/air diffusion flame (Garo et al. [6])
(Reproduced with the permission of the Combustion Institute.)

reactants, after which there is a steady rise in N, f_v and d up to 43 mm above the burner. A *possible* mechanism of initial solid carbon formation is:

$$2CO \rightarrow CO_2 + C.$$

At 43 mm, d and f_v continue to rise, indicating that solid carbon continues to be formed. However, N decreases because of aggregate formation. Attainment of the maximum in f_v, 45–50 mm above the burner, signals the end of the regime where the profiles are determined solely by soot formation and aggregation, and that some soot destruction by oxidation is occurring. The fact that after the maximum in f_v the mean diameter d continues to rise suggests that even though some soot is being oxidized, aggregation continues. Also, destruction is likely to be of the smallest particles first, causing mean particle size d to increase. The region of the flame axis where N, f_v and d are all decreasing can clearly be identified with soot destruction but the highly interesting eventual *increase* in N should be noted. It has also been observed in studies of premixed flames and is thought to be due to disintegration in this part of the flame of aggregates formed closer to the burner. Particles so formed are subsequently oxidized and a point is reached above the burner where there are no particles detectable by the optical technique. This height above the flame was called the flame front (FF).

The path of a fluid element after entry to the burner up to the FF is termed a streamline, and such paths were calculated at various radial positions. The results are shown in Figure 10.5 for fluid elements entering the system at the axis and at various radial distances from the axis. Each dashed line is a streamline for a fluid element admitted at a particular value of the radial coordinate. An element entering the system at the axis travels further before encountering the FF than an element entering along the radius.

Considering the line farthest to the left, which is the streamline for a fluid element entering at the axis, we see by comparison with Figure 10.4 that the maximum in f_v, which lies between 45 and 50 mm from the burner, will occur about two-thirds of the way along the streamline. At FF the value of f_v is zero, so a profile of f_v along this or one of the other streamlines drawn will have a maximum and then drop to zero at FF. Profiles for N and d along a streamline will have the same general shape. The distance coordinate defining a streamline can be converted to time by means of calculated estimates of velocity and it is predicted that the entire formation–aggregation–destruction sequence lasts approximately 10 ms.

10.1.3 Sooting in premixed propane flames

Propane is of practical importance because of its inclusion in certain manufactured gases (see Chapter 5) as well as the fact that it is a major constituent of LPG. Sooting in the combustion of premixed propane–oxygen mixtures under laminar conditions with various amounts of inert gas across a range of initial pressures was studied in a combustion bomb using laser techniques to follow particle formation [7]. The bomb was cylindrical with eight spark plugs at 45°

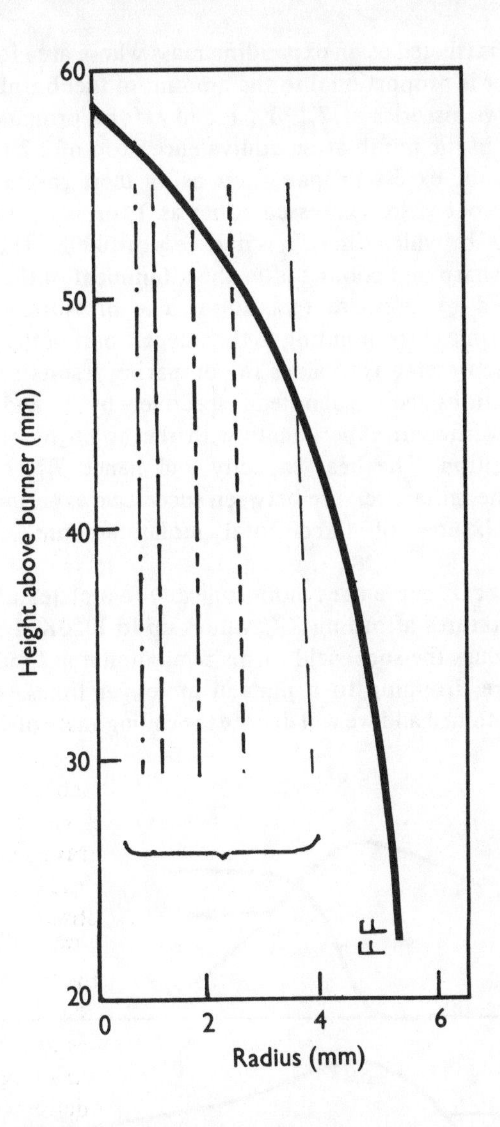

Figure 10.5 Streamlines for fluid elements admitted to a burner at its axis (line farthest left) and at various radial distances from the axis (Garo et al. [5])
(Reproduced with the permission of the Combustion Institute.)

intervals along its circumference. Once a particular propane–oxygen–inert gas mixture was ignited by means of the sparks, the yield of soot (Y_s, percent of the carbon originally in the propane present as soot) and the volume fraction f_v of the soot were followed, and concurrently the pressure P inside the bomb was measured. The burnt gas temperature T_{bg} at various times after ignition was calculated and details of the calculation procedure include the following:

(a) T_{bg} has a single value at any time.

(b) The burnt gas is treated as an expanding mass whose area for the purpose of heat transfer is proportional to the amount of fuel burnt.

Figure 10.6 shows histories of T_{bg}, Y_s, P and f_v for a propane–oxygen (no inert gas) mixture in the bomb at an equivalence ratio of 2.2 (that is, a rich mixture, in which the excess propane *acts* as an inert gas in lowering the temperature). The soot yield, expressed either as Y_s or as f_v, starts to ascend only after $\approx$ 100 ms, by which time T_{bg} is in excess of 1000K. The pressure rise during reaction is sharp and cooling after the attainment of the maximum in T_{bg} is accompanied by pressure relaxation. The pressure and burnt-gas temperature at the time corresponding to the steepest part of the f_v, t curve can be used as single representative values in comparing results under different experimental conditions and are denoted respectively by P^* and T^*_{bg}. T^*_{bg} can be varied between different experiments with the bomb by controlling the initial gas composition. The heat capacity and hence T^*_{bg} can be varied independently of the initial pressure between successive experiments by using argon/nitrogen mixtures of fixed total molar amount, but different proportions.

In Figure 10.6, the Y_s curve rises monotonically to a plateau, as it does for other initial gas mixtures affording T^*_{bg} values up to 1350K. In experiments with higher T^*_{bg} values the soot yield forms a maximum at times between 50 and 100 ms, before dropping to a plateau at longer times. A plateau is, however, always obtained and we will denote the ceiling value of Y_s by Y^*_s. This

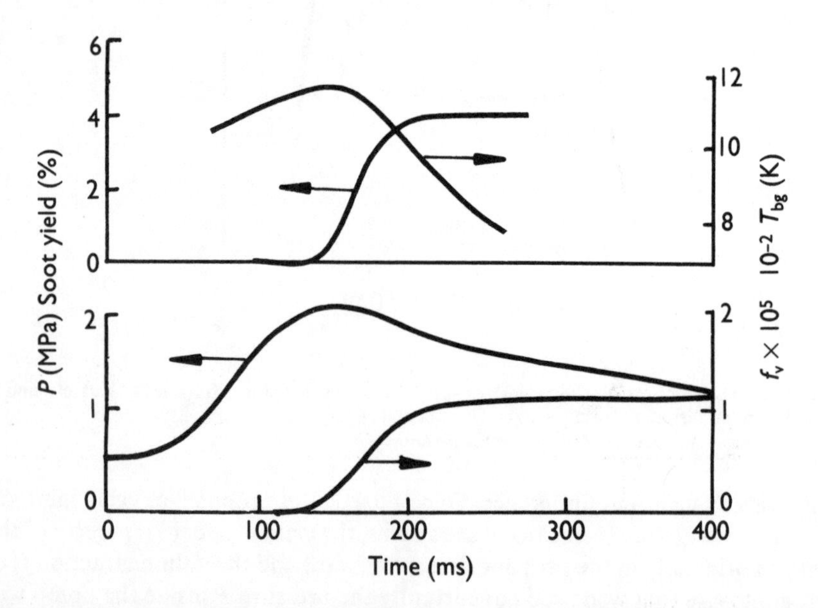

Figure 10.6 Burnt gas temperature T_{bg}, pressure P, soot yield Y_s and volume fraction of soot f_v histories for a propane–oxygen mixture with equivalence ratio 2.2 and total initial pressure 0.58 MPa in a bomb (Kamimoto et al. [7])
(Reproduced with the permission of the Combustion Institute.)

contrasts with the diffusion-flame experiments using methane discussed in Section 10.1.2, in which all the soot is eventually oxidized.

The quantity Y_s^* increases with P^* for fixed stoichiometric ratio and with stoichiometric ratio for a particular P^*. For a stoichiometric ratio of 2.1, values of Y_s^* against P^* for various T_{bg}^* are plotted in Figure 10.7. Clearly there is a very steep dependence of soot formation on pressure at low T_{bg}^* and a diminishing dependence as T_{bg}^* increases. In some experiments, turbulence was created by use of a screen containing multiple holes through which the flames had to pass. Other things being equal, turbulence causes T_{bg}^* values to be 50–150K higher than under the laminar conditions prevailing without the screen, but for particular values of T_{bg}^* the pressure dependence of Y_s^* is the same for laminar or turbulent conditions as shown. The curves can perhaps be described as isotherms since each is for a particular T_{bg}^* value at different P^*, some points on each isotherm being for laminar conditions and some for turbulent conditions. When Y_s^*, P^* isotherms are plotted for the same equivalence ratio

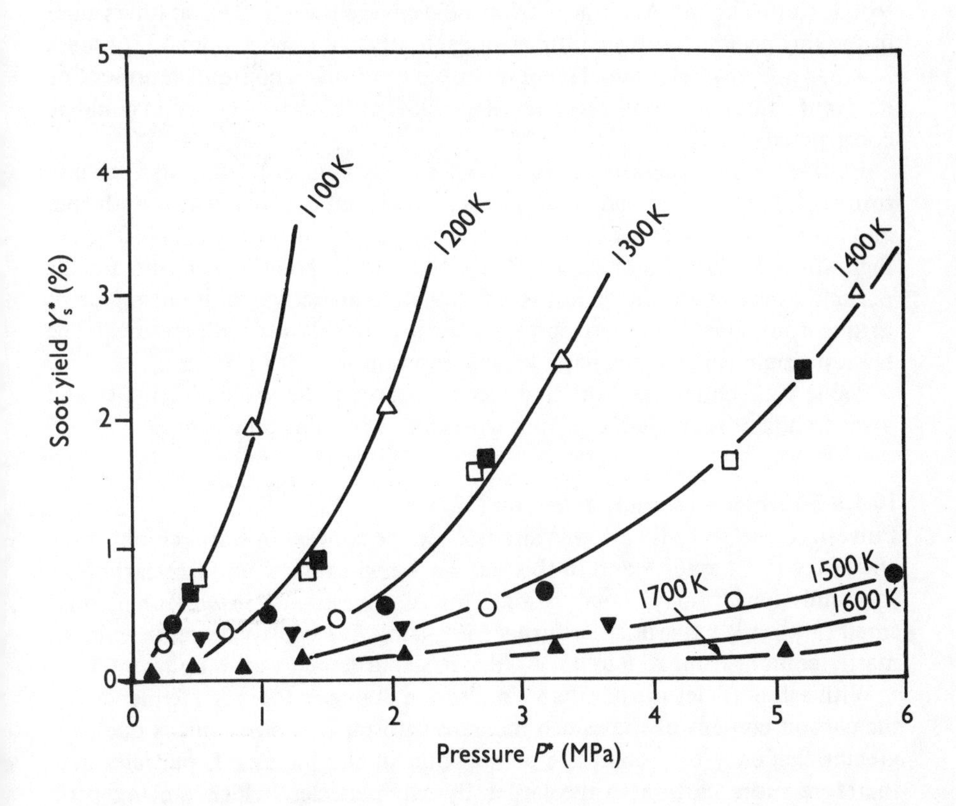

Figure 10.7 Soot yield Y_s^* against pressure P^* at various temperatures T_{bg}^* for propane–oxygen–N_2/Ar mixtures with stoichiometric ratio 2.1. All points are for laminar conditions, except those denoted by filled squares or filled inverted triangles which are for turbulent conditions (Kamimoto et al. [7])

(Reproduced with the permission of the Combustion Institute.)

and laminar conditions with differing amounts of inert gas, they mirror very closely those in Figure 10.7: strong pressure dependence at lower temperatures and much weaker dependence at higher temperatures. The indication is therefore that in practical systems using propane (or LPG), pressure and temperature rather than turbulence or amounts of inerts are the factors most strongly affecting sooting.

10.1.4 Particles from diesel combustion

Particle emission was studied in work utilizing a single-cylinder diesel engine [8]. The particles themselves, once trapped in a filter, were characterized in various ways. In the work cited, the term *gas oil* describes the fuel; petroleum-derived materials used in compression-ignition engines and those used in some gasification processes occur at about the same temperature range in fractionation, so the terms *diesel* and *gas oil* are sometimes used synonymously. The engine was operated at various speeds (rpm) and with various values of applied torque (units, kg.m). At each rpm value the engine was operated at four values of the torque which were multiples of each other, termed 1/4 load, 2/4 load, 3/4 load and 4/4 (full) load. Hence in torque–rpm space, contours representing different values of the various quantities relevant to particle emission could be constructed.

Exhaust-gas temperatures are shown as contours according to rpm and torque in Figure 10.8, and exhaust-gas temperatures increase with both rpm and load. While the contour of highest temperature corresponds to 650° C the highest single value observed was 722° C. The corresponding contours for the particle emission are shown in Figure 10.9, with higher values in the region of torque–rpm space also corresponding to higher exhaust temperatures. The highest single value of the particle emission rate was 10.7 g kWh^{-1}.

Table 10.2 summarizes the findings as regards particle characteristics and gives trends of such characteristics with engine running conditions.

10.1.5 Submicron unburnt carbon (SUC)

Pulverized fuel (p.f.) fly ash was discussed in the context of sulfur combustion chemistry in Chapter 8 and in this section we will review unburnt carbon in solid emissions from p.f. combustion. This was investigated in work on p.f. in a small (5 kg h^{-1}) experimental furnace [9] using five coals varying in volatile matter content from 28.6 to 45.3% (dry basis) at temperatures up to 1400° C.

With ash particles greater than a micron in diameter there is a tendency for the carbon content of the ash to increase with particle size. This is due to a greater tendency to incomplete combustion in the larger p.f. particles and therefore more carbon in the larger fly ash particles, which are taken to originate from p.f. at the higher end of the particle-size range. However, with submicron particles there is a very steep *increase* in carbon content with *decrease* in diameter, the opposite effect from that displayed by the larger particles. For example, 1 μm ash particles from one of the coals had a carbon

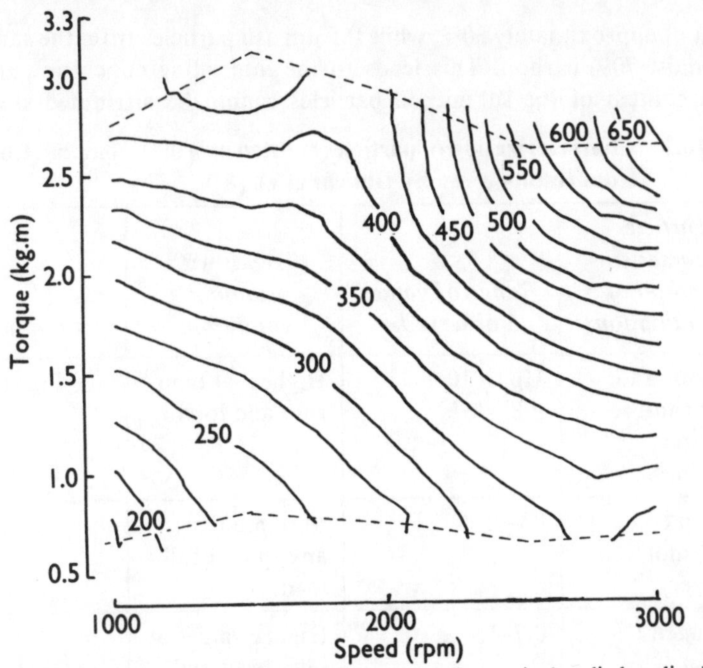

Figure 10.8 Exhaust gas temperature (°C) contours in a single-cylinder diesel engine (Obuchi et al. [8])
(Reproduced with the permission of the Combustion Institute.)

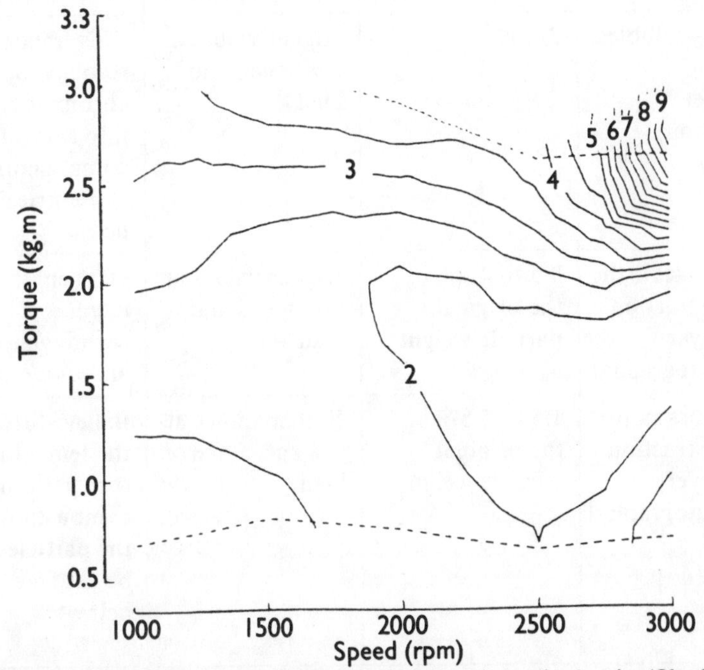

Figure 10.9 Particle emission (g kWh^{-1}) contours in single-cylinder diesel engine (Obuchi et al. [8])
(Reproduced with the permission of the Combustion Institute.)

content of approximately 30%, while 0.5 μm ash particles from the same coal had almost 70% carbon. This leads to the interesting conclusion that the carbon content of the submicron particles cannot be attributed simply to

Table 10.2 Characterization of particles emitted by a diesel engine (Compiled from results given by Obuchi et al. [8])

Particle characteristic (method of determination)	*Range of values observed*	*Trends with running conditions*	*Comments*
Emission rate (filter capture and direct weighing)	Up to 10.7 g kWh^{-1}	Highest at high rpm and load	See Fig. 10.9
Carbon content (elemental analyzer)	83–95%	Maximum at any rpm at full load	
Hydrogen content (elemental analyzer)	0.7–1.7%	Higher values at low speed and load	
CH_2Cl_2-soluble fraction (Soxhlet extraction)	2–14%	Higher values at low speed and load	This fraction resolved by chromatography into several components (see entries below)
Oxygenates in the soluble fraction (thin layer chromatography)	0.5 to 4.1% of the original particle weight	Higher values at low speed and load	Oxygenates include carboxylates and quinones
Aliphatics in the soluble fraction (thin layer chromatography)	1.5 to 5.5% of the original particle weight	Higher values at low speed and load	Believed from the temperature pattern of combustion of the particles to be C_{7+} (see below)

(Continued p. 203)

Table 10.2 (Cont.)

Particle characteristic (method of determination)	*Range of values observed*	*Trends with running conditions*	*Comments*
Benzo-(a)-pyrene in the soluble fraction (fluorescence spectrometry)	0.01 to 5.9 ppm of the original particle weight	Lowest at the higher speeds	This and other polycyclic aromatics discussed in Section 10.3.
Fraction of the particle weight combustible at ≈ 400° C (differential scanning calorimetry (DSC))	Up to 6%	Higher at the lower speeds, reflecting the dependences of the hydrogen and aliphatic contents on running conditions	Largely combustion of the CH_2Cl_2-soluble fraction and believed to include aliphatics of C_{7+}, since this temperature is characteristic of C_{7+} burning in DSC
Burnout temperature of the particles other than the fraction combustible at ≈ 400° C (differential scanning calorimetry and thermogravimetric analysis)	673 to 707° C	Higher at the higher loads	Trend suggests a positive correlation between the burnout temperature and the degree of carbonization of the particles

incomplete combustion. In fact, it is believed to arise from sooting in the combustion of the coal volatiles. Two observations support this view:

(a) A H/C mass ratio of 0.02 for the submicron carbon, typical of the values for soot.

(b) Electron-micrographs of the submicron fly ash particles, which reveal chain-like carbon structures on the surface of the ash. Such structures would not be expected to be present in the original fuel but are characteristic of soot.

10.2 Hydrocarbons

10.2.1 Introduction

Sources of hydrocarbon contamination of the atmosphere include gas turbines utilizing petroleum-based fuels as well as engines using gasoline or diesel. There is evidence that in combustion engines the fuel and the lubricating oil are both involved in hydrocarbon emission.

10.2.2 Experiments with different fuels and oils in a combustion bomb

Relatively recent investigations by research personnel of the Ford Motor Company, Michigan USA, have contributed significantly to our understanding of hydrocarbon emission from combustion processes and among their seminal publications is one describing combustion of certain fuel blends in an oil-lined combustion bomb [10]. The fuel blend/air mixture was admitted to a small (0.11 litre volume) bomb with 0.2 g of oil forming a thin layer at its base. Time was allowed for phase equilibrium between the fuel and the oil to be established before spark ignition of the mixture, and the combustion products were analyzed.

That the various oils do affect hydrocarbon formation substantially can be seen from the selected results presented in Table 10.3 (a) and (b). The alkane and the alcohol carbon structures both appear in the post-combustion gases, although the latter exists partly as the nitrite, as indicated in the table. The less polar oils cause the greatest increase in ethane emission, while the opposite is true for the alcohol (methanol or ethanol) emissions, which are higher in the presence of the more polar oils. The mechanism of the observed effect is likely to involve the dissolution of some of the fuel in the oil, relevant to which is Henry's Law, which applies to a dilute solution with the liquid and vapor phases in equilibrium [11] as follows:

$$P_s = HX_s, \tag{10.4}$$

where P_s = partial pressure of solute in the vapor phase
X_s = mole fraction of solute in the liquid phase
H = Henry's Law constant, units pressure, dependent upon the identities of solvent and solute.

The Henry's Law constants for the alcohols and the alkane in oils A, B, C and D are known, and the value for methanol in oil A is about an order of magnitude higher than that for methanol in oil D. On this basis we expect a lower mole fraction of methanol in oil A than in oil D, given equal or very close values of the equilibrium methanol pressure in the vapor phase before firing. This appears to correlate directly with the significantly smaller increase in unburnt methanol for oil A. The way in which oils cause the unburnt carbon to increase is dissolution of the fuel vapor in the oil, enabling all or part of the dissolved oil to resist oxidation. At ignition, the phase equilibrium is lost and the dissolved fuel returns to the gas phase as unburnt fuel in the post-combustion mixture.

As we have seen, although there is an order of magnitude spread of values of *H* for methanol in the oils, there is a much smaller spread of values of *H* for ethane in the oils. This is reflected in the absence of large variations between the oils in the increased percentage of unburnt ethane. However, the *H*-value for ethanol in oil B is 15 times the value for ethanol in oil C and this is reflected in

Table 10.3 (a) Products of the combustion of fuel/air* mixtures in the presence of various oils in a bomb (Compiled from data given by Adamczyk et al. [10])

		Partial post-combustion gas composition				
Fuel	*Oil*	CO_2 %	N_2 %	C_2H_6 ppm (*% initial unburnt*)	CH_3OH** ppm (*% initial unburnt*)	*Increase in % unburnt due to oil*
CH_3OH, 38%, 22 400 ppm in initial fuel/O_2/N_2 mixture C_2H_6, 62%, 37 100 ppm in initial fuel/O_2/N_2 mixture	None	11.7	85.4	85 ± 6 (0.2)	1170 (5.2)	—
As above	Oil A, a synthetic motor oil, moderately polar, m.w. 445	11.9	84.5	318 ± 8 (0.9)	2360 (10.5)	Alcohol, 5.3 Alkane, 0.7
As above	Oil B, a petroleum-derived motor oil, m.w. 470	11.8	84.3	257 ± 17 (0.7)	1500 (6.7)	Alcohol, 1.5 Alkane, 0.5
As above	Oil C, a polar and highly oxygenated oil, m.w. 1040	11.2	85.5	208 ± 7 (0.6)	3980 (17.8)	Alcohol, 12.6 Alkane, 0.4
As above	Oil D, also highly oxygenated, m.w. 1030	10.9	84.7	181 ± 1 (0.5)	4500 (20.1)	Alcohol, 14.9 Alkane, 0.3

Table 10.3 (b)

Fuel	*Oil*	*Partial post-combustion gas composition*				
		CO_2 %	N_2 %	C_2H_6 ppm (% initial unburnt)	C_2H_5OH** ppm (% initial unburnt)	*Increase in % unburnt due to oil*
C_2H_5OH, 24%, 12 100 ppm in initial fuel/O_2/N_2 mixture C_2H_6, 76%, 38 000 ppm in initial fuel/O_2/N_2 mixture	None	11.4	84.5	87 ± 4 (0.2)	345 (2.9)	—
As above	Oil A	11.7	84.5	315 ± 12 (0.8)	1630 (13.5)	Alcohol, 10.6 Alkane, 0.6
	Oil B	11.5	84.7	274 ± 3 (0.7)	990 (8.1)	Alcohol, 5.2 Alkane, 0.5
As above	Oil C	10.5	84.0	254 ± 6 (0.7)	2880 (23.8)	Alcohol, 20.9 Alkane, 0.5
As above	Oil D	10.3	85.6	204 ± 31 (0.5)	2500 (20.7)	Alcohol, 17.8 Alkane, 0.3

* 21% O_2, balance N_2 used instead of atmospheric air. Fuel:oxidant equivalence ratio 0.9 for each entry in the table. All percentages molar basis. m.w. denotes average molecular weight in g per Avogadro number of molecules.

** Actually the alcohol ROH + the nitrite derivative RONO, formed by reaction with NO_2 in the bomb.

the very significant difference in unburnt ethanol between the two. Small (ppm) quantities of acetaldehyde were also observed in the ethanol/ethane fuel emission.

10.3 Polycyclic aromatic hydrocarbons (PAH)

10.3.1 Introduction

The simplest compound in this category is naphthalene which contains two benzene rings and this can occur in the combustion products of some petroleum-based materials, for example, kerosene. Other examples of PAH occurring in combustion processes include anthracene and phenanthrene (three rings), pyrene (four rings) chrysene (four rings) and 3,4 benzpyrene (five rings). To these we must add the many possible alkylated derivatives. Some of these compounds are known to have serious effects on health so attention has to be given to their formation even in trace amounts. One of the earliest reported identifications of a substance believed to cause cancer was soot, in 1775, and it is now known that PAH deposited on soot is largely responsible for the carcinogenic activity.

Some liquid fuels contain PAH compounds, for example, gasoline at the high end of its boiling range, so one source of PAH is unburnt fuel. Another is what is sometimes possibly erroneously referred to as fuel pyrolysis, which would perhaps be better called PAH synthesis. For example, acetylene, when heated to 973K under conditions precluding combustion, produces pyrene at a yield of 6.5 percent as well as smaller quantities of other PAH including chrysene [12]. Toluene, butadiene and styrene also yield significant amounts of PAH when heated in this way. The term *pyrolysis*, though widely used to describe the formation of PAH from simple organics, is inappropriate since pyrolysis is the breaking down or degradation of an organic structure by heat. PAH production from acetylene or other simple organics is the opposite of this, actually a building up of the carbon structure. However, some routes to PAH are genuinely pyrolytic, for example, PAH from coal pyrolysis.

What has been described above as PAH synthesis can occur in flames and we will outline studies of the combustion of benzene as a model fuel in which PAH formation is studied.

10.3.2 PAH formation in laboratory combustion of benzene [13]

PAH formation is *often* accompanied by sooting and in an experiment a small laboratory burner used to study benzene combustion was operated just below the minimum equivalence ratio for sooting (see Section 10.1.1). The influx mixture was 13.5% (molar basis) benzene vapor, 56.5% oxygen, and 30.0% argon. The flame was subatmospheric and laminar. Gas samples from along the axis of the flame were drawn into a molecular-beam mass spectrometer system for analysis. Analyses were made of stable species such as unreacted fuel, unreacted oxygen, acetylene, product water and oxides of carbon, as well as H, OH and HO_2. The mass spectrometry was also extended to higher mass numbers signifying PAH. Distances from the burner were converted into times using direct estimates of velocities at different heights above the flame. This was by admitting to the burner a dummy cold gas with the same Reynolds number as the flame, and using an anemometer. The temperature profile along the axis of the flame was measured with a coated rare-metal thermocouple and was found to have a peak value of approximately 1800K at 10 mm above the burner entry.

Figure 10.10 shows profiles in the flame for species of mass numbers 166, 176, 178 and 202. These are attributed respectively to $C_{13}H_{10}$, $C_{14}H_8$, $C_{14}H_{10}$, and $C_{16}H_{10}$. Obviously each structure could represent a number of isomers and the mass spectra do not enable identification of the precise structures. $C_{16}H_{10}$ for example is the molecular formula of pyrene, which comprises four fused aromatic nuclei, and $C_{14}H_{10}$ is the molecular formula of anthracene or phenanthrene, each tricyclic as already noted. All the structures fitting these formulae are examples of PAH so PAH synthesis in the flame is confirmed in spite of the lack of precise structural details. As can be seen from Figure 10.10, the PAH mole fractions peak at about 8 mm above the burner. In this part of the flame the CO mole fraction is 0.4 and the CO_2 fraction just below 0.1. There is also a significant amount of C_2H_2. The fact that there is a high degree of partial oxidation to CO is consistent with the fact that although the initial mixture is fuel-rich, the benzene and oxygen mole fractions have *both* dropped to low values by 8 mm into the flame. C_2H_2 continues to have a significant mole fraction at points in the flame beyond those at which the benzene mole fraction has become relatively very small. The benzene in fact has a mole fraction of the order of 10^{-3} in the region of the flame corresponding to maxima in PAH and there are other species present at about the same level including C_6H_6O, C_6H_5, and C_6H_8, which are intermediates in the oxidative destruction of benzene.

In Figure 10.10 the disappearance of PAH, say at 12 mm above the burner, is apparent rather than real, since there continues to be PAH but of higher mass numbers than those in the figure. This is manifest as a signal representing mass numbers > 200. There is also a signal due to material of mass number > 700. The maximum in the mole fraction of mass number > 200 occurs at about 8.5 mm above the burner, close to the maxima in Figure 10.10. The maximum

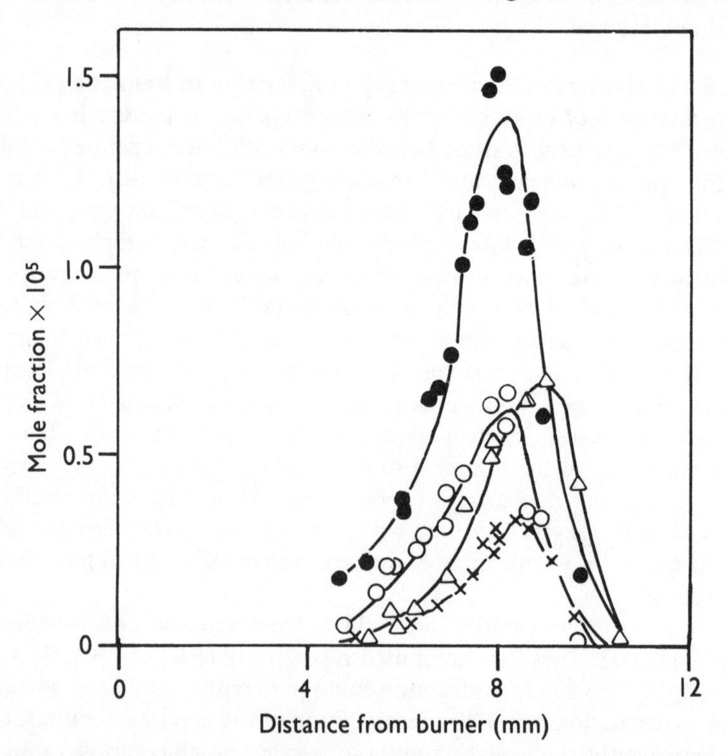

Figure 10.10 Profiles of mass number characteristic of PAH in a benzene/oxygen/argon laminar subatmospheric flame. Symbols: ×, $C_{16}H_{10}$; Δ, $C_{14}H_8$; ○, $C_{14}H_{10}$; ●, $C_{13}H_{10}$ (Bittner and Howard [13])

(Reproduced with the permission of the Combustion Institute.)

in the fraction of mass number > 700 is 10.5 mm above the burner. Hence there is evidence that the PAH, up to the position corresponding to the maximum in the signal for mass number > 700, is simply adding on carbon during its progression along the flame. Beyond this point there is a drop due to burning of the carbon structures so formed, but not to zero. The signal for the mass number > 200 is about one-hundredth of the intensity at the maximum, even 15 mm from the burner and quite measurable, hence the potential for PAH emission in the burnt gas is established. The signal for higher mass number also retains a measurable intensity at relatively large distances from the burner.

When the burner is operated with an influx above the minimum equivalence ratio for sooting, the maximum in the profile for carbon structures of greater mass number than 700 coincides exactly with the position of a yellow region of the otherwise blue flame due to soot burning. Therefore the results suggest that in this flame, PAH is the precursor of soot.

In the region of the maximum in the material of mass number > 700, the benzene profile drops very rapidly whether the equivalence ratio is just above or just below that required for sooting. This indicates that benzene is involved in stabilization of incipient carbon structures. Mass growth in this region of the flame is due primarily to hydrocarbons other than aromatics, including acetylene, which in the regions of the flame approaching this maximum are present in much more abundant amounts than benzene. The role of benzene is to provide a mechanism for ring closure. There are several feasible ways in which an existing aromatic could assist ring closure and aromatization, and they include:

$$\underset{\text{phenyl radical}}{C_6H_5} + \underset{\text{acetylene}}{HC{\equiv}CH} \rightarrow \underset{\text{vinyl radical}}{C_6H_5\text{-}CH{=}CH}$$

$$\downarrow HC{\equiv}CH$$

$$C_6H_5\text{-}CH{=}CH\text{-}CH{=}CH$$

$$\downarrow \text{cyclization}$$

$$C_{10}\ \text{PAH}$$

Once a PAH structure is synthesized by a scheme such as this it develops into a larger structure farther into the flame, as we have seen when discussing the profiles of the various mass number ranges.

References

[1] Calcote H. F., Manos D. M., 'Effect of molecular structure on incipient soot formation', *Combustion and Flame* **49** 289 (1983)

[2] Wark K., Warner C. F., *Air Pollution*, Second Edition, New York: Harper and Row (1981)

[3] Hunt R. A., 'Relation of smoke point to molecular structure', *Industrial and Engineering Chemistry* **45** 602 (1953)

[4] Clarke A. E., Hunter T. G., Garner F. H., 'The tendency to smoke of organic substances on burning', *Institute of Petroleum Journal* **32** 627 (1946)

[5] Garo A., Lahaye J., Prado G., 'Mechanisms of formation and destruction of soot particles in a laminar methane–air diffusion flame', *Twenty-first Symposium (International) on Combustion*, 1023, Pittsburgh: The Combustion Institute (1987)
[6] Garo A., Prado G., Lahaye J., 'Chemical aspects of soot particles oxidation in a laminar methane–air diffusion flame', *Combustion and Flame* **79** 226 (1990)
[7] Kamimoto T., Bae M-H., Kobayashi H., 'A study on soot formation in premixed constant volume propane combustion', *Combustion and Flame* **75** 221 (1989)
[8] Obuchi A., Ohi A., Aoyama H., Ohuchi H., 'Evaluation of gaseous and particulate emission characteristics of a single-cylinder diesel engine', *Combustion and Flame* **70** 215 (1987)
[9] Sadakata M., Kurosawa Y., Sakai T., 'Emission of submicron carbon from a pulverised coal combustion system', *Combustion and Flame* **56** 245 (1984)
[10] Adamczyk A. A., Rothschild W. G., Kaiser E. W., 'The effect of fuel and oil structure on hydrocarbon emission from oil layers during closed vessel combustion', *Combustion Science and Technology* **44** 113 (1985)
[11] Moore W. J., *Physical Chemistry*, Fourth Edition, London: Longman (1970)
[12] Badger G. M., Kimber R. W. L., Spotswood T. M., 'Mode of formation of 3,4 benzopyrene in human environment', *Nature* **187** 663 (1960)
[13] Bittner J. D., Howard J. B., 'Composition profiles and reaction mechanisms in a near-sooting premixed benzene/oxygen/argon flame', *18th Symposium (International) on Combustion*, 1105, Pittsburgh: The Combustion Institute (1981)

CHAPTER 11

Modes of Burning of Liquids and Solids

Abstract

Spray combustion of liquids is discussed with special reference to kerosene and heavy fuel oils. Mechanisms of different spray flames, for example that by which the flame is due to droplet evaporation followed by a hydrocarbon diffusion flame, are outlined. Pool fires are discussed with special reference to polymers and the role of free convection in pool fires is fully discussed. Flaming combustion of solids is given fundamental treatment and the concept of thermal thickness is discussed, giving a suitable experimental example. Smoldering combustion of solids is also outlined.

11.1 Spray combustion

11.1.1 Introduction

For liquids we will consider combustion as sprays, and in pool fires, and for solids, flaming combustion and smoldering combustion. Sprays have obvious importance in, for example, fuel oil combustion where the fuel is dispersed into droplets (atomized) before burning. The principles of droplet and spray combustion can be understood by reference to selected research literature, beginning with an experimental treatment of kerosene spray combustion. Heavy fuel-oil spray combustion is then discussed.

11.1.2 Spray combustion of kerosene [1]

Kerosene is distilled from crude oil in the approximate temperature range 150° C to 300° C. A particular kerosene was examined [1] in a spray burner at various flow rates of fuel and air/fuel ratios. The kerosene was admitted at the base of the system and droplets travelled vertically upward. The combustion air and fuel droplets were not premixed. Atomization was brought about by air, however the quantity of atomizing air was equivalent to not more than 3 percent of the combustion air. The number of droplets was measured at various distances from the point of entry by rapid sampling onto a glass plate coated with magnesium oxide, followed by microphotographic analysis. The droplet distribution was found to be bell-shaped with respect to the axis of the flame. Size distributions of droplets 15 cm from the inlet at various distances from the flame axis had modal values of about 15 μm but not a normal distribution, most droplets being larger than this and some larger than 100 μm. Analyses were made of total hydrocarbons (HC), oxygen, and oxides of carbon

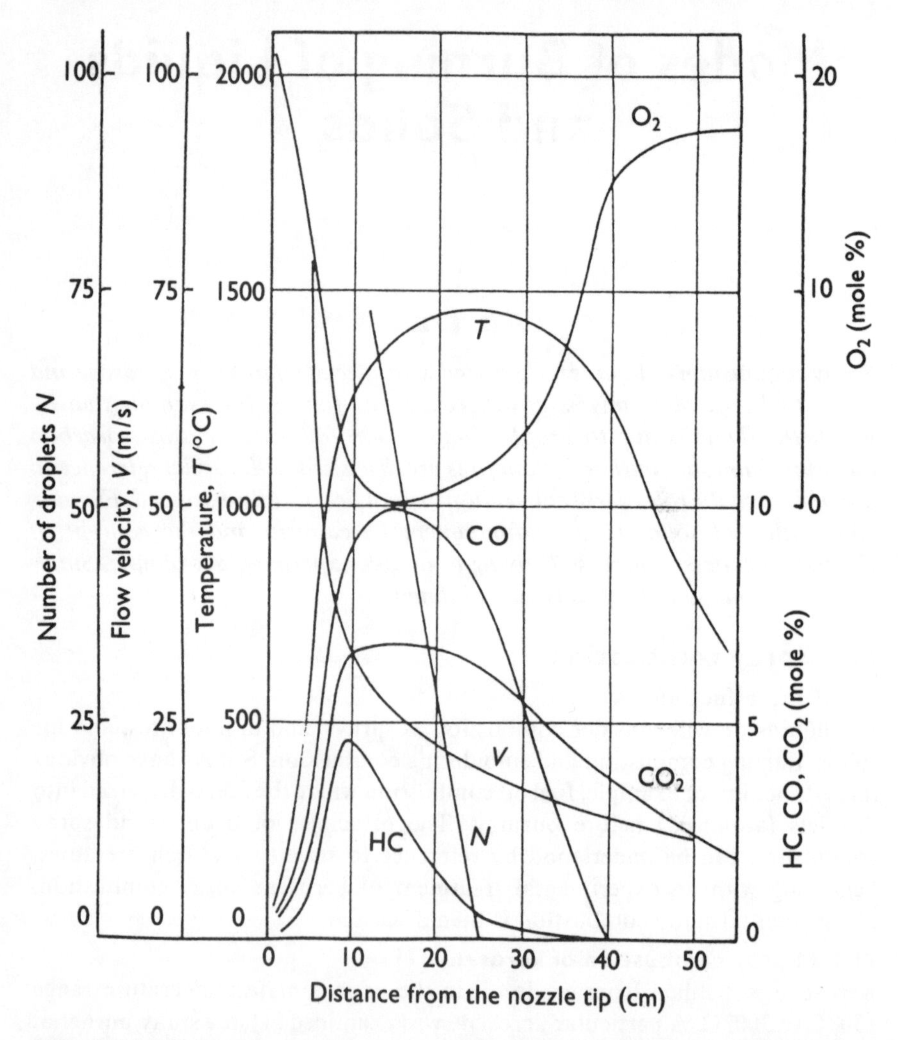

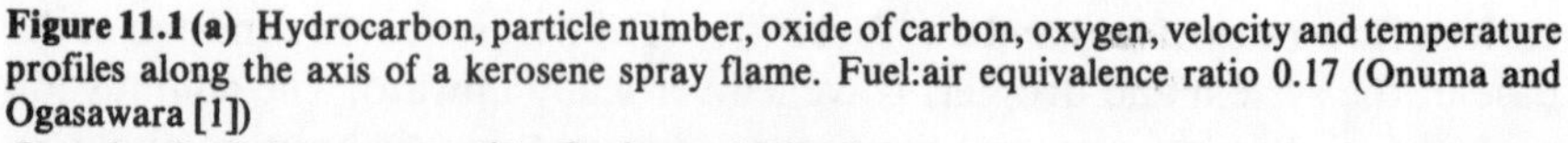

Figure 11.1 (a) Hydrocarbon, particle number, oxide of carbon, oxygen, velocity and temperature profiles along the axis of a kerosene spray flame. Fuel:air equivalence ratio 0.17 (Onuma and Ogasawara [1])

along the axis of the flame and, for one axial distance, along a line normal to the axis. At each point the gas velocity was determined with a Pitot tube and the temperature was measured using a thermocouple technique.

Figure 11.1(a) shows the various profiles along the axis of the flame and Figure 11.1(b) the profiles along a line normal to the axis intersecting it 20 cm from the inlet. In Figure 11.1(b), the number of particles shows an approximately Gaussian distribution about the axis and the hydrocarbons, carbon dioxide and carbon monoxide each show maxima at progressively greater distances from the nozzle. The height of the flame was 53 cm. The flame

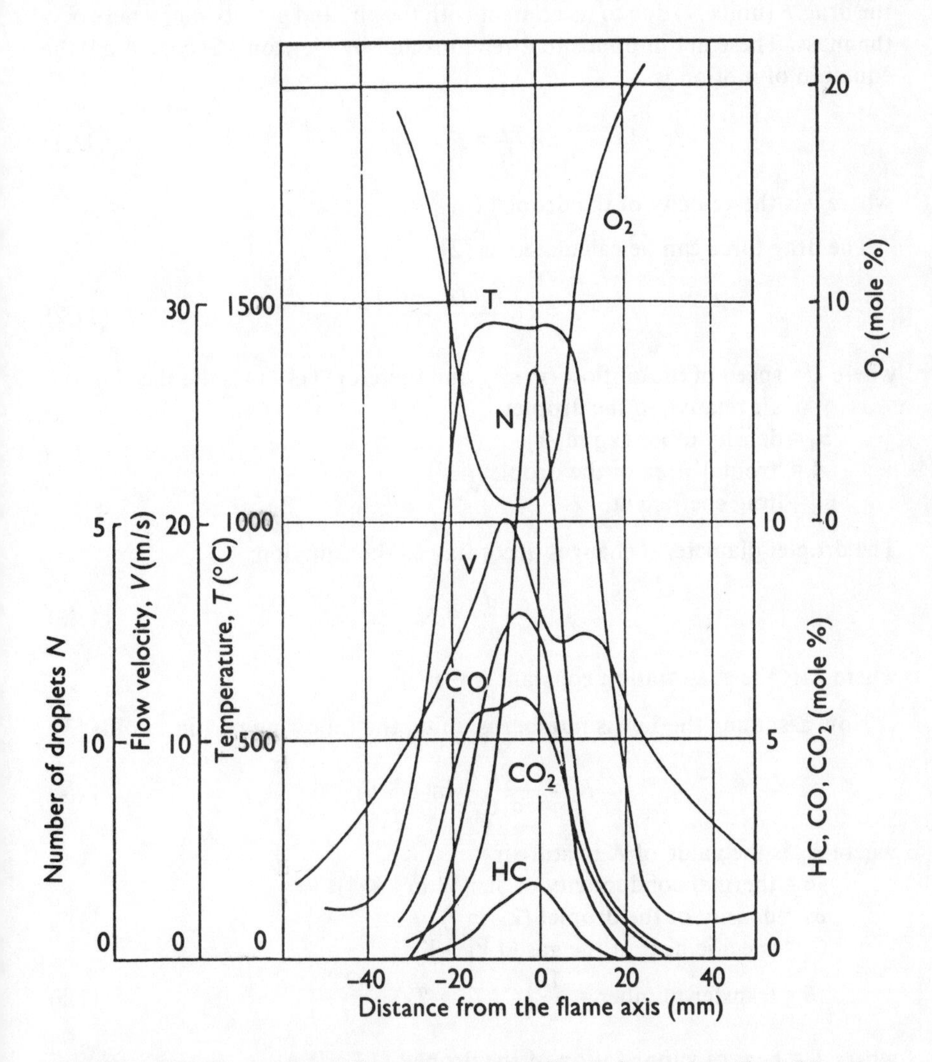

Figure 11.1 (b) Corresponding profiles along a line normal to the flame axis, cutting at 20 cm from the inlet point of the spray (Onuma and Ogasawara [1])
(Reproduced with the permission of the Combustion Institute.)

structure indicates that the fuel is being vaporized and pyrolyzed before entering the hot region closer to the boundary, where it goes first to carbon monoxide and then to carbon dioxide. Thus the combustion is more like that of a gaseous diffusion flame than an ensemble of droplets. This was found for all the equivalence ratios used. When propane gas was substituted for kerosene spray, profiles of temperature and the chemical species both along the axis and normal to it were very similar to those obtained with the spray. The following analysis therefore applies to the kerosene spray combustion.

Two mechanical forces act on a kerosene droplet ascending from the nozzle: the drag F (units N) due to its contact with the air, and gravity mg, where m is the mass. These act in opposite directions and by Newton's Second Law the equation of motion is:

$$m\frac{dv}{dt} = F - mg, \tag{11.1}$$

where v is the velocity of the droplet (m s^{-1}).

The drag force can be calculated as [2]:

$$F = C_d' \frac{\sigma_a (U - v)^2 A}{2}, \tag{11.2}$$

where U = speed of the air flow (m s^{-1}), and therefore $(U - v)$ is the speed of the air relative to the droplet
σ_a = density of air (kg m^{-3})
A = 'frontal' area of the droplet (m^2)
C_d' = drag coefficient.

The droplet diameter d changes according to the equation:

$$-\frac{dd^2}{dt} = K, \tag{11.3}$$

where K is the evaporation constant (m^2 s^{-1}).

Now assuming the Lewis number is unity, the following relation holds [3]:

$$K_o = \frac{8k}{\sigma_l C_g} \ln(1 + B), \tag{11.4}$$

where K_o is the value of K in still air
k = thermal conductivity of air (W m^{-1} K^{-1})
σ_l = density of the droplet (kg m^{-3})
C_g = specific heat of the gas (J kg^{-1} K^{-1})
B = transfer number $= \frac{1}{L}\left\{C_g(T_\infty - T_L) + \frac{QY_o}{i}\right\}$, (11.5)

where L = heat of vaporization of the droplet (J kg^{-1})
T_∞ = air temperature (K) in isolation from the droplet
T_L = boiling temperature of the drop (K)

Q = heat of combustion of the drop (J kg^{-1})
Y_o = weight fraction of oxygen
i = stoichiometric weight ratio of oxygen to fuel.

By means of suitable correlations the evaporation constant K in moving air can be calculated, the Reynolds number (Equation 4.1) being required, and the drag coefficient can also be calculated from the Reynolds number and B. A convection equation for heat transfer from the air to the inflowing particle is also developed using an established correlation and the time taken for the droplet to reach its boiling temperature is calculated. Equation 11.1 is solved from time zero, that is, the time of exit of the particle from the nozzle, and equation 11.3 from the time taken for the droplet to reach boiling point. The velocity solution can be converted to distance by integration, and for particles of different initial sizes the numerical results are shown in Figure 11.2. In a flame of height 53 cm, even droplets initially larger than 100 μm have shrunk to 10 μm or less by 20 cm into the flame. This calculated result is consistent with the hydrocarbon peak in Figure 11.1(a), this being from the evaporated and possibly partly decomposed droplets. This, and the similarity of profiles with those for a propane flame in the same system, indicate that the kerosene droplet combustion in these experiments can be understood as droplet evaporation, followed by a hydrocarbon diffusion flame.

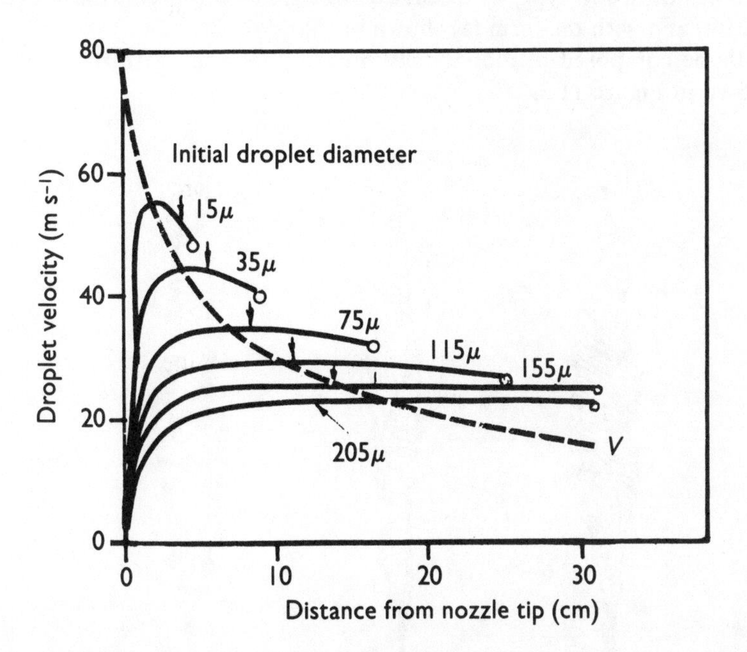

Figure 11.2 Calculated velocity and evaporation behavior of kerosene droplets in spray combustion. Arrows denote the point at which evaporation begins. Circles indicate the point where the diameter becomes 10 microns. Dashed curve = measured flow velocity profile (Onuma and Ogasawara [1])
(Reproduced with the permission of the Combustion Institute.)

11.1.3 Spray combustion of heavy fuel oils

Whereas kerosene burns essentially as a gaseous diffusion flame in spray combustion, different ways of burning are possible with materials such as heavy fuel oils. These include individual droplet burning with an enveloping flame. Any particular fuel oil might display one or another way of burning depending on conditions such as droplet size and droplet momentum on entry from the atomizer. This is shown by work on the spray combustion of a heavy fuel oil [4] in which these conditions were varied from one experimental trial to another. We will consider in detail the conditions for one of the fuel oil spray flames examined.

Atomization was by air and the rate M_0 of momentum passage (units N) from the atomizer into the burner was measured by a transducer. The droplet sizes were measured using a laser technique and were found to fit the well-known Rosin–Rammler (R–R) distribution:

Percent of droplets with diameter greater than $d = 100 \exp[-(d/d')^N]$, (11.6)

where d' = R–R mean diameter and N the R–R exponent. These and other data for the fuel oil spray flame under consideration are given in Table 11.1. Nine spray flames were examined, with different values of M_0, and the fuel flow rate m_F, etc. and according to the conditions, displayed one of three types of flame:

(a) The gas-diffusion type, as discussed for kerosene spray in the previous section and with the form (a) shown in Figure 11.3.

(b) A flame composed of individually burning droplets, with the form (b) shown in Figure 11.3.

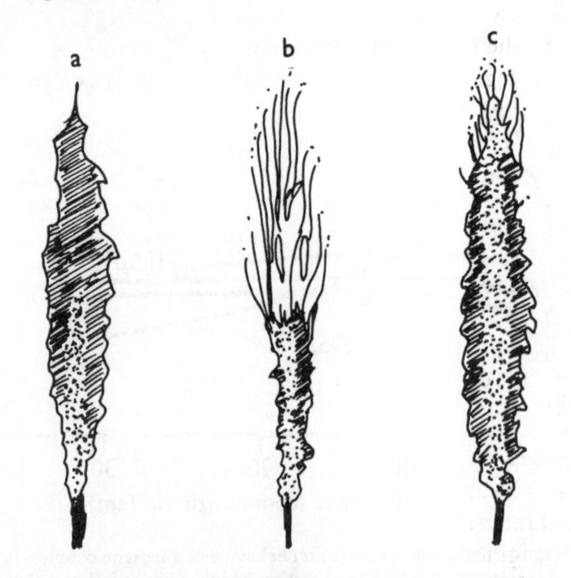

Figure 11.3 Shapes of fuel-oil spray flames (a) Gas-diffusion regime, (b) Individual-droplet regime (c) Mixed regime (Yule and Bolado [4])
(Reproduced with the permission of the Combustion Institute.)

Table 11.1 Fuel-oil spray flame data [4]

Rate of momentum delivery M_o (N)	0.207
Fuel flow rate m_F (kg s^{-1})	3.5×10^{-3}
Atomizing air flow rate m_{A_o} (kg s^{-1})	2.24×10^{-3}
Initial fuel/air ratio ϕ	25
R–R mean droplet diameter d' (μm)	107.5
R–R exponent N	0.98
Ratio $\dfrac{\text{flame length}}{\text{orifice diameter}}$	217
Spray vaporization parameter $\mathfrak{b}$: $\mathfrak{b} = M_o^{0.5}\, d'^2$ ($N^{0.5}$ μm^2)	5260
Spray entrainment parameter $\mathfrak{e}$: $\mathfrak{e} = m_F M_o^{-0.5}$ ($N^{0.5}$m^{-1})	7.69×10^{-3}
Burning regime parameter $\mathfrak{b}/\mathfrak{e}$	6.84×10^5

(c) A flame with some of the characteristics of each of the above, called a flame in the mixed regime and shown as (c) in Figure 11.3. This is predominantly the gas-diffusion type flame but with individually burning droplets evident.

The flame for which details are given in Table 11.1 burnt according to the gas-diffusion regime and across the nine flames the regime was found to correlate with the burning regime parameter $\mathfrak{b}/\mathfrak{e}$. Values of $\mathfrak{b}/\mathfrak{e}$ varied from 1.47×10^5 to 30.94×10^5, with values at the low end signifying the gas-diffusion regime, values at the high end the individual-droplet regime and intermediate values the mixed regime.

Whether a spray flame is gas-diffusion type or individual-droplet type depends on whether droplet evaporation is complete by the time of entrainment of the stoichiometric amount of air. If it is, then the gas-diffusion regime is the probable one. By treating the spray issuing from a nozzle by the same principles that apply to passage of gas through a nozzle, it is shown that:

Time taken for complete evaporation of a drop $\propto M_o^{0.5} d'^2$ ($= \mathfrak{b}$),

and also that:

Time taken for entrainment of the stoichiometric requirement of air $\propto m_F M_o^{-0.5}$ ($= \mathfrak{e}$).

Hence the burning regime parameter $\mathfrak{b}/\mathfrak{e}$ is an index of the relative values of these times. Thus it can be understood why low $\mathfrak{b}/\mathfrak{e}$ signifies rapid evaporation relative to oxidant entrainment, and hence the gas-diffusion type of flame, while high $\mathfrak{b}/\mathfrak{e}$ signifies relatively rapid entrainment, and hence individual-droplet burning. Progressive increase of $\mathfrak{b}/\mathfrak{e}$ in different experiments with the same fuel oil will show a transition from gas-diffusion to individual-droplet burning and from there, at sufficiently high values, to incomplete combustion.

Recalling that:

$$\frac{\mathfrak{h}}{\mathfrak{e}} = \frac{M_0 d'^2}{m_F}, \tag{11.7}$$

The incomplete combustion at sufficiently high values can be identified physically with high M_0, requiring high velocities of the particles. This causes passage through the system so rapid that there is not time for complete combustion and/or convective cooling, which inhibits reaction.

11.2 Pool fires

11.2.1 Introduction

A fire above a horizontal surface of liquid is called a pool fire and the liquid may be either a substance which is a liquid at room temperature, or the melting/degradation products of a polymer. The temperature gradients in a pool fire lead to free convection effects and the characteristics of pool fires are often discussed in terms of the Grashof number Gr, defined as:

$$\mathrm{Gr} = \frac{g\beta\delta T\sigma^2 d^3}{\mu^2}, \tag{11.8}$$

where β (K^{-1}) is a reciprocal temperature, d in the case of a pool fire is the diameter and δT (K) is the temperature difference between the surface of the hot body and the surrounding air. Free-convection heat transfer coefficients are normally expressed in terms of this dimensionless group. In pool fires there is also a need to incorporate radiation heat transfer into the burning rate equations.

Smaller pool fires tend to be laminar, and larger ones turbulent. This is consistent with the form of the Reynolds number (equation 4.1). Laboratory-scale fires which would be laminar at atmospheric pressure are turbulent at sufficiently high pressures above atmospheric, and this also could be predicted from the Reynolds number.

11.2.2 Analysis of pool fires

The transfer number B introduced in Section 11.1.2 in the discussion of spray combustion can also be used in describing pool fires. An approximate expression for the burning rate [5] which separates the convective and radiative heat transfer is:

$$m'' = \frac{h}{C_g} \ln(1 + B) + \frac{q_R}{L \ln(B + 1)}, \tag{11.9}$$

where m'' = mass flux from the pool (kg m^{-2} s^{-1})
q_R = radiative flux from the flame to the pool surface (W m^{-2})
h = coefficient of natural convection (W m^{-2} K^{-1}).

Alternatively the mass flux m'' can be expressed as:

$$m'' = \frac{h}{C_g} \ln(B' + 1), \tag{11.10}$$

where $B' = \mathrm{B} + \left[\frac{B}{\ln(B+1)}\right]\left[\frac{q_{\mathrm{R}}\, C_{\mathrm{g}}}{h\, L}\right]$. (11.11)

Clearly B′ → B in the limit of nil radiation flux. The convection heat transfer coefficient h can be estimated from the well-known correlation:

$$\mathrm{Nu} = 0.14\mathrm{Gr}^{0.33}\,\mathrm{Pr}^{0.33}, \tag{11.12}$$

where Nu = Nusselt number = $\frac{hd}{k}$

$$\mathrm{Pr} = \text{Prandtl number} = \frac{\text{kinematic viscosity}}{\text{thermal diffusivity}}. \tag{11.13}$$

Combining equations 11.10 and 11.12, with Pr expressed in terms of its constituent quantities:

$$\left[\frac{m'}{(\pi d^2/4)}\,\frac{d}{\mu}\,\frac{1}{\ln(B'+1)}\right]\left[\frac{\mu C_{\mathrm{g}}}{k}\right]^{0.66} = 0.14\mathrm{Gr}^{0.33}, \tag{11.14}$$

where m' is the steady mass loss rate in kg s^{-1}. For a Prandtl number of unity this simplifies to:

$$\left[\frac{m'}{(\pi d^2/4)}\,\frac{d}{\mu}\,\frac{1}{\ln(B'+1)}\right] = 0.14\mathrm{Gr}^{0.33}. \tag{11.15}$$

This expression was applied to laboratory-scale pool burning of several polymers in air [5]. In each experiment the temperature T_{L} was taken to be 400° C and the pool diameter was 2.875 inches. The sample fitted tightly into a holder and only the surface had contact with the air. Steady mass-loss rate m' was measured for combustion under air at different pressures, giving different values of Gr, which in each experiment could be calculated. The viscosity μ of air was assigned a single value in all the experiments and B' was determined from B together with q_{R} which was calculated for each flame. For polymers, values of B of the order of unity are expected, as the following example shows.

A typical value of the stoichiometric weight ratio i of oxygen to fuel is ≈ 3; of the heat of combustion Q, 40 MJ kg^{-1}; and of the latent heat L, 2.5 MJ kg^{-1}. The weight fraction of oxygen in air is 0.233. Putting $C_{\mathrm{g}} \approx 1200$ J kg^{-1} K^{-1}, $T_{\mathrm{L}} = 400$° C and $T_\infty = 25$° C, and substituting in equation 11.5, gives $B = 1.1$.

Figure 11.4(a) gives results for many polymers at different Grashof numbers, plotted according to equation 11.15. The conformity indicates:

(a) that the mass flux from the polymer can be described by an equation of the form of (11.10);
(b) that the free convection resulting from the temperature difference depends on the third root of the Grashof number.

Since m' is a steady burning rate:

$$m' = -\frac{\mathrm{d}m}{\mathrm{d}t} = \mathcal{Z}\mathrm{Gr}^{0.33}, \tag{11.16}$$

where $\mathcal{Z} = 0.14 \ln(B'+1)\,\frac{\mu}{d}\,\frac{\pi d^2}{4}$ (kg s^{-1}).

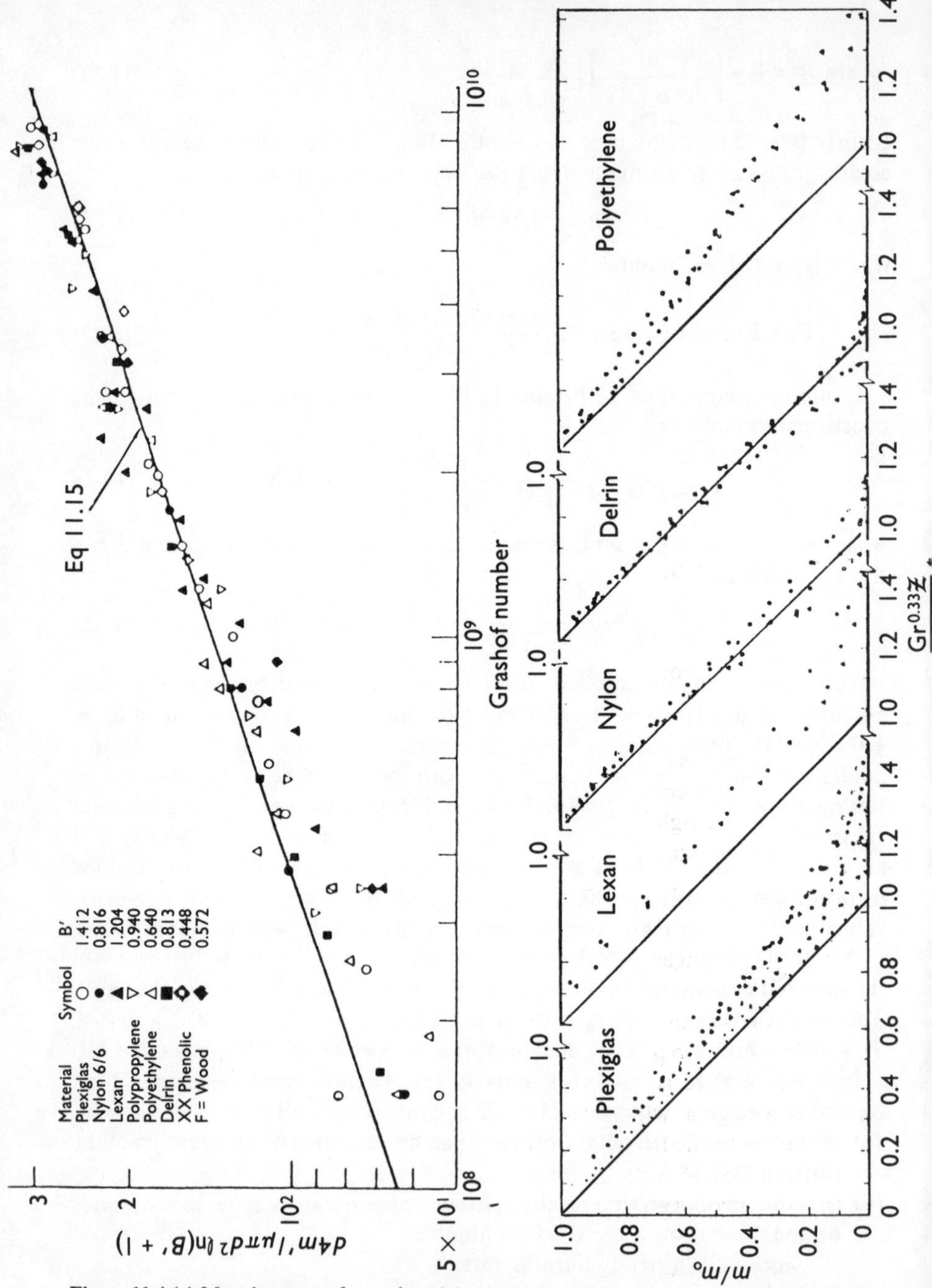

Figure 11.4 (a) Mass-loss rates for various laboratory-scale polymer-pool fires against Grashof number. Line drawn = equation 11.15 (Kanury [5])

(Reproduced with the permission of the Combustion Institute.)

Figure 11.4 (b) Fraction weight loss m/m_o against $\frac{Gr^{0.33}}{m_o}t$ for five polymers (Kanury [5])

(Reproduced with the permission of the Combustion Institute.)

Integrating equation 11.16 and letting the initial mass = m_o:

$$\frac{m}{m_o} = 1 - \frac{\mathrm{Gr}^{0.33}\mathcal{Z}}{m_o}\,t. \tag{11.17}$$

The quantity $\mathcal{Z}$, containing B', is characteristic of a particular polymer and in Figure 11.4(b) the fractional weight m/m_o is plotted according to equation 11.17 for five of the polymers. The line drawn through each set of points has a slope of unity, that is, it is equation 11.17 for each polymer.

The total radiative heat flux from the flame q_T (W) was calculated from thermopile measurements and the ratio $\mathcal{X}$ of this to the total heat release rate is clearly:

$$\mathcal{X} = \frac{q_T}{m'Q}. \tag{11.18}$$

It was found that any particular polymer had a constant value of $\mathcal{X}$ across the range of Grashof numbers used in the experiments. Values ranged from ≈ 0.04 to ≈ 0.22, the higher values being associated with sooty flames.

11.3 Flaming combustion of solids

11.3.1 Effect of sample thickness

Thermal thickness has a conceptual similarity to thermal resistance. For a plane surface, the conductive thermal resistance is:

$$R_{\text{thermal}} = \frac{\tau}{kA} \qquad \text{K W}^{-1}, \tag{11.19}$$

where τ = thickness (m)
k = thermal conductivity (W m^{-1} K^{-1})
A = area at right angles to the direction of heat flux over which conduction is occurring,

and we expect intuitively that the thermal thickness, like the thermal resistance, will contain k in the denominator. We also expect that it might contain V, the combustion propagation velocity, in the numerator since the less time the flame spends occupying a specific portion of the body, the less time there is for heat transfer from the flame to the surface and from there to the interior of the body. We might also predict intuitively that a thermal thickness will be reached above which there is no effect of thickness on rate of flame propagation, signalling a stage at which the thermal thickness is sufficient for the temperature response to the flame to be nil at some depth in the body.

For a flat bed of fuel, a criterion for the value of the thermal thickness at which flame propagation becomes independent of it has been developed as follows [6]:

Let $T' = T(y,\ t) - T_o$,

where T_o = ambient temperature (K)
$T(y,\ t)$ = temperature at time t and position y (K).

The quantity y is the coordinate normal to the direction of flame propagation, so solution of a heat-balance equation in this coordinate gives a temperature profile in the bed, with $y = 0$ denoting an insulated surface at the base of the medium.

The heat-balance equation is:

$$\sigma c \frac{\partial T'}{\partial t} = k \frac{\partial^2 T'}{\partial y^2}, \tag{11.20}$$

where σ = density (kg m^{-3})

c = heat capacity (J kg^{-1} K^{-1}),

and the physical quantities σ, c and k relate, of course, to the solid fuel itself. Conditions applying are:

(a) For all y, $t = 0$:

$$T' = 0. \tag{11.21}$$

(b) For $y = \tau$, where τ = bed thickness, and $t > 0$:

$$q = k \frac{dT'}{dy}, \tag{11.22}$$

that is, constant conductive heat flux at the surface.

(c) The temperature gradient is considered to become small close to the base of the bed, so to a good approximation, at $y = 0$:

$$\frac{dT'}{dy} = 0. \tag{11.23}$$

Although this comparison is not made by the developers of this treatment, support for the boundary condition expressed as equation 11.23 can be obtained from the well-documented solution [7] of the heat-balance equation (11.20) for a *semi-infinite solid* subject to a step change in temperature at its surface, that is:

$$T(y = \tau) = T_s \text{ at } t > 0, \tag{11.24}$$

with equation 11.21 also applying. This can perhaps be viewed as a very simple model of a fuel bed, and graphical solutions show that after quite a steep drop close to the surface, the temperature profiles at various times after the step change take on very gentle slopes farther into the medium. We will return to the semi-infinite medium as a basis for understanding fuel beds in Chapter 12.

Equation 11.20, subject to the conditions indicated, has an analytical solution which in the example under consideration can be simplified by putting t in the solution as the ignition time. This can be equated to δ / V where δ is the length of the ignition zone. When the product $q\delta$, which appears in the solution as a result of the above simplification, is amalgamated into a single quantity Q_s, the solution of the heat-balance equation becomes:

$$T' = T_b - T_o = Q_s (k\sigma c\delta V)^{-0.5} f(\tau'), \tag{11.25}$$

where T_b = temperature of the surface at $t = \delta / V$. The product $kc\sigma$ is known as the thermal inertia and will be further discussed in Chapter 12. Also:

$$\tau' = \frac{\tau}{(k\delta/\sigma c V)^{0.5}} \qquad \text{(no units)}, \tag{11.26}$$

$$\text{and } f(\tau') = 1.128 + 2\sum_{n=0}^{\infty} 2\,\text{ierfc}(n+1)\tau'. \tag{11.27}$$

The error function complement (erfc) was introduced in Chapter 3 and ierfc [7] is an integrated form:

$$\text{ierfc } x = \frac{1}{\sqrt{\pi}} \exp(-x^2) - x \text{ erfc x}. \tag{11.28}$$

Two important points arising from solution of heat-balance equation 11.20 are:

(a) The quantity τ' is a natural choice of definition of *thermal thickness* and has the advantage of being dimensionless. It has the expected dependences on τ, V, k.
(b) A graph of $f(\tau')$ against τ' can be generated and will indicate the dependence of $f(\tau')$, and hence T_b on the thermal thickness.

This is shown as Figure 11.5. At low thermal thickness τ' the function $f(\tau')$ has values up to 10 on the scale shown, corresponding to $\tau' = 0.1$. The value of $f(\tau')$, and hence T_b, drops with τ' up to $\tau' \approx 1$ where $f(\tau')$ becomes independent of τ' and has a constant value of 1.128, meaning that the second term in equation 11.27 has diminished to zero. Therefore, on the basis of this fundamental treatment:

(a) $\tau' < 1$, material thermally thin, and T_b is influenced by sample dimension in the direction normal to that of flame spread;

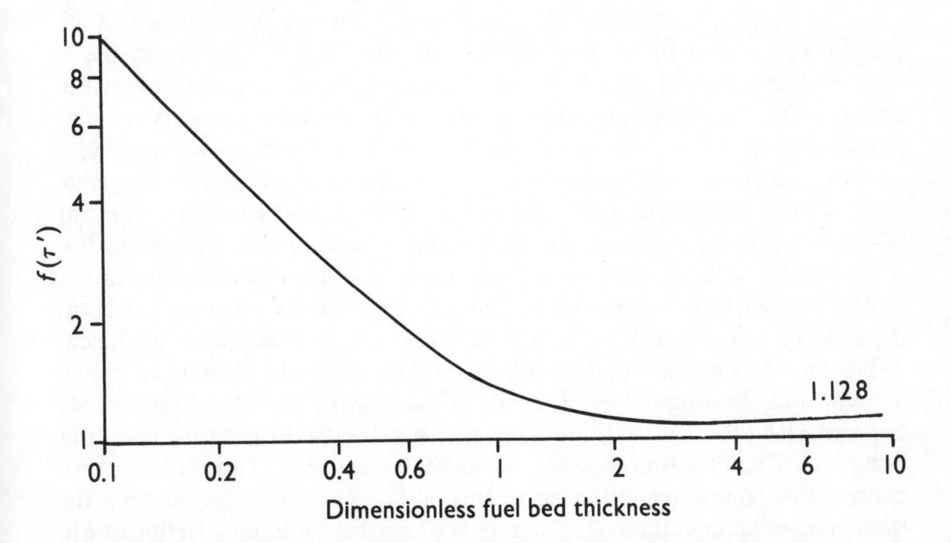

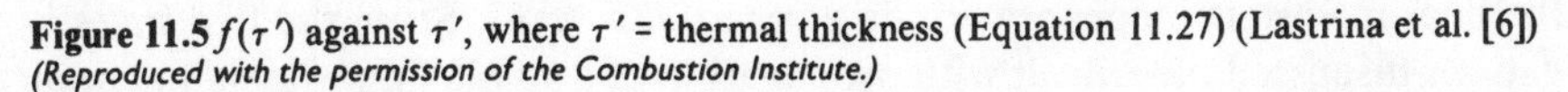
Figure 11.5 $f(\tau')$ against τ', where τ' = thermal thickness (Equation 11.27) (Lastrina et al. [6]) *(Reproduced with the permission of the Combustion Institute.)*

(b) $\tau' > 1$, material thermally thick, and T_b simply depends on quantities other than thickness, according to:

$$T_b - T_o = 1.128 Q_s (k\sigma c \delta V)^{-0.5}. \tag{11.29}$$

Thus a clear demarcation has been deduced.

The criterion for thermal thickness can be expressed very simply in terms of the time t spent by the flame along a length δ. Putting $\delta / V = t$ as before, the criterion for thermal thickness becomes:

$$\frac{\tau}{(\alpha t)^{0.5}} > 1,$$

where $\alpha = k / c\sigma$ = thermal diffusivity ($m^2\ s^{-1}$),

or $$\tau > (\alpha t)^{0.5}.$$

11.3.2 Experimental tests with PMMA rods

Burning tests using cylindrical rods of polymethyl methacrylate (PMMA) were carried out [8] and for this shape, as opposed to a plane surface, the thermal thickness is expressed as:

$$\tau'_{rod} = \frac{d}{4(k\delta / \sigma c V)^{0.5}}, \tag{11.30}$$

where d = diameter (m). With τ' so defined, the treatment predicts a critical value of τ'_{rod} (value below which sample thickness affects T_b, and above which it does not) exactly the same as that for τ' in the case of a plane sample.

Vertical cylindrical rods of PMMA [8] were ignited using a Bunsen burner and after allowing the flame time to stabilize, the pyrolysis zone became conical. We should note the difference between these conditions and those of the polymer pool fires discussed in Section 11.2.2 where the only part of the sample with access to air was the flat surface. Here the entire sample is suspended in air. The length δ of the burning zone was estimated from surface temperatures. The flame propagation rate V was determined visually and was plotted against the cross-sectional area A_c of the rods. From knowledge of k, c, σ, V and δ, a value of the thermal thickness τ' for each size of rod is obtainable and it is to be expected that the value of τ' at which T_b becomes independent of thickness will be approximately 1. The result is shown in Figure 11.6 in which V is used as the ordinate instead of $f(\tau')$. There is a clear reproduction of the shape of the curve in Figure 11.5 and, noting the τ'-values, the transition from dependence on rod size to independence occurs at $\tau' \approx 1$, exactly as predicted.

The heat-flux model for burning of PMMA therefore appears to match experimental data quite well. It is worth mentioning that the model simply imposes a heat flux in the formulation and therefore takes no explicit account of the role of combustion chemistry in the heat release rate. This contrasts with some of the flame and ignition treatments in earlier chapters, for example, the discussion of laminar flames in Chapter 4, where the role of the reaction rate in the heat release is incorporated via a reaction rate K_i. Since the two investigations discussed have dealt with propagation of a stable flame, assignment of

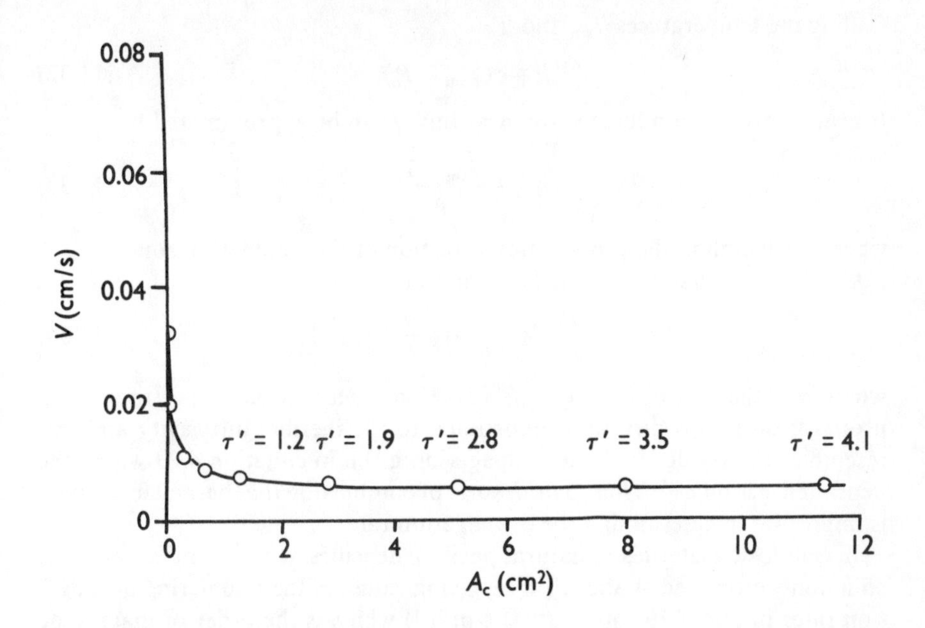

Figure 11.6 V against A_c, with values of τ' for ignited horizontal PMMA rods (Lee [8])
(Reproduced with the permission of the Technomic Publishing Company.)

a flux is correct but it should be remembered that the magnitude of the flux will depend on the reaction rate and this dependence will sometimes cause smoldering, rather than flaming combustion, as discussed below.

11.4 Smoldering combustion

11.4.1 A simple model for smoldering

Smoldering is a much slower phenomenon than flaming combustion and cellulosic materials are among those capable of displaying this mode of burning. Upholstered furniture, for example, is susceptible to this type of combustion. We will therefore examine studies of cellulose smoldering as a means of developing the principles.

We will first consider a very basic model [9] for smoldering which links the key quantities with the phenomenology. The combustion propagation rate V of a smoldering layer can be estimated as:

$$V = \frac{q}{\sigma h}, \tag{11.31}$$

where q = heat flux in the direction of propagation (W m^{-2})
σ = density (kg m^{-3})
h = enthalpy required to raise the fuel from ambient temperature to smoldering combustion temperature (J kg^{-1}).

The heat flux is from an already smoldering region to an unaffected one.

Calling the temperatures T_{sm} and T_o:

$$h = c(T_{sm} - T_o). \tag{11.32}$$

If heat transfer is conductive, the heat flux q can be approximated by:

$$q = k\frac{T_{sm} - T_o}{x}, \tag{11.33}$$

where x = length in the propagation direction of the smoldering zone.

Combining equations 11.31, 11.32 and 11.33:

$$V = \frac{k}{\sigma c x} = \frac{\alpha}{x}, \tag{11.34}$$

where α = thermal diffusivity ($m^2\ s^{-1}$). Hence on this basis the smoldering propagation rate is directly proportional to the thermal diffusivity and this resembles the result for flame propagation given in equation 4.29 where the dependence is on $\alpha^{0.5}$. Thus in *both* sorts of combustion the thermal diffusivity is important in determining the propagation rate.

In cellulosic materials at natural packing densities, $\alpha \approx 10^{-7}\ m^2\ s^{-1}$ and x is commonly estimated as about 1 mm, giving values of the smoldering propagation rates of about $10^{-4}\ m\ s^{-1}$ ($\approx 0.4\ m\ h^{-1}$) which is the order of magnitude expected in simple experiments to measure V for cellulosic materials. It is perhaps instructive to reflect on this figure in terms of a very common form of household fire: the accidentally ignited mattress. According to this treatment, a 2 m mattress filled with coir—a cellulosic material—ignited at one end will require of the order of five hours for the smoldering combustion to spread to the other end.

11.4.2 Transition from smoldering to flames

Having thus placed smoldering combustion on a semiquantitative basis it is necessary to identify factors causing smoldering rather than flaming combustion. Air supply is central and in situations where still air promotes smoldering, flowing air can sometimes cause a transition to flames. This was observed [10] with beds of grain dust (principal constituent starch) 0.1 m deep. Beds of the dust in still air gave smoldering propagation rates of the order of $10^{-5}\ m\ s^{-1}$, with a small dependence on the depth in the bed at which ignition, caused by a heated wire, occurred. Temperatures when the propagation actually crossed a thermocouple site were recorded as 700–800K. Both propagation rates and temperature maxima increased when air flow across the free surface was up to $4\ m\ s^{-1}$. This air-flow rate, the highest used, was capable of causing a transition from smoldering to flaming combustion but, interestingly, only if the ignition source was 2.5 cm or more below the free surface. When the ignition source was closer to the surface there was no transition. The transition is preceded by formation of a cavity due to substantial local heating at the depth of the ignition source and, once formed, the cavity provides a reserve of air enabling subsequent flaming combustion to occur. With the ignition source too close to the surface this mechanism does not exist.

11.4.3 Detailed treatment of smoldering in beds of powder

A somewhat more detailed but still quite straightforward treatment of smoldering combustion was reported [11] together with experimental data for a variety of materials including coal dust and cocoa. These gave respective propagation rates of 1.7×10^{-5} and 2.7×10^{-5} m s^{-1}. Sawdust gave a propagation rate an order of magnitude higher than these. In this treatment the velocity V_a of the air due to natural convection promoted by the temperature gradients is incorporated in the heat-balance equation. It is expected that V_a can be estimated fairly accurately from free convection correlations and the rather unusual approach was taken of estimating it both in this way and from heat balance, comparing the two for consistency. The heat-balance equation is similar to that derived from equation 11.31ff, except that heating is considered to occur over a very small distance d in the x-direction, resulting in almost a step change in temperature. Heat uptake by the air in the voids is considered separately from that by the particles.

When, in unit time, the propagation front moves a distance V, as it must from the definition of V, the volume swept out per unit area is also numerically equal to V. The mass of dust in the volume is $(1 - \epsilon)\sigma_s V$,

where ϵ = voidage (fraction of the apparent volume occupied by voids)
σ_s = particle density of the dust (kg m^{-3}).

Air moves by free convection, as noted, with velocity V_a m s^{-1}, creating in unit time across unit area a volume numerically equal to V_a. The mass of air in the volume is $\epsilon\sigma_a V$. The rate of heat uptake is therefore:

$$\begin{aligned} R &= (1 - \epsilon)c_s\sigma_s (T_{sm} - T_o)V + \epsilon c_a\sigma_a (T_{sm} - T_o)V_a \\ &= (T_{sm} - T_o)[(1 - \epsilon)c_s\sigma_s V + \epsilon c_a\sigma_a V_a] \quad \text{W m}^{-2}. \end{aligned} \tag{11.35}$$

This can be rewritten as:

$$R = (T_{sm} - T_o)(1 - \epsilon)c_s\sigma_s V\left[1 + \frac{\sigma_a c_a V_a \epsilon}{\sigma_s c_s V(1 - \epsilon)}\right]. \tag{11.36}$$

The quotient in the square brackets has the nature of air:fuel weight flux ratio in the propagation front, multiplied by the ratio of the heat capacities, and is assigned the symbol ϕ. Equating the heat uptake to the linearized conductive heat flux gives:

$$(T_{sm} - T_o)(1 - \epsilon)c_s\sigma_s V(1 + \phi) = k\,\frac{T_{sm} - T_o}{d}, \tag{11.37}$$

where k is the thermal conductivity of the composite particle/air assembly (W m^{-1} K^{-1}) and d the length over which the flux occurs. This can be rearranged to:

$$V = \frac{\alpha}{d(1 + \phi)(1 - \epsilon)}. \tag{11.38}$$

This is clearly the analogue of equation 11.34 for the alternative model.

The chemistry of the process can be schematized:

$$Z + \frac{n}{2}(O_2 + 3.76N_2) \rightarrow ZO_n + 1.88nN_2,$$

and for coal, Z is its carbon content, forming CO, that is, $n = 1$. The quantity of heat required to raise a sample of coal to T_{sm} is only 1–2 percent of the quantity of heat that the sample can release when fully burnt. However, the rise to T_{sm} will be sufficient to cause appreciable weight loss by pyrolysis. For extent of reaction y, the weight of air required is 2.38 ynM_a per mole equivalent of fuel reacting according to the above scheme, therefore we can write ϕ as:

$$\phi = \frac{2.38ynM_a}{M_s} \times \frac{c_a}{c_s}, \qquad (11.39)$$

where M denotes molar weight (kg mol^{-1}) which in the case of a dust sample means the quantity capable of releasing on combustion one mole of the combustion product ZO_n and can be calculated from elemental analysis. The quantity y is easily estimated from the experimentally measured temperature rise:

$$T_{sm} - T_o \approx \frac{yQ}{1.88nM_a c_a + M_s c_s}, \qquad (11.40)$$

where Q is the molar heat of reaction. This rise was about 440K for the coal and considerably lower for the cocoa and the sawdust. Combining equations 11.39 and 11.40, together with our earlier definition of ϕ, we obtain:

$$V_a = V\frac{\sigma_s(1-\epsilon)}{\epsilon\sigma_a} \times 2.38n\frac{M_a}{M_s} \times \frac{(T_{sm} - T_o)(M_s c_s + 1.88nM_a c_a)}{Q}, \qquad (11.41)$$

enabling V_a to be calculated via the measured propagation rate and temperature rise ($T_{sm} - T_o$). The quantity V_a can be obtained independently from the following expression for free convection [12]:

$$V_a = \frac{g\sigma_a(T_{sm} - T_o)d_p^2\epsilon^3}{\mu T_o(1-\epsilon)^2}, \qquad (11.42)$$

where g = acceleration due to gravity (9.81 m s^{-2})
d_p = average particle diameter (m)
μ = dynamic viscosity of air (kg m^{-1} s^{-1}).

Table 11.2 gives values of V_a determined in the experiments on smoldering from equation 11.41, and from equation 11.42, for coal, cocoa powder and sawdust as fuel. In every case the agreement is acceptable. In each case, for the purposes of equation 11.41, n was taken as 1 with ZO_n as CO.

In concluding this discussion of smoldering, we should perhaps reconsider the question of what causes some fires to display smoldering and not flaming combustion. The two can be described by essentially the same equations: compare equations 11.22 (flame) and 11.33 (smoldering), each expressing

Table 11.2 Air-flow velocities at the propagation front of smoldering materials (Cohen and Luft [11])

	V_a (m s^{-1})	
Material	*From equation 11.41*	*From equation 11.42*
Coal dust, particle size < 104 μm	1.2×10^{-2}	1.4×10^{-2}
Cocoa, particle size < 40 μm	1.6×10^{-3}	1.0×10^{-3}
Sawdust, particle size 76–152 μm	5.0×10^{-3}	4.0×10^{-3}

Fourier's Law with the heat for the flux having its origin in the combustion reaction. The difference is therefore in the magnitude of this flux, depending on the rate of heat release by the fuel. In at least some cases, for example, combustion of grain dust, it is the dependence of this rate on air supply that can cause smoldering, rather than flames.

References

[1] Onuma Y., Ogasawara M., 'Studies on the structure of a spray combustion flame', *Fifteenth Symposium (International) on Combustion*, 453, Pittsburgh: The Combustion Institute (1975)
[2] Welty J. R., Wilson R. E., Wicks C. E., *Fundamentals of Momentum, Heat and Mass Transfer*, 2nd Edition, New York: John Wiley (1976)
[3] Williams A., *Combustion of Liquid Fuel Sprays*, Sevenoaks: Butterworths (1990)
[4] Yule A. J., Bolado R., 'Fuel spray burning regime and initial conditions', *Combustion and Flame* **55** 1 (1984)
[5] Kanury A. M., 'Modelling of pool fires with a variety of polymers', *Fifteenth Symposium (International) on Combustion*, 193, Pittsburgh: The Combustion Institute (1975)
[6] Lastrina F. A., Magee R. S., McAlevy R. F., 'Flame spread over fuel beds: solid-phase energy considerations', *Thirteenth Symposium (International) on Combustion*, 935, Pittsburgh: The Combustion Institute (1971)
[7] Carslaw H. S., Jaeger J. C., *Conduction of Heat in Solids*, Oxford: The University Press (1947)
[8] Lee C. K., 'Flame propagation characteristics of cylindrical PMMA rods', *Journal of Fire and Flammability* **7** 104 (1976)
[9] Drysdale D. D., *Introduction to Fire Dynamics*, New York: John Wiley (1985)
[10] Leisch S. O., Kauffman C. W., Sichel M., 'Smoldering combustion in horizontal dust layers', *Twentieth Symposium (International) on Combustion*, 1601, Pittsburgh: The Combustion Institute (1985)
[11] Cohen L., Luft N. W., 'Combustion of dust layers in still air', *Fuel* **34** 154 (1955)
[12] Prandtl L., cited by Cohen and Luft [11]

CHAPTER 12

Examples of Fire Modelling

Abstract

In the first part of the chapter the semi-infinite solid, with different boundary conditions, is outlined as an aid in understanding fire behavior. A formulation for room fires based on a thermal diagram is given and flashover, that is, transition from a localized fire to full room involvement, is fully discussed. Radiation effects are incorporated into the treatment and the well-known effect radiation enhancement *is put on a firm basis. The final part of the chapter gives a recent fundamental treatment of the interaction of chemically and electrically generated heat, for example, where a current-bearing cable is buried in a spontaneously ignitable material.*

12.1 The semi-infinite solid as a basis for modelling fire*

12.1.1 Introduction

When discussing flaming combustion of solids in Chapter 11, reference was made to the *semi-infinite solid* model, with a constant temperature condition at the surface, as being a rough approximation to a fuel bed. In modelling fires, the semi-infinite solid concept is often useful and an alternative condition to constant surface temperature is convective heat flux at the surface. The solution is obtainable analytically [2] and is discussed more fully below.

At the surface the solution is:

$$\frac{T(t) - T_i}{T_\infty - T_i} = 1 - \exp\left\{\frac{\alpha t}{(k/h)^2}\right\} \operatorname{erfc}\left\{\frac{\sqrt{(\alpha t)}}{k/h}\right\}, \tag{12.1}$$

*Most of this section is developed from material in a paper by Drysdale [1].

where $T(t)$ = surface temperature at time t
T_i = temperature of the body before application of the convective flux, assigned a value of 20° C in what follows
T_∞ = temperature of the fluid in isolation from the surface
α = thermal diffusivity of the body ($m^2\ s^{-1}$)
h = convection coefficient ($W\ m^{-2}\ K^{-1}$)
k = thermal conductivity of the body ($W\ m^{-1}\ K^{-1}$).

The quotient k^2/α (= $kc\sigma$, units $J^2\ s^{-1}\ m^{-4}\ K^{-2}$) is known as the thermal inertia and the response of the temperature of the semi-infinite body to the heat flux depends strongly on this quantity. It is of course a characteristic of a particular material. Figure 12.1 shows the graphical solution of equation 12.1 for several materials of different thermal inertia with h set arbitrarily at 20 $W\ m^{-2}\ K^{-1}$. A difference of the order of 10^5 between the thermal inertia values of polyurethane foam and steel is reflected in the rapid response of the former and the very sluggish response of the latter. The semi-infinite formalism applies to bodies which are thermally thick in the sense discussed in Chapter 11, where we saw that the criterion for thermal thickness was:

$$\tau > \sqrt{(\alpha t)},$$

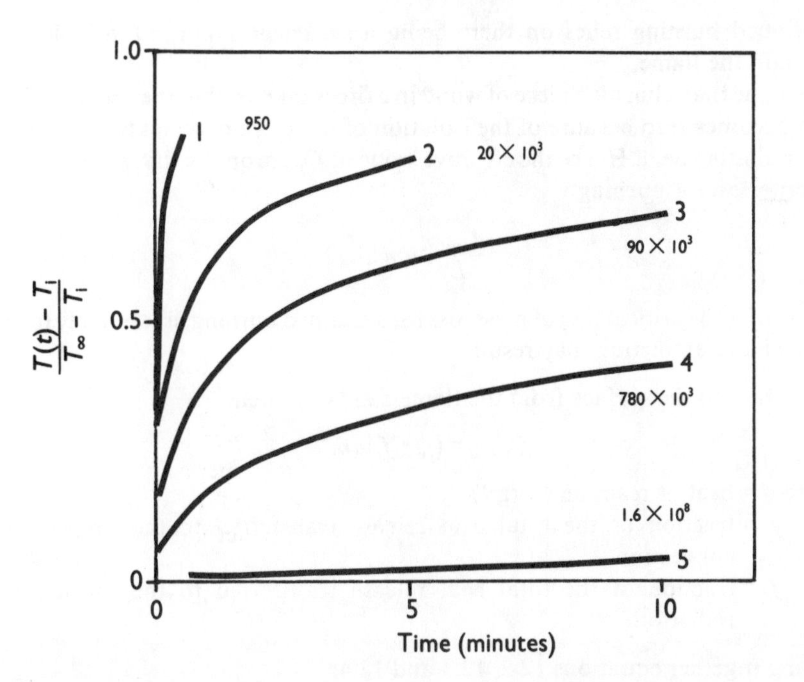

Figure 12.1 Graphical solutions of Equation 12.1 for materials of various thermal inertia. h = 20 $W\ m^{-2}\ K^{-1}$. Curve 1, polyurethane foam. Curve 2, fibre insulating board, Curve 3, asbestos, Curve 4, oak, Curve 5, steel. Value of the thermal inertia for each material ($J^2\ s^{-1}\ m^{-4}\ K^{-2}$) given on the curves (Drysdale [1])
(Reproduced with the permission of the Society of Fire Protection Engineers.)

in the case of a propagating flame spending time t in the ignition zone. For a body whose surface is receiving a constant external flux, the approximate criterion is usually stated:

$$\tau > 2\sqrt{(\alpha t)},$$

where t is now identified with the time of exposure of the surface to the flux.

Considering the surface of a burning material, there will be three components of the heat flux: one from the flame at the surface, one from sources away from the surface, and the flux *from* the surface. Giving these, respectively, the symbols Q_{flame}, Q_{ext} and Q_{loss}, then clearly the total flux is:

$$Q_{tot} = Q_{flame} + Q_{ext} - Q_{loss} \qquad \text{W m}^{-2}. \qquad (12.2)$$

Now weight loss is due primarily to release of volatiles whose production is endothermic by an amount L J kg^{-1}. The volatiles burn in the vapor phase and the heat so released promotes further volatile loss from the solid via Q_{flame}. The rate m' of mass loss is therefore:

$$m' = \frac{Q_{tot}}{L} \qquad \text{kg m}^{-2}\text{ s}^{-1}. \qquad (12.3)$$

Continued burning relies on there being a sufficient supply of volatiles to maintain the flame.

Imagine that a burning piece of wood in a fire is taken out of the fire. The Q_{ext} term becomes zero because of the isolation of the piece of wood from nearby ones radiating heat. Hence the effective value of Q_{tot} drops by that amount and the criterion for burning:

$$\frac{Q_{tot}}{L} > m'_c,$$

where m'_c is the critical rate of mass loss for sustained burning, is less likely to be fulfilled and extinction may result.

The flux to the surface from the flame can be written:

$$Q_{flame} = (f_c + f_r)qm', \qquad (12.4)$$

where q = heat of reaction (J kg^{-1})
f_c = fraction of the total heat release transferred to the surface by convection
f_r = fraction of the total heat release transferred to the surface by radiation.

Putting together equations 12.2, 12.3 and 12.4:

$$m'[q(f_c + f_r) - L] + Q_{ext} - Q_{loss} = 0. \qquad (12.5)$$

This is the criterion for steady burning and the extinction criterion is that the left-hand side should drop below zero through insufficient heat transfer from the flame and the surroundings to the surface. A left-hand side above zero will result in adjustment of conditions via Q_{loss} until the conditions of equation 12.5 are fulfilled, and steady burning ensues. Close to extinction, f_r is expected to be small in comparison with f_c, so the burning criterion is:

$$m'[qf_c - L] + Q_{ext} - Q_{loss} = 0. \quad (12.6)$$

The quantity f_c has a value for many combustible solids of about 0.3.

12.1.2 Detailed treatment of PMMA

The predictive power of this approach can be understood by reference to polymethylmethacrylate (PMMA), the same material that was discussed in Chapter 11 in relation to thermal thickness. PMMA is known from tests to burn if its surface temperature is maintained for long enough at or above 270° C. Imagine a large piece of PMMA approximating a semi-infinite medium which experiences impingement by a non-luminous flame ($f_r \approx 0$). The thermal inertia of PMMA is $3.2 \times 10^5\ J^2\ s^{-1}\ m^{-4}\ K^{-2}$. A flame temperature of 1300° C will be assumed, and a convection heat transfer coefficient from flame to PMMA surface of $50\ W\ m^{-2}\ K^{-1}$. The time taken for the surface temperature of the PMMA to rise from room temperature to 270° C will be $\approx$ 6 s. The interested reader can easily verify this by substitution into equation 12.1.

To predict whether steady burning will occur if external heating ceases at 6 s ($Q_{ext} \to 0$) it is necessary to estimate Q_{loss} which is the sum of radiative and conductive components:

$$Q_{loss} = \epsilon \Omega T^4 - k\frac{dT}{dx}, \quad (12.7)$$

where ϵ = emissivity
Ω = Stefan's constant
x = 0 at the surface and has positive values going into the medium and hence the differential is negative, since temperature obviously decreases in this direction.

In the literature cited [1] the temperature gradient is estimated from the approximation that the depth into the PMMA over which the heating extends is $\sqrt{(\alpha t)}$*. It is, however, possible to improve on this by reference to the full solutions of the time-dependent equation for the semi-infinite medium with the surface convection condition. These are reproduced as Figure 12.2 from which it can be seen that the solutions are in the form of curves of:

$$\frac{T(t) - T_i}{T_\infty - T_i} \text{ versus } \frac{x}{2\sqrt{\alpha t}}, \text{ for various } \frac{h\sqrt{(\alpha t)}}{k}.$$

---

*For the semi-infinite model with the constant surface temperature condition, $\frac{dT}{dx}$ at the surface is $\frac{T_{surface} - T_i}{\sqrt{(\pi \alpha t)}}$.

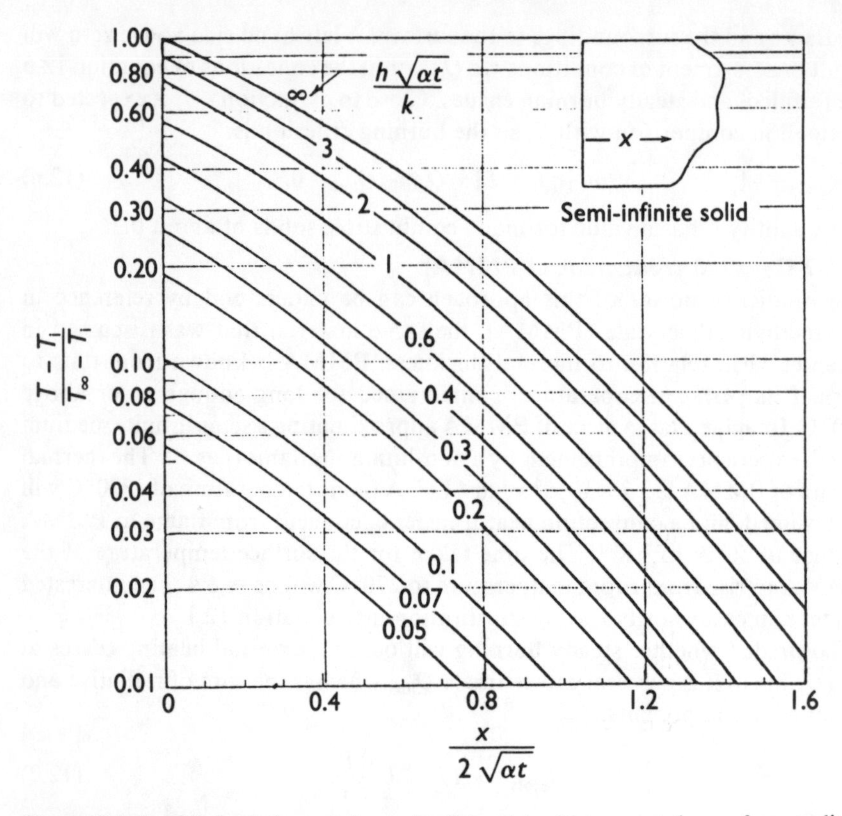

Figure 12.2 Graphical solutions of the semi-infinite slab with a convection surface condition (Incropera and De Witt [2])
(Reprinted with the permission of Prentice-Hall, Englewood Cliffs, New Jersey from Geankoplis C.J., 'Transport Processes and Unit Operations', 1993.)

In the present example with $t = 6$ s, $h = 50$ W m^{-2} K^{-1}, $k = 0.19$ W m^{-1} K^{-1}, and $\alpha = 1.1 \times 10^{-7}$ m^2 s^{-1}, we have:

$$\frac{h\sqrt{(\alpha t)}}{k} = 0.2. \tag{12.8}$$

This determines the curve in Figure 12.2 to which we have to refer and we should perhaps note the consistency whereby the surface temperature 270° C with the flame temperature 1300° C gives the value of the ordinate at $x = 0$ as 0.20, exactly as on the curve.

To calculate $\frac{dT}{dx}$ we note that on the relevant curve, the ordinate is 0.1 when the abscissa is 0.32. Substitution gives the result that after 6 s of heating, $T = 148$° C at $x = 5.2 \times 10^{-4}$ m (≈ 0.5 mm) and this serves as a control distance for calculating the temperature gradient very close to the surface. The temperature profile will be taken as linear over this very small distance. The temperature gradient can therefore be estimated as:

$$\frac{dT}{dx} = -\frac{270 - 148}{5.2 \times 10^{-4}} = -2.3 \times 10^5 \text{ K m}^{-1}. \tag{12.9}$$

With an emissivity of 0.8, the value of Q_{loss} can be calculated from equation 12.7 as 47.6 kW m^{-2}, conduction being by far the larger component. The value of m' for this material at 270°C is known from experiment to be 4×10^{-3} kg m^{-2} s^{-1} and f_c is 0.3 for many materials, as previously noted. With q = 26 MJ kg^{-1} and L = 1.6 MJ kg^{-1} we can therefore write, with $Q_{ext} = 0$:

$$m'[qf_c - L] - Q_{loss} = -22.8 \text{ W m}^{-2} < 0. \tag{12.10}$$

The fact that the calculated balance of the flux from the flame to the surface and that from the surface to the surroundings is negative indicates that the condition for steady burning expressed as equation 12.6 is not fulfilled and the material will not continue to burn if the applied heat stops at 6 s.

If the flame is maintained in contact with the polymer surface beyond 6 s it can be removed without extinction of burning, provided that equation 12.6 can be fulfilled when $Q_{ext} \to 0$. However, continued application of the convective flux causes the surface temperature to continue to rise, and hence the radiative part of Q_{loss}. The rise in surface temperature will also affect m'. Hence it is not straightforward to address the time of application of the flame necessary for subsequent sustained burning on the basis of the convective flux model, but it is not difficult to do so on the basis of the semi-infinite medium concept with the alternative surface conditions:

$$T_{surface} = T_i \text{ at } t = 0, \tag{12.11}$$

$$T_{surface} = T_f \text{ at } t > 0. \tag{12.12}$$

The solution to this is [2]:

$$\frac{T(x,t) - T_f}{T_i - T_f} = \text{erf} \frac{x}{2\sqrt{(\alpha t)}}, \tag{12.13}$$

where the x-coordinate, as before, denotes distance from the surface into the medium. Imagine that the PMMA surface is heated very rapidly to 270° C and maintained there for one minute before withdrawal of the applied heat. Will the PMMA continue to burn? Using our control distance of 5.2×10^{-4} m as a means of estimating the temperature gradient, the argument of the erf function is easily calculated as 0.10 and the function itself is found from tables [3] to be 0.11. This gives, by substitution:

$$T(5.2 \times 10^{-4} \text{ m, 1 minute}) = 242.5°\text{C}, \tag{12.14}$$

and

$$\frac{dT}{dx} = \frac{242.5 - 270}{5.2 \times 10^{-4}} = -0.53 \times 10^5 \text{ K m}^{-1}. \tag{12.15}$$

The temperature gradient is smaller than that calculated after 6 s of convection flux exposure (equation 12.9) because of the longer time for heat transfer into the PMMA. The radiative part of Q_{loss} is the same as in the previous example and, added to the newly calculated conductive part, this gives Q_{loss} = 14.0 kW m^{-2}. For the sustained one-minute heating:

$$m'[qf_c - L] - Q_{loss} = +\ 10.8\ \mathrm{kW\ m^{-2}} > 0. \quad (12.16)$$

Therefore according to the constant surface temperature model the PMMA rod will continue to burn once the external heat is removed. We have therefore shown that 6 s of heating to the surface by convective flux to take the surface of the material to 270° C will not bring about steady burning, but maintenance of the surface at that temperature for one minute will. Of course there is some margin of uncertainty in the numerical values of m' and f_c used and the calculated results have corresponding limitations.

12.2 Room fires [4]

12.2.1 A basic formulation

Room fires usually begin when one object in a room ignites. This is followed by fire growth and consequent involvement of the whole room in the fire. The transition from localized burning to full room involvement is rapid and is termed flashover. Very useful insights into room fires can be obtained from analysis of a fairly simple model which approximates fire growth to a gas layer covering a fuel surface and subject to the equation:

$$c\frac{\mathrm{d}(mT)}{\mathrm{d}t} = \mathbf{r}(T, t) - \boldsymbol{L}(T, t), \quad (12.17)$$

where m = total (fuel + air) mass of gas in the layer (kg)
c = heat capacity of the gas (J kg^{-1} K^{-1})
$\mathbf{r}$ = heat-release rate (W)
$\boldsymbol{L}$ = heat-loss rate (W).

We will consider a fairly stable gas layer with a low value of $\frac{\mathrm{d}(mT)}{\mathrm{d}t}$ in which case quasi-steady solutions of equation 12.17 are obtainable, given functional forms of $\mathbf{r}$ and $\boldsymbol{L}$. For the former:

$$\mathbf{r} = (1 - f)m'_{fb}Q \qquad \mathrm{W}, \quad (12.18)$$

where $(1 - f)$ = fraction of the total energy released entering the layer under consideration
Q = exothermicity (J kg^{-1})
m'_{fb} = rate of fuel burning (kg s^{-1}).

The precise form of m'_{fb} depends on whether fuel is in excess (ventilation control) or air is in excess (fuel control). If air is in excess, as it will be in the earlier stages of the fire, the burning rate can be equated with the rate m'_f of release of fuel into the gas layer by vaporization/pyrolysis.

This is:

$$m'_f = \frac{A_f q''(T)}{q_{vap}} = m'_{fb} \qquad \text{kg s}^{-1}, \qquad (12.19)$$

where $q''(T)$ = heat flux to the undecomposed fuel (W m^{-2})
A_f = area experiencing the flux (m^2)
q_{vap} = heat of vaporization/pyrolysis (J kg^{-1}).

Now $q''(T)$ is the heat flux from the hot layer, subject to equation 12.17, to the fuel surface which it surrounds and is at least partly radiative. This means that $q''(T)$ and hence **r** will have a stronger dependence on T than a linear one. When the stage is reached where the fire is ventilation controlled:

$$m'_{fb} = \frac{m'_a}{\beta} \text{ kg s}^{-1}, \qquad (12.20)$$

where m'_a = entry rate of air into the layer (kg s^{-1})
β = ratio (in stoichiometric terms) of air to fuel.

No equation analogous to (12.19) applies to m'_a, and if m'_a is taken to be temperature independent, so is the burning rate. If for the purposes of preliminary discussion L is taken to be linear, and recalling that $L = 0$ at the ambient temperature T_o, Figure 12.3 describes the state of affairs. The strong similarity between this graph and that for the Semenov model of thermal ignition (Figure 1.1) should be noted*. Quasi-steady solutions of equation

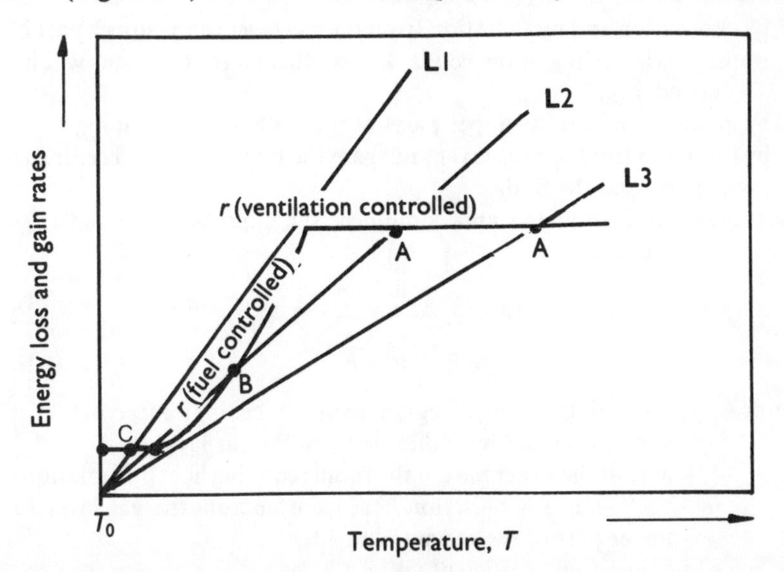

Figure 12.3 Rates of heat release and heat loss versus temperature for a layer of gas causing spread of a room fire (Thomas et al. [4])
(Reproduced with the permission of the Combustion Institute.)

*This type of diagram is also used to analyze the stability of steady states in a stirred-tank reactor, where it is known as a van Heerden diagram [5].

12.17 are signified by intersections of the **r** curve with the $\boldsymbol{L}$ line and those denoted C represent small fuel-controlled fires. In fires of this type, **r** and $\boldsymbol{L}$ *do* change slowly with time because of slow changes in T, and intersections of the C type are followed by those of the B type which are not physically stable because of the steeper slope of the **r** curve than of the $\boldsymbol{L}$ line. The system will therefore jump to a steady state of the A type, with two accompanying physical changes:

(a) A sharp increase in T.
(b) Transition from fuel control to ventilation control.

This jump can therefore be identified with flashover, and against this generalized background which interprets the phenomenology, a more detailed treatment is now given.

12.2.2 Detailed treatment

The heat loss rate $\boldsymbol{L}$ from the gas layer can be written:

$$\boldsymbol{L} = (m'_a + m'_f)c(T - T_o) + UA(T - T_o) \qquad \text{W}, \qquad (12.21)$$

where T_o = ambient temperature
U = heat transfer coefficient (W m^{-2} K^{-1})
A = area over which heat transfer from the gas layer to the surroundings occurs.

In detailed analysis, the effective value of UA has a temperature dependence. Heat is transferred in the following ways:

(a) by convection and by radiation from the gas layer to the upper part of the walls and the ceiling of the room. The coefficient for the latter will have a T^4 dependence;
(b) by conduction into the upper parts of the walls and the ceiling;
(c) by radiation from the gas layer and from the upper walls and ceiling to the lower parts and the floor.

When these are formulated and combined, the expression for the effective value of UA becomes:

$$UA = \frac{A_u}{\dfrac{1}{h_k} + \dfrac{1}{h_c + h_r}} + A_l h'_r, \qquad (12.22)$$

where A_u = area of the upper region (walls + ceiling) receiving heat by convection and by radiation from the gas layer (m^2)
A_l = area of the lower part of the room receiving heat by radiation (m^2)
h_r = coefficient for radiation heat transfer from the gas layer to the upper part of the room (W m^{-2} K^{-1})
h_c = convection coefficient (W m^{-2} K^{-1})
h'_r = coefficient for radiation heat transfer from the gas layer and the upper parts of the room to the lower parts (W m^{-2} K^{-1})
$h_k = \dfrac{\text{thermal conductivity of wall/floor material}}{\text{wall/floor thickness}}$ (W m^{-2} K^{-1}).

Equation 12.21, when the temperature dependences are put in place, predicts upward curvature for $L(T)$.

Whereas under fuel control, **r** rises steeply with T and has in the preliminary treatment above been taken to be independent of temperature under ventilation control, a widely used functional form for fire-induced air passage through doors and windows [4] has a dependence of m'_a on:

$$\left[\frac{T_o}{T}\left(1-\frac{T_o}{T}\right)\right]^{0.5},$$

and predicts that at the higher temperatures associated with ventilation control, m'_a will decline with temperature. Figure 12.3 can therefore be modified to incorporate the temperature dependences of $\mathbf{L}(T)$ and of $\mathbf{r}(T)$ in the ventilation control regime and in Figure 12.4 this is shown for three different graphs of $\mathbf{r}(T)$ in the fuel control regime, describing three different kinds of fire. The intersection at C is a quasi-steady fire with fuel control which can subsequently become tangential to the **L** curve, as at B, where there is flashover and therefore transformation to ventilation control. This is essentially as already outlined for the simpler treatment in Figure 12.3. The other possibility is a fire initially very powerfully heat releasing, called type 3 in the diagram. Such a fire might occur through spillage of a flammable liquid, or a gas leak. This does not display flashover but only ever intersects the **r** curve in the ventilation control branch.

12.2.3 Interpretation of flashover

This approach has considerable interpretive capability. Consider a room fire under ventilation control. A door to the room is then opened (or burnt

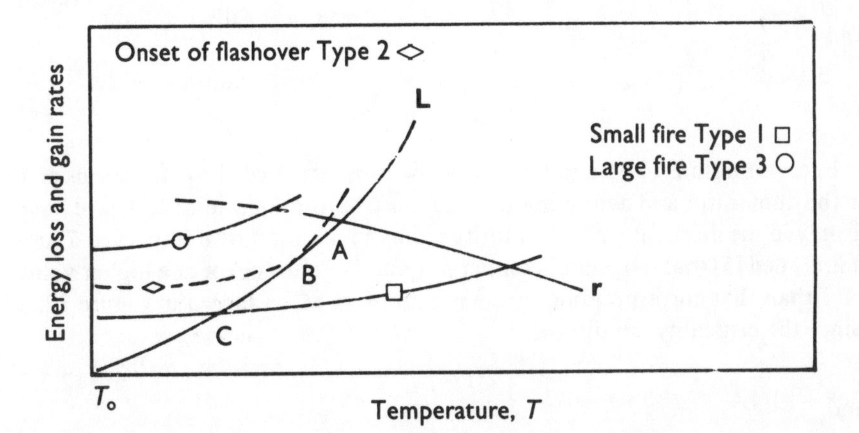

Figure 12.4 Rates of heat release and heat loss versus temperature for a layer of gas, causing spread of a room fire with non-linear heat loss rate with temperature, and declining heat release rate with temperature when ventilation controlled (Thomas et al. [4])
(Reproduced with the permission of the Combustion Institute.)

through) and the consequences in terms of the interplay of **r** and L are as follows:

The L curve takes on a steeper slope and consequently can become tangential to the **r** curve at a point on the fuel control part of the curve beyond the ventilation control regime before the door opening. There can then be flashover to a ventilation control branch higher up the curve, so conditions continue to be ventilation controlled but at a higher release rate than before the door opening.

On the other hand, imagine a fire under fuel control conditions and a sudden reduction in the slope of the heat-loss curve due to a restriction on heat transfer. This can cause a tangency with the **r** curve and hence flashover to ventilation control conditions. Again, considering fuel control conditions, an increase in A_f might be caused by disintegration or leakage during the fire and, by equations 12.18 and 12.19, this will cause the system to move on to a different **r** curve. Tangential contact between this and the L curve will result in flashover. Soot deposition increases the emissivity and hence the radiation heat transfer which influences **r** and L, with a possible transition from an intersection of the C type in Figure 12.4 to tangency and hence flashover. All these occurrences are consistent with practical experience and can be understood qualitatively by means of Figure 12.4.

12.2.4 Calculation of the critical ignition condition

Clearly the tangency condition in Figure 12.4 is central to this approach. However, with a temperature-dependent L, this tangency condition cannot be calculated in the simple way that it was in Chapter 1 for the Semenov model. There is, however, a bound that applies and uses the functions:

$$\frac{\mathbf{r}}{(T-T_o)} = \mathbf{r}^{\#} \tag{12.23}$$

and

$$\frac{L}{(T-T_o)} = L^{\#}. \tag{12.24}$$

From equations 12.18 and 12.19 it can be seen that $\mathbf{r}^{\#}$ will have functions of T in the numerator and denominator, whereas $L^{\#}$, from equation 12.21, will have T only in the numerator. The quantity $\mathbf{r}^{\#}$ has a minimum with respect to T and it is argued [5] that tangential contact of $L^{\#}$ and $\mathbf{r}^{\#}$ must occur at a higher value of T than that corresponding to the minimum in $\mathbf{r}^{\#}$, as shown in Figure 12.5. Since the criticality conditions:

$$\mathbf{r} = L, \tag{12.25}$$

and

$$\frac{d\mathbf{r}}{dT} = \frac{dL}{dT}, \tag{12.26}$$

also imply

$$\mathbf{r}^{\#} = L^{\#}, \tag{12.27}$$

and

$$\frac{d\mathbf{r}^{\#}}{dT} = \frac{dL^{\#}}{dT}, \qquad (12.28)$$

the temperature corresponding to the minimum in $\mathbf{r}^{\#}$ is a lower bound on the critical temperature for flashover, T_c. The condition at the minimum:

$$\frac{d\mathbf{r}^{\#}}{dT} = 0, \qquad (12.29)$$

also implies:

$$\frac{d}{dT}\left\{\frac{q''(T)}{T - T_o}\right\} = 0, \qquad (12.30)$$

as the condition establishing an approximate value of T_c. Of course, values of T_c so obtained will depend on the form of $q''(T)$ and whereas previously we considered fuel close to the gas layer receiving combined convective and radiative flux $q''(T)$, it is possible to modify the treatment to consider fuel more distant from the gas layer receiving convective flux $q''(T_o)$, where T_o is the average temperature of the surroundings. This will be supplemented by radiation from the gas layer at T, hence:

$$q''(T) = q''(T_o) + \phi(T)\Omega[T^4 - T_o^4], \qquad (12.31)$$

where $\phi(T)$ is the product of the temperature-dependent emissivity and the

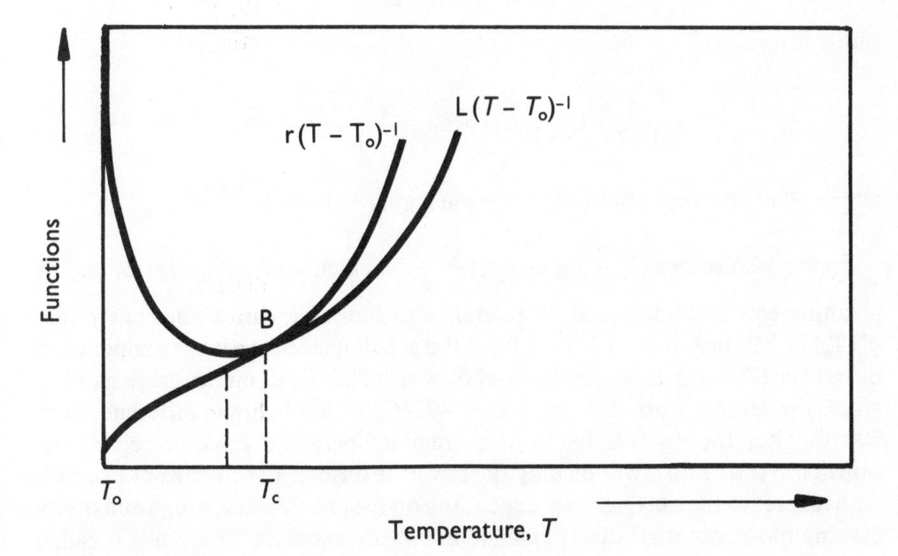

Figure 12.5 Tangency condition of $\frac{r}{T - T_o}$ and $\frac{L}{T - T_o}$, establishing an approximate value of the critical temperature for flashover T_c (Thomas et al. [4])
(Reproduced with the permission of the Combustion Institute.)

fraction of heat radiated from the gas layer actually incident upon the fuel, that is, the view factor. (The emissivity and the view factor each have a maximum value of 1, therefore ϕ has a maximum value of 1.) Substitution of this form for $q''(T)$ into equation 12.30 will, in principle, yield T_c, comparison of which with T_o will indicate the temperature rise necessary for flashover under the conditions stated. This substitution leads to:

$$\frac{T_c q''(T_o)}{\phi\Omega(T_c^4 - T_o^4)(T_c - T_o)} = \frac{\phi'}{\phi} T_c + \frac{3 + 2x + x^2}{(1 + x)(1 + x^2)}, \qquad (12.32)$$

where $x = \frac{T_o}{T_c}$ and $\phi' = \frac{d\phi(T)}{dT}$.

Now if ϕ' is taken as positive, that is, the emissivity increases with temperature , then:

$$\frac{T_c q''(T_o)}{\phi\Omega(T_c^4 - T_o^4)(T_c - T_o)} \geq \frac{3 + 2x + x^2}{(1 + x)(1 + x^2)},$$

the equality denoting $\phi' = 0$, that is, constant emissivity. The equality therefore gives T_c in the limiting case of constant emissivity. Solution gives $(T_c - T_o)$ for a particular value of $\frac{q''(T_o)}{\phi}$ and since this is for constant ϕ, this is a lower limit. At the upper limit of T_c the emissivity is greater than at T_o and it is shown that as ϕ rises with T:

$$\phi'(T_c - T_o) < \phi T_c.$$

Substituting for $\frac{\phi'}{\phi}$ gives:

$$\frac{q''(T_o)}{\phi\Omega(T_c^4 - T_o^4)} < 1 + \frac{(3 + 2x + x^2)(1 - x)}{(1 + x)(1 + x^2)},$$

which gives the upper limit of T_c for particular values of $\frac{q''(T_o)}{\phi}$.

Figure 12.6 shows $(T_c - T_o)$ against $\frac{q''(T_o)}{\phi}$, calculated as indicated, and its meaning can be understood by reference to fuel receiving a convective flux $q''(T_o)$ 10 kW m^{-2}. If $\phi = 1.0$, then from the graph it is seen that the temperature difference $(T_c - T_o)$ necessary for flashover is $\approx 300°$ C. If, on the other hand, $\phi = 0.3$, the temperature difference is $\approx 400°$C. If $\phi = 0.1$, the difference is $\approx 600°$C. Therefore fuels at the same average temperature T_o and receiving the same convective flux will display flashover at different temperature excesses with respect to the hot-gas layer depending on the emissivities, a high emissivity causing flashover at relatively small temperature excesses. This effect is called radiation enhancement and is well known to fire investigators. We should note, however, that the temperature excesses are always larger than those characterizing criticality in the Semenov model, which uses a very similar heat-balance diagram.

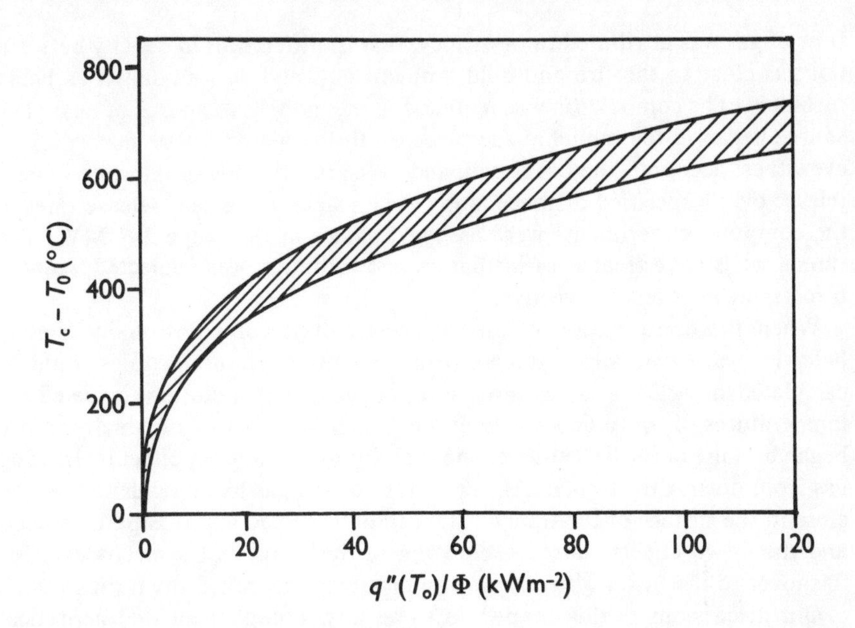

Figure 12.6 Calculated upper and lower limits of temperature excess ($T_c - T_o$) (Thomas et al. [4]) *(Reproduced with the permission of the Combustion Institute.)*

All that remains is for extinction to be discussed in terms of this treatment. Improved heat loss, for example by spraying water, can bring about the reverse of flashover: transformation of a large ventilation-controlled fire to a small fuel-controlled one. Imagine a fire on the ventilation control part of the curve in Figure 12.3. Such a fire is dependent on there being an intersection between the **r** and L curves and as the L curve becomes steeper, that intersection will move along the flat part of the **r** curve until it reaches the low-temperature limit of this branch, whereupon there is return to a C-type intersection, a small fuel-controlled fire.

12.3 Aerodynamic interpretation of flashover

A fatal fire at King's Cross Underground Station, London, in 1987 began in an escalator tunnel. The tunnel housed three escalators constructed largely of wood and the fire originated in one of them. The sudden and rapid transformation of this fire from a small one in the escalator tunnel to a much larger one in the ticket hall and surrounding subways, that is, its flashover, was simulated in aerodynamic terms by detailed computer modelling [6]. We now outline the principles incorporated in the model, and its predictions.

The escalator tunnel in which the fire began was approximated to a tube of hemispherical cross-section, and the ticket office area at the top of it was treated as having flow inlets and outlets according to the configuration of the passageways. Gas originating at the position in the tunnel where the fire began was the subject of mass, energy and momentum conservation equations. The flow of gas was attributed to buoyancy, that is, differences in density between hot gas close to the fire and cold ambient gas, and was modelled as being turbulent. The combustion was included in the model as a source of heat. The source began in the model at the place on the escalator where, according to eyewitness accounts, the fire began, and then grew to a maximum rate of heat release over a specified distance along the escalator. The heat-release rates in the computer experiments were assigned values in the range 2–7 MW. The tunnel walls were treated as insulators and radiation was neglected, so heat transfer was entirely convective.

When the conservation equations were solved computationally, steady behavior was eventually obtained, from which temperature profiles could be calculated, as well as a steady flow pattern of gas. The simulations revealed gas temperatures of up to 660°C along the trench of escalator in which the fire began because of the heat source, and that this had a chimney effect in drawing gas from outside the trench. Also, the hot flowing gas had a tendency to stay close to the surface of the trench rather than to spread and dissipate its heat, and these two effects caused preheating of the wood and a mechanism for flashover to the ticket office area. The behavior was called the trench effect.

Our discussions in this chapter in some ways complement the theoretical parts of Chapter 11, which deal with flaming and smoldering combustion of solids. We conclude this chapter with a discussion of a rather specialized, though highly interesting, treatment of fire due to the interaction of electrically generated heat with chemically generated heat.

12.4 Heating of materials by proximity to electrical cables

12.4.1 Introduction

The plastic coating around an electrical cable has to be certified as able to withstand the heat effect of the current in the wire that it surrounds. This is done by assessing the temperature rise in prolonged operation of the cable with the maximum current for which it is rated. Coiling or contact with a surface, or with other cables, can inhibit heat dissipation and therefore cause temperature rises greater than those in the certification tests.

A further interesting case, the theory of which was recently addressed [7], is where combustible powder or dust has deposited on the outside of a cable, in which case it can, by its oxidative self-heating, exacerbate the electrical heating. For example, a conduit holding a cable might become full of dust, and this scenario is attractive from the theoretical standpoint since the system would then comprise three concentric cylinders long enough for all heat transfer to be taken to be radial. These can be summarized as follows:

(a) the metal wire itself, undergoing uniform internal heat generation and radial heat conduction;
(b) the insulation around the wire, undergoing radial heat transfer but no heat generation;
(c) the dust layer, undergoing radial heat transfer and heat generation which,

unlike that in the wire, is temperature-dependent and therefore nonuniform.

12.4.2 Formulation

The steady-state heat-balance equation for the wire is a well-known standard one [3]:

$$\frac{d^2T}{dr'^2} + \frac{1}{r'}\frac{dT}{dr'} + \frac{q}{k} = 0, \tag{12.33}$$

where T = temperature (K)
r' = radial coordinate (m)
q = heat-release rate (W m^{-3})
k = thermal conductivity (W m^{-1} K^{-1}).

Putting $\theta = \frac{E}{RT_o^2}(T - T_o)$ as in earlier chapters, and also:

$x = \frac{r'}{a}$, where a = radius of the wire, gives:

$$\frac{RT_o^2}{Ea^2}\frac{d^2\theta}{dx^2} + \frac{RT_o^2}{Ea^2x}\frac{d\theta}{dx} + \frac{q}{k} = 0. \tag{12.34}$$

Now if R = wire resistance/unit length (ohm m^{-1}) and I = current (amperes), then:

$$q = \frac{I^2 R}{\pi a^2} \qquad \text{W m}^{-3}. \tag{12.35}$$

Combining equations 12.34 and 12.35:

$$\frac{d^2\theta}{dx^2} + \frac{1}{x}\frac{d\theta}{dx} + \frac{E I^2 R}{RT_o^2 \pi k} = 0 \qquad (0 < x < 1). \tag{12.36}$$

Equation 12.36 then is the heat-balance equation for the wire itself, and for the insulation:

$$\frac{d^2\theta}{dx^2} + \frac{1}{x}\frac{d\theta}{dx} = 0 \qquad (b/a > x > 1), \tag{12.37}$$

where b = radius (m) of the outer layer of the insulation.

For the outer layer, where there is heat release by oxidation, the dimensionless form of this equation (see Chapter 1) is incorporated:

$$\frac{d^2\theta}{dx^2} + \frac{1}{x}\frac{d\theta}{dx} + \delta \exp\theta = 0 \qquad (\ell/a > x > b/a). \tag{12.38}$$

In (12.38) δ is as defined in equation 1.31, with r_o identified with the outer radius ℓ of the dust layer.

Equations 12.36, 12.37 and 12.38 therefore define the problem. Analytical integration is possible and conditions fixing the constants include:

$$\frac{dT}{dx} = 0 \text{ at } x = 0, \tag{12.39}$$

that is, there is a temperature maximum at the center of the wire and hence of the assembly, with dimensionless temperature excess θ_o at the center. There are continuity conditions: if the entire assembly is in a steady state, the heat passing through any component of it in unit time is the same, and also $T = T_o$ at the outside surface of the dust layer.

When the heat-balance equation 12.36 for the wire is solved, subject to the conditions imposed on the wire by the presence of the other two media, and $x = 0$ is inserted, corresponding to $\theta = \theta_o$, the solution is:

$$\exp\theta_o = \frac{[(F + k' + 2) + b'^{-F}(F - k' - 2)]^2 b'^{F-2}}{4F^2 k''}, \tag{12.40}$$

where F is a constant of integration, and:

$$k' = -\frac{EI^2R}{2RT_o^2 k_i \pi}, \tag{12.41}$$

where k_i = thermal conductivity of the insulating material (W m^{-1} K^{-1}) and:

$$k'' = \exp\left[-\frac{EI^2R}{2RT_o^2 k_i \pi}\ln\frac{b'}{a'} - \frac{EI^2R}{4RT_o^2 k\pi}\right], \tag{12.42}$$

where $a' = \dfrac{a}{b + h'}$ and $b' = \dfrac{b}{b + h'}$, the quantity h' being the thickness of the dust layer, that is, $b + h' = \ell$.

With the composite system in a steady state, δ can be obtained from the three heat-balance equations solved according to the conditions indicated as:

$$\delta = \frac{2F^2(F - k' - 2)(F + k' + 2)b'^{-F}}{[(F + k' + 2) + (F + k' - 2)b'^{-F}]^2} \tag{12.43}$$

$$= \frac{\sigma Q A \ell^2 E \exp(-E/RT_o)}{k_d T_o^2 R}, \qquad \text{(Equation 1.31)},$$

where k_d = thermal conductivity of the dust coating the insulation material (W m^{-1} K^{-1}) and the other quantities (Q, A, E, σ) also relate to this material, since this is the only one of the three layers with chemical activity.

For chosen values of k' and k'' (and hence of E, I, R, etc.) and combination of equations 12.40 and 12.43, θ_o can be plotted against δ, and an example for a particular set of the constants is shown in Figure 12.7. The value of θ_o rises with δ until $\delta = \delta_{crit}$ and $\theta_o = \theta_o$ (crit), beyond which there is no steady state, and ignition of the dust layer results. The consequence is probable melting of the insulation layer below the dust, leading to very grave electrical and fire

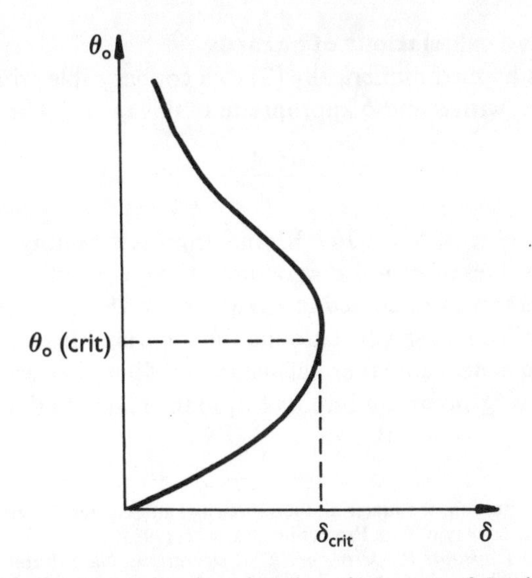

Figure 12.7 Dimensionless temperature θ_o against dimensionless rate of chemical heat release for a system comprising a current-bearing wire with its outer insulation coated with dust (Gray et al. [7]) *(Reproduced with the permission of Elsevier Science Publishers.)*

hazards. Setting $k'' = 0.5$, θ_o and δ can be calculated for various k' and h', and δ_{crit} and θ_o (crit), can be calculated from the graph as shown in the example outlined in Table 12.1. The h' values are expressed as multiples of b, and of course δ and θ_o depend on h' because of their dependence on b'. However, k'' does not depend on h' because the scaling factor $(b + h')$ cancels.

Table 12.1 Calculated critical conditions for dust coating an electric power cable (Gray et al. [7])
$k'' = 0.5$, $k' = -5$

h' =	$9b$	$4b$	$3b/2$	b	$b/4$	$b/9$
θ_o (crit)	13.0	9.8	6.3	5.3	3.0	2.4
δ_{crit}	0.0006	0.0075	0.13	0.39	11	63

Before discussing these calculated results quantitatively, the following trends require comment. First, most of the θ_o (crit) values are much higher than those encountered in Chapter 1 which had values of about 1. The explanation is, of course, that the temperature excess θ_o at the center of the wire is electrical in origin, influenced by heat transfer rates to the outside, but certainly not limited to values of ≈ 1 for a steady state such as where the heat release is by combustion. Secondly, since δ_{crit} is the minimum value of the rate of heat release by the dust necessary for ignition, we expect larger values of δ_{crit} for smaller values of h' and consequently lower thermal resistances of the dust layer. The figures in Table 12.1 also support this prediction.

12.4.3 Predictive calculations of hazards

This model was applied numerically [7] to a copper cable with PVC coating bearing 14 amps, with a and b appropriate to this rating. The group:

$$\frac{\sigma QA}{k_d},$$

was assigned a value of 5.3×10^{13} K cm^{-2} from self-heating experiments on cellulosics, and values of k'' and k' were respectively 0.5 and –5. A δ_{crit} of 0.0006 therefore applies to a cable coated to an extent $h' = 9b$ and when this is solved, the result is that 15 cm of sawdust around the cable is sufficient to bring δ to its critical value and hence cause thermal instability. That a cable should be buried in 15 cm of dust is by no means impossible, so the reality of the hazard is clear.

References

[1] Drysdale D. D., 'Ignition: the material, the source and subsequent fire growth', *Technology Report 83-5*, Boston: Society of Fire Protection Engineers (1983)
[2] Geankoplis C. J., *Transport Properties and Unit Operations*, 2nd Edition, Englewood Cliffs N. J.: Prentice Hall (1993)
[3] Holman J. P., *Heat Transfer*, SI Metric Edition, New York: McGraw-Hill (1989)
[4] Thomas P. H., Bullen M. L., Quintiere J. G., McCaffrey B. J., 'Flashover and instabilities in fire behavior', *Combustion and Flame* **38** 159 (1980)
[5] van Heerden C., 'Autothermic processes', *Industrial and Engineering Chemistry* **45** 1242 (1953)
[6] Simcox S., Wilkes N. S., Jones I.P., 'Computer simulation of the flows of hot gases from the fire at King's Cross Underground Station', *The King's Cross Underground Fire: Fire Dynamics and the Organisation of Safety*, 19, London: Institution of Mechanical Engineers (1989)
[7] Gray B. F., Dewynne J., Hood M., Wake G. C., Weber R., 'Effect of deposition of combustible powder on to electric power cables', *Fire Safety Journal* **16** 459 (1990)

CHAPTER 13

Industrial Explosion Hazards

Abstract

Flash fires and vapor cloud explosions are discussed and two approaches to understanding them are given, each from relatively recent research literature. One approach is essentially a systematic compilation of details from documented case histories and application of damage/injury analysis, and the other is concerned with pressure effects at various distances into an exploding cloud of vapor. Fireballs are treated quantitatively and the chapter concludes with a treatment of dust explosions. Selected case histories are discussed in detail throughout the chapter.

13.1 Vapor cloud explosions (VCE)

13.1.1 Introduction

Loss of life and property can attend the explosion of an accidentally discharged gas or vapor, for example, LPG or cyclohexane. A gas or vapor having leaked can be ignited by an energy source of the order of mJ, as we saw in Chapter 3, and this can be supplied by a frictional or electrical spark. The result might, at first consideration, be expected to be a flame (deflagration) with great destructive potential by reason of its heat release, but without a blast due to the pressure, at least if the vapor is unconfined or weakly confined. That a leaked gas or vapor *does* often result in a blast is well known from case histories and this signifies that flame speeds at least an order of magnitude higher than those characterizing hydrocarbon deflagrations in laboratory measurements such as were discussed in Chapter 4 are being generated. The term *vapor cloud explosion* (VCE) applies where there are significant pressure effects, and the term *flash fire* is used otherwise.

The need to avoid leakage of hydrocarbon gases and vapors and to eliminate ignition sources anywhere close to storage of these materials hardly needs stating. However, when there is accidental leakage, the initial absence of a spark or a hot surface to cause ignition makes the accident all the more serious if ignition does eventually occur. It has happened [1] that there has been abundant leakage of a hydrocarbon for 15 minutes before ignition, during which time the amount of leaked material accumulated has become very large and the consequences of ignition correspondingly grave. It is this possibility of long ignition delays which in phenomenological terms distinguishes a flash fire from a boiling liquid expanding vapor explosion (BLEVE), discussed in detail in Section 13.2. A particular accident may involve both VCE and flash-fire types of behavior. The research literature for the last decade contains details of two approaches to understanding the combustion behavior of vapor clouds and we will consider them separately.

13.1.2 Scientific evaluation of vapor leak accident case histories [2], [3]

In this work, information regarding 165 accidents due to gas or vapor leak, spanning the period 1921–80, was collated according to the characteristics listed in Table 13.1. For example, the 1974 Flixborough, UK, disaster has descriptors M5, R3, I1, D3, E2, L2, Dy2, F5, W5, and Do3. This means that there was leakage of between 10^4 and 10^5 kg of a substance of medium reactivity (it was actually cyclohexane). It is unknown whether ignition was by a continuous ignition source or a transient one and the cloud drifted 10^2–10^3 m. There was a VCE and, as noted in the table, this does not necessarily mean that there was not also a flash fire. (In fact there was and their combined occurrence is discussed in Section 13.1.3.) The cloud encountered obstacles such as buildings. The delay time to occurrence of the VCE was < 1 minute, and numbers of dead and injured were each between 16 and 50. There was a domino effect.

The 165 accidents contributing to this statistical examination are believed to account for about 2 percent of the total number during the period, that is, there was a total of about 8000. Many of the 8000 were too insignificant in damage terms to have been the subject of discussion in the open literature so the 2 percent sample is believed to contain more very serious accidents than would a strictly representative sample. There was not a complete set of information available for all the 165 accidents, so trends in M, R, I, etc. are based on fewer than 165 data. M was known for 82 of the 165 vapor cloud accidents and when plotted as a histogram there was a peak at M4 to M5. There was only one recorded accident with M2, leaked quantity $< 10^2$ kg, which suggests that leakages that small do not often lead to a VCE or flash fire. In 56 of the 165 accidents, the ignition source was identified and of these, somewhat more than half were I2, a continuous ignition source known to be present. Drift had been estimated and reported in as many as 87 cases and the histogram shows that of these, 60 percent were D2, a drift of less than 100 m. Most of the remainder were D3, a drift up to 1 km. Only two cases of a drift exceeding 1 km were recorded.

The classification either E2, signifying VCE, or E3, signifying flash fire, was known for 150 of the accidents considered and nearly 60 percent were VCE. Whether there was significant confinement due to buildings was known in 81 cases, and for 80 percent of them there was such confinement (L2 according to the scheme in Table 13.1). The Dy histogram is very spread out, with 20 percent of the 55 cases for which this information was available being Dy2, <1 minute delay. However, 60 percent were Dy 2 or 3, with up to 5 minutes delay between release of the cloud and VCE or flash fire as the case may be. A significant proportion (≈ 10 percent) were Dy6, with over half an hour delay. While F and W were known for respectively 143 and 114 of the 165 accidents, the value of this information in statistical analysis is diminished by the obvious possibility that some potentially fatal leaks occurred when, by good fortune, there were no persons at the scene.

Table 13.1 Classification of gas or vapor leak accidents

Property	*Physical basis*	*Categories*
Mass M	Mass of gas or vapor having leaked	1. Unknown 2. $< 10^2$ kg 3. 10^2–10^3 kg 4. 10^3–10^4 kg 5. 10^4–10^5 kg 6. $> 10^5$ kg
Reactivity R	Relative combustion reactivity of the leaked gas or vapor according to laboratory measurements	1. Unknown 2. High (e.g., H_2. Only 4 cases of the 165 cases known to involve substances in this category) 3. Medium (e.g., LPG. 70% of the cases known to involve substances in this category) 4. Low (e.g., methane)
Ignition source I	Whether a continuous ignition source such as an open fire was present or not. Where there is no such source, ignition is assumed to be due to an accidental source which is not continuous, e.g. a spark.	1. Unknown 2. Continuous source present 3. No continuous source present

Continued p. 252

Table 13.1 Continued

Drift distance D	Distance that the vapor cloud drifted from the source of the leak	1. Unknown 2. $< 10^2$ m 3. 10^2–10^3 m 4. $> 10^3$ m
Explosion E*	Whether the ignition resulted in a significant pressure blast or not	1. Unknown 2. VCE 3. Flash fire only, no VCE
Location L	Whether obstacles such as buildings were present in the path of the cloud. Obstacles cause a vapor cloud to behave not entirely as an unconfined one.	1. Unknown 2. Obstacles 3. No obstacles
Delay Dy	Time between release and occurrence of the VCE or flash fire. (The BLEVE type of accident, in which ignition occurs immediately at release, is not included in this analysis.)	1. Unknown 2. < 1 min 3. 1–5 min 4. 6–15 min 5. 16–30 min 6. > 30 min
Fatalities F	Number of persons killed	1. Unknown 2. 0 3. 1–5 4. 6–15 5. 16–50 6. > 50
Wounded W	Number of persons injured	As for Fatalities F
Domino Do	Whether other hazardous materials were released because of damage caused by the primary occurrence. No such effect is possible if there are no susceptible materials nearby.	1. Unknown 2. No effect possible 3. Domino effect 4. No domino effect in spite of the presence of nearby susceptible materials

*Category 2 means that a VCE occurred and a flash fire may or may not have occurred during the same accident. Classification 3 means that there was a flash fire but no VCE.

Whether there was a domino effect was known in 83 cases, 40 percent of which were Do1, with no effect possible for the reason noted in Table 13.1. There was a domino effect in about two-thirds of the remaining cases.

Of the 150 cases for which the E descriptor was known, 115 were R3, with medium reactivity substance. Of these, 62 were VCE and 53 flash fire. This group of data was subjected to further appraisal to provide a basis for predicting from the M, I, D, L and Dy descriptors for a particular leakage whether a VCE or a flash fire is more probable. The following trends emerged:

Up to 10^5 kg, the quantity leaked has no bearing on whether a VCE or a flash fire will result. While a VCE is possible with larger leaks than 10^5 kg (M6) there is a somewhat greater likelihood of a flash fire. While there was no significant connection between the behavior type and the drift distance, there was a very marked tendency for the presence of obstacles (L2) to promote VCE behavior (although flash fire was not excluded in the presence of obstacles). Short delay times also favored VCE occurrence rather than flash fire. Clearly, a degree of confinement due to obstacles increases the turbulence and, as we saw in Chapter 4, this increases flame propagation rates. The trend with delay times can also be understood on this basis. With short delay times the leaked material will have retained much of the momentum with which it was ejected from the leaking vessel, thus promoting turbulence.

According to the 115 sets of accident data appraised, for a VCE the number of injured persons tended to be about an order of magnitude higher than the number of fatalities, whereas for flash fires the numbers of injured and dead were about the same. Injury or death due to a flash fire is by burning only. Since there is no significant pressure blast, it is not expected that persons too far from the flash fire to be affected by its heat flux will be harmed in any other way. Injury *actually in* a VCE will be due primarily to the heat, quite sufficient to be fatal, so the pressure blast is in a sense irrelevant. However, the pressure blast also causes danger to persons distant from the cloud. Hence in flash fires there appears to be essentially a single class of casualties, people in or close to the cloud at the time of ignition. On the other hand, in a VCE there are two distinct classes of casualty: those in the cloud and affected mainly by the heat, and those distant from the cloud and affected only by the blast. This is reflected in the order of magnitude difference between fatal and nonfatal injuries in a typical VCE.

The injuries experienced by victims other than those affected by the heat flux of a VCE are unquestionably due to the blast, but in an indirect way. The danger to people arises from damage to buildings, walls and other structures and consequent collapse or movement of building materials including glass. There is no evidence that a person standing outside the actual explosion zone has ever been killed or seriously injured directly by a VCE, although it has been known for a person to be knocked over without being seriously hurt. Nor is there evidence that a motor vehicle outside the explosion zone has ever been overturned by the blast. By means of a rather novel scientific approach, these observations enable the checking of independently obtained upper bounds on the quantities characterizing the VCE blast as felt outside the actual explosion area. These quantities are overpressure P (units Pa), the maximum amount by which the pressure exceeds atmospheric, and the impulse I (units Pa.s), defined as follows:

$$I = \int_{0}^{t_p} P(t)\,dt, \tag{13.1}$$

where t_p, the positive phased duration, is the time during which the overpressure is above zero (actual pressure above atmospheric). Overpressure and impact obviously depend on the distance from the explosion zone. From calculations based on flame speeds for M3 (medium reactivity) gases and vapors the following bounds are estimated for the blast immediately outside the explosion zone [2]:

$$\text{Upper bound: } P = 3 \times 10^4 \text{ Pa} \tag{13.2}$$
$$\text{Lower bound: } P = 1.3 \times 10^4 \text{ Pa}, \tag{13.3}$$

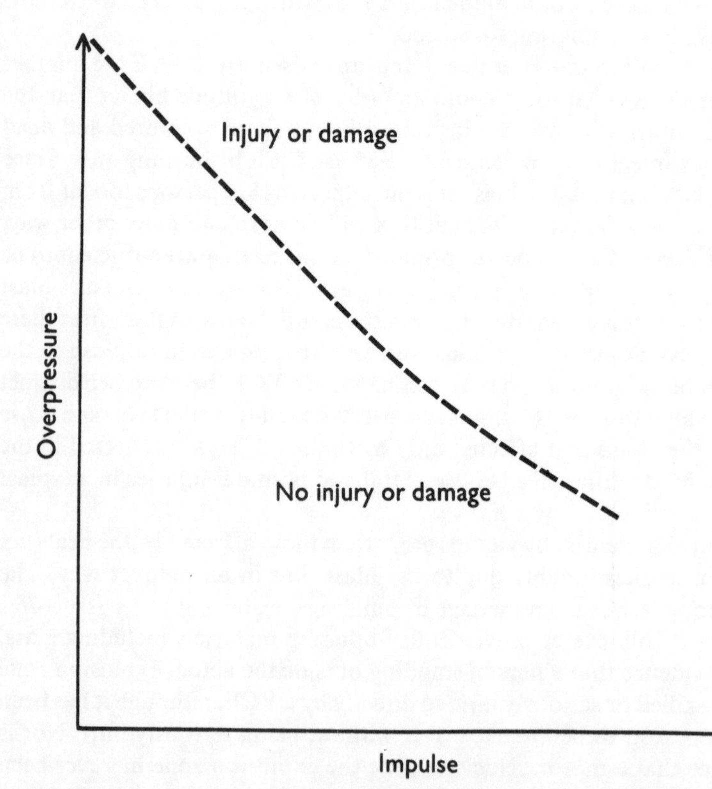

Figure 13.1 Generalized form of a damage/injury criteria diagram

and the corresponding values of I were also estimated, each being of the order of 10^4 Pa.s.

The conventional way to express the criteria of damage (in the case of property) or injury (in the case of people) is by a diagram of the type schematised in Figure 13.1: a line or curve in overpressure–impulse space below which there is no damage or injury, as the case may be, and above which there is. In injury terms, reliable graphs of the form of Figure 13.1 are available [4] for death or injury by lung damage, by skull damage and by *whole body translation* (that is, being thrown by the blast). In the specific versions of Figure 13.1 for each of these, the area corresponding to the above bounds lies well inside the part of the diagram denoting no injury and this is consistent with there being no reports of death or injury by the direct effects of a VCE outside the explosion zone itself.

Figure 13.1 in terms of building damage actually consists of three curves: minor structural damage, major structural damage and partial demolition. The area represented by the bounds on P and corresponding values of I envelops the region signifying minor damage and goes well into the region signifying major damage, coming close to the threshold for partial demolition. This is consistent with records of building damage outside the explosion zone of a VCE and with the view that it is this that causes injury to persons, not, in locations away from the cloud, the explosion itself.

The form of Figure 13.1 for building damage can, for each extent of damage, be drawn to a reasonable approximation as two lines at right angles as shown schematically in Figure 13.2. Clearly for any value of the impulse higher than

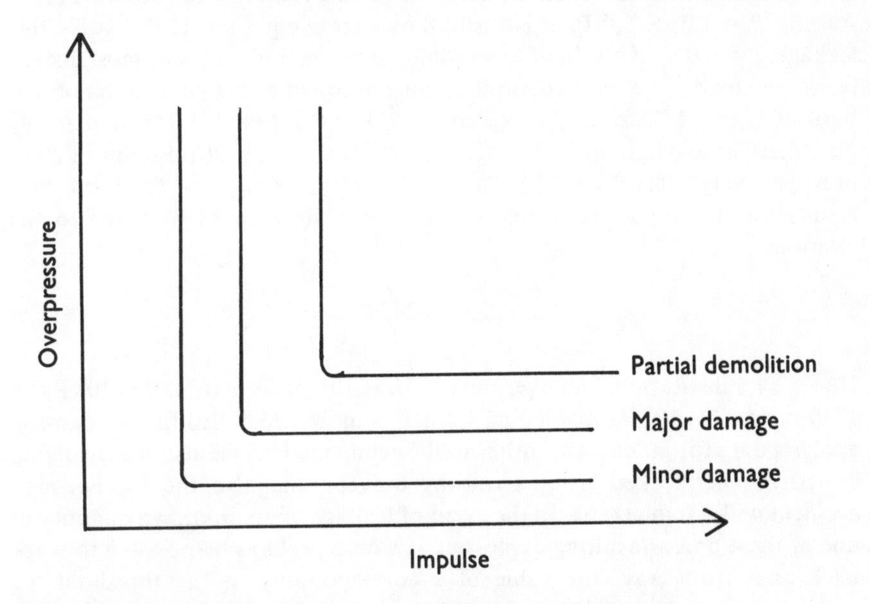

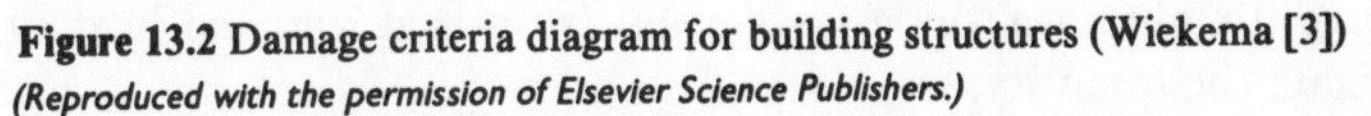
Figure 13.2 Damage criteria diagram for building structures (Wiekema [3])
(Reproduced with the permission of Elsevier Science Publishers.)

that corresponding to the horizontal part of the minor damage curve the extent of damage can be predicted from knowledge of the overpressure only, without precise knowledge of the impulse. The practical value of this is actually retrospective rather than predictive; from the damage pattern after an accident it is possible to estimate the overpressure profile during the blast. Before detailed consideration of a case history in which these ideas are brought out it is necessary to introduce the term *explosion length*, defined as:

$$L = \left[\frac{\boldsymbol{E}}{P_0}\right]^{0.33}, \tag{13.4}$$

where L = explosion length (m)
P_0 = atmospheric pressure (Pa)
$\boldsymbol{E}$ = energy that the leaked gas or vapor releases when fully reacted with oxygen (J).

The quantity $\boldsymbol{E}$, and therefore L, is dependent on the quantity of substance leaked. In damage pattern analysis, distances from the actual explosion are usually expressed as multiples of L, that is, as y/L, where y is the actual distance in m, and the overpressure P as a fraction of P_0, that is, as P/P_0. The overpressure will decrease with distance from the exploding cloud, hence an overpressure profile deduced from damage inspection is expected to express P/P_0 as a direct function of L/y.

The case history we will examine [3] involved leakage from a railway tankcar of approximately 18 tonnes of butadiene (m.w. 54 g mol^{-1}, heat of combustion 2539 kJ mol^{-1}, medium reactivity according to the classification in Table 13.1). Putting $P_0 = 1.01 \times 10^5$ Pa it is easily shown from equation 13.4 that for this leakage, $L = 200$ m. Now, in practical damage inspection the scene most distant from the cloud showing structural damage to buildings is considered on the basis of Figure 13.2 to have experienced $P/P_0 = 0.1$, ($P = 10^4$ Pa) while heavy glass damage and light glass damage are indicated by value thresholds of P/P_0 of respectively 0.03 ($P = 3 \times 10^3$ Pa) and 0.01 (P = 1×10^3 Pa). The inspection results for the butadiene accident along these lines could be fitted to the relation:

$$\frac{P}{P_0} = 0.2\frac{L}{y}. \tag{13.5}$$

Hence by substitution, the overpressure was 10^4 Pa at 400 m, 3×10^3 Pa at 1330 m and 1×10^3 Pa at 4000 m. In fact, equation 13.5 also fits the damage analysis data for at least two other well-documented accidents: one involving dimethyl ether leaked from a railway tankcar and also the Flixborough accident with cyclohexane. In the event of leakage of an unknown quantity of one of these and a resulting explosion, it would perhaps be possible to work back and, from, say, the value of y corresponding to the threshold for structural damage, to obtain a value for L, and from that and the heat of reaction to estimate the quantity.

This approach to understanding gas or vapor cloud accidents is certainly soundly based. One of the trends revealed in the statistical analysis was that confinement by buildings increases the likelihood of a pressure blast and this is a point deserving further attention and is, in fact, the keynote of a complementary treatment to be described in Section 13.1.3.

13.1.3 The multi-energy approach [5]

It was mentioned in Chapter 4 that grids can be used in experimental flame investigations to bring about turbulence, and with leaked gas or vapor, obstacles such as buildings and large plant will act in the same way. Three quantities are considered in the multi-energy method: peak static pressure, acting in all directions; peak dynamic pressure, acting only in the direction of flow; and positive phase duration. Each will have a dependence on distance from the exploding cloud and the essentials of the method can be understood by reference chiefly to the static pressure. Numerical simulation of a stoichiometric hydrocarbon explosion with uniform propagation rate and arbitrarily hemispherical geometry provides Figure 13.3 which can be understood as follows: the vertical axis is the dimensionless overpressure P/P_0 which was discussed in Section 13.1.2, as was the dimensionless distance y/L on the horizontal axis. The numerical prediction encapsulated in Figure 13.3 is that there are varying severities of overpressure at the edge of the cloud, depending on the extent of generation of turbulence by obstacles. In the numerical work, different boundary conditions lead to different conditions of flow. In all cases there is a decline, eventually to zero, of overpressure with distance.

The numbers on the curves signify the severity of the overpressure at the edge of the cloud, the highest value of 10 denoting an actual detonation with a very strong pressure effect. Profile 1, for example, shows feeble pressure effects while profiles 6–9 take on overpressure values equivalent to those of a detonation at the larger values of the dimensionless distance. Therefore, the treatment predicts that even where the overpressure at the edge of the area initially occupied by the cloud is an order of magnitude or more lower than it would be for a detonation, the profile for affected areas outside the cloud merges with the detonation profile. It is not only the magnitude of the overpressure that is relevant, however, and profiles 4–6 show significant blast in the regions denoted by the solid line. To understand this we have to consider another aspect of the numerical work underlying Figure 13.3: the *temporal* dependence of the pressure at any point, which can be expressed as a graph of pressure against time throughout the positive phase duration. In regions of the profiles shown as solid lines, the pressure rise occurs over a time which is an extremely small fraction of the positive phase duration and therefore appears vertical on the pressure–time graph. In other words there is a shock wave and concomitant blast. In the parts of the profiles shown as dashed lines there is a relatively gradual rise of the overpressure to the maximum value which forms the ordinate of Figure 13.3. A shock wave therefore occurs at all points in a detonation as shown, and at regions of some of the other profiles, depending on

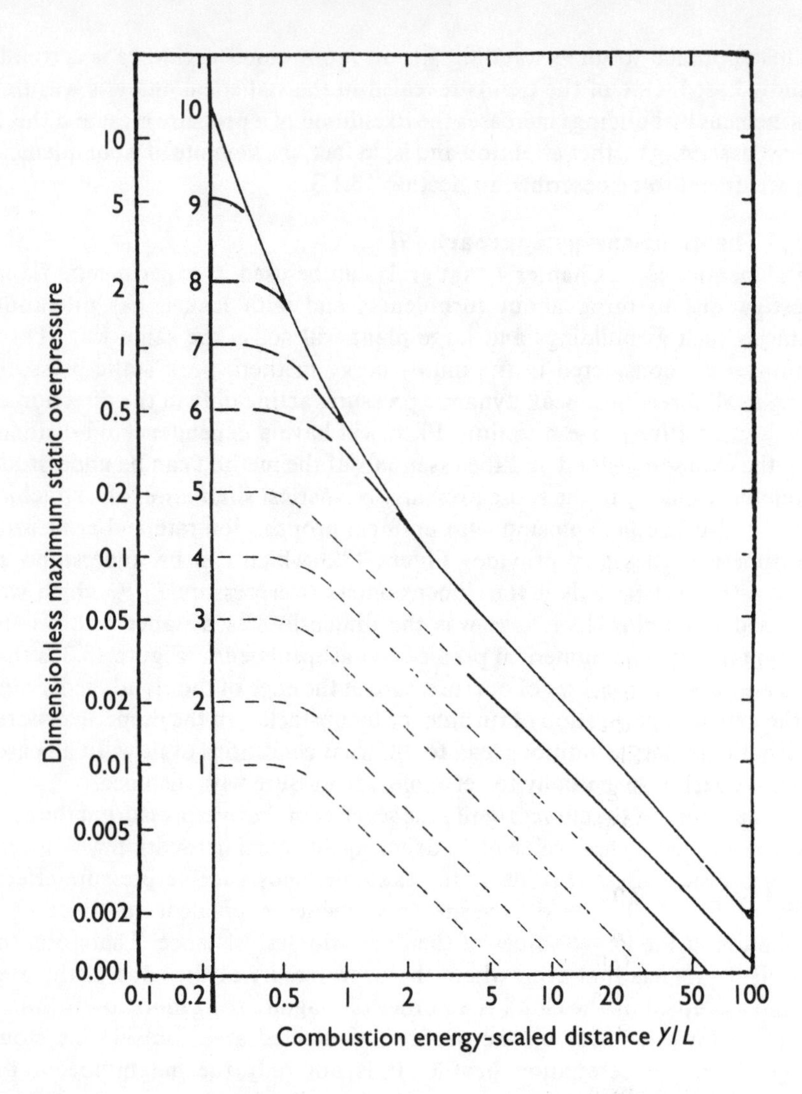

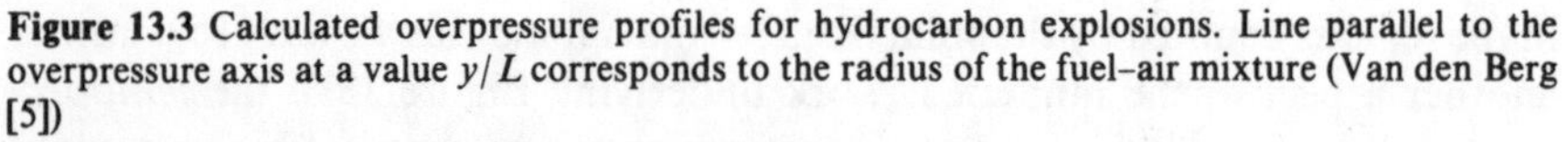

Figure 13.3 Calculated overpressure profiles for hydrocarbon explosions. Line parallel to the overpressure axis at a value y/L corresponds to the radius of the fuel–air mixture (Van den Berg [5])

(Reproduced with the permission of Elsevier Science Publishers.)

the overpressure at the edge of the cloud which, in turn, depends on the turbulence.

To apply the method to a particular explosion it is necessary to know ***E***, as in the previous treatment, in order to convert the dimensionless distances to actual ones, as well as to have an estimate for the initial cloud radius. According to the extent of confinement, it is then necessary to judge whether the relevant curve is 1, 2, 3, etc. and from all this information it can be deduced whether there will be a blast, and if so, at what position relative to the initial cloud. Of course, a particular accident may involve, in effect, two or more clouds of the same material and the strength of this approach is that such an occurrence can be treated as separate explosions of different ***E*** values depending on the amounts in the different clouds (hence the term multi-energy). The Flixborough explosion, already referred to, can be treated in this way since it is believed to have developed from two vapor clouds: one involving about 1×10^4 kg of the leaked cyclohexane confined by the process equipment adjacent to the leak, and another of about 3×10^4 kg of cyclohexane which was unconfined. In view of the confinement, the smaller leak is placed on a number 7 overpressure profile and the larger leak where there is no confinement on a number 2 profile. From knowledge of ***E*** for each explosion, an actual distance can be substituted for the dimensionless one on the horizontal axis. The results are shown in Figure 13.4, the higher curve representing the profile for the confined explosion displaying blast. There is a shock pressure rise at distances beyond approximately 100 m. The points represent the profile calculated independently from damage analysis along the lines indicated in Section 13.1.2 and agreement is very convincing. The lower curve represents the unconfined explosion, which hardly contributes to the blast damage.

A concluding comment on flash fires and VCEs, and factors causing one or the other, relates to the precise value of the energy of the ignition source. We saw in Chapter 3 that an energy source will impose an initial temperature profile. If that profile is very steep it is possible that the resulting pressure effect will be sufficient for generation of a shock wave. Under these conditions a blast will accompany the explosion.

13.2 BLEVEs and fireballs

13.2.1 Introduction

A BLEVE has the following features:

(a) The leaked substance is well above its normal boiling point at ordinary temperatures, having been stored under pressure before escape, as with LPG.
(b) Ignition is immediate.
(c) Propagation is deflagrative and there is no significant blast, so injury and damage are due to the heat.

In much of the research literature the term *fireball* is used to describe accidents conforming to the above definition of a BLEVE. The term *BLEVE fireball* is also used. We will examine a descriptive treatment of combustion of this kind with LPG and follow with a case study.

13.2.2 Model of an LPG fireball [6]

Combustion in a fireball is very rapid and the released energy is transferred as a radiant flux. The duration of flux of a particular magnitude required to cause burn injuries is known, so calculation of fluxes from fireballs enables the potential for injury at various distances from a fireball to be predicted.

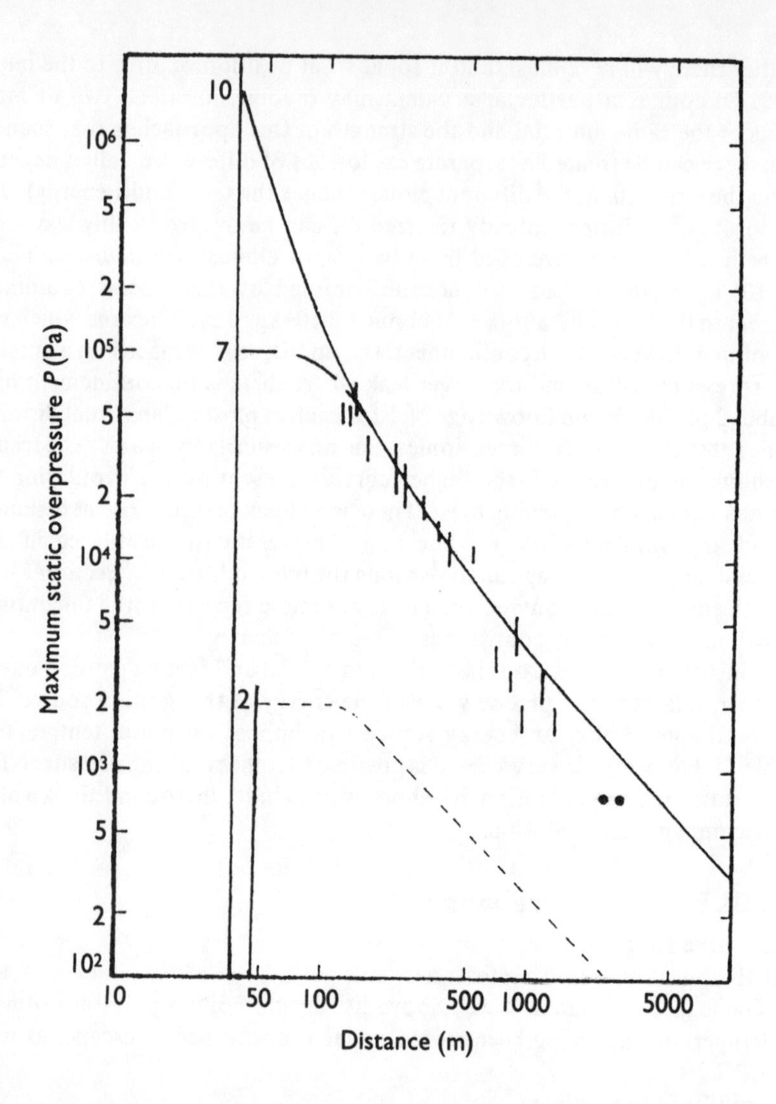

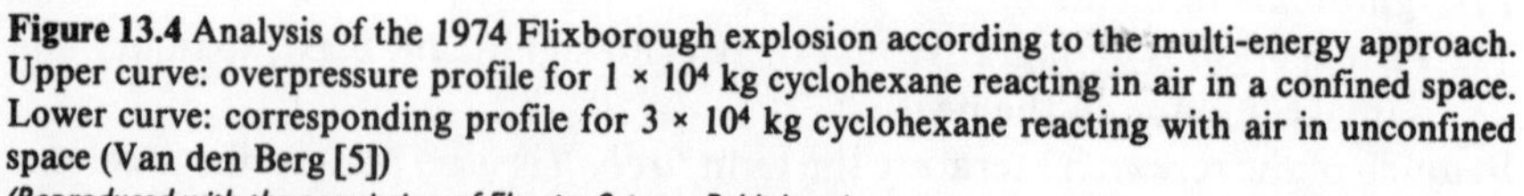

Figure 13.4 Analysis of the 1974 Flixborough explosion according to the multi-energy approach. Upper curve: overpressure profile for 1 × 10^4 kg cyclohexane reacting in air in a confined space. Lower curve: corresponding profile for 3 × 10^4 kg cyclohexane reacting with air in unconfined space (Van den Berg [5])

(Reproduced with the permission of Elsevier Science Publishers.)

We consider a fireball receiving influx of fuel from a leak. This supply is sustained for a period of time t_b after ignition, so t_b is the effective lifetime of the fireball and is sometimes called the burnout time. Assigning a fireball an emissivity of unity, the flux q_o from the surface of a fireball to the surroundings is:

$$q_o = \Omega(T^4 - T_o^4) \qquad \text{W m}^{-2}, \qquad (13.6)$$

where Ω = Stefan's constant = 5.7×10^{-8} W m^{-2} K^{-4}
T = surface temperature of the fireball (K)
T_o = ambient temperature (K).

The flux q at a distance d from the center of the fireball is, when the appropriate view factor is used:

$$q = q_o \frac{r^2}{d^2} \qquad \text{W m}^{-2}, \tag{13.7}$$

where r = radius of the fireball (m) and unlike T, which can be identified with the adiabatic flame temperature of the mixture, is assigned a time dependence.

The quantity Q of heat transferred per unit area during the lifetime of the fireball is:

$$Q = \int_0^{t_b} q \, dt = t_b \int_0^1 q \, d\tau \qquad \text{J m}^{-2}, \tag{13.8}$$

where the dimensionless time $\tau = t / t_b$. The following radius–time relationship is characteristic of fireballs.

$$r = r_1 \tau^{1/3}, \tag{13.9}$$

where r_1 = radius at burnout. An empirical expression for the lifetime of fireballs expressed for the specific case of a stoichiometric mixture of propane and air is:

$$t_b = 1.09 \, W_{prop}^{1/6} \qquad \text{s}, \tag{13.10}$$

where W_{prop} = weight of propane (kg), and equations 13.7 and 13.10 inclusive give the result:

$$Q = 0.654 \frac{q_o r_1^2}{d^2} W_{prop}^{1/6} \qquad \text{J m}^{-2}. \tag{13.11}$$

Now, for a propane/air stoichiometric mixture the adiabatic flame temperature (see Chapter 5) is approximately 2200K, so with T_o = 298K this gives q_o as 1320 kW m^{-2}, so:

$$Q = 860 \frac{r_1^2}{d^2} W_{prop}^{1/6} \qquad \text{kJ m}^{-2}. \tag{13.12}$$

This analysis is clearly leading to ability to predict from the amount leaked the heat received from the fireball by objects at various distances d, but first we have to substitute for r_1. Thus we define a rate $\boldsymbol{R}$ of total (fuel and air) mass influx into the fireball as follows:

$$\boldsymbol{R} = \frac{W_{prop} + W_{air}}{t_b} \qquad \text{kg s}^{-1}, \tag{13.13}$$

where W_{air} is the mass of air (kg) stoichiometrically equivalent to W_{prop} kg of propane. The expression of which equation 13.10 is a special case for a stoichiometric mixture is:

$$t_b = 0.68(W_{prop} + W_{air})^{1/6}. \quad (13.14)$$

Hence

$$R = \frac{(W_{prop} + W_{air})^{5/6}}{0.68}. \quad (13.15)$$

For a fireball approximated to spherical geometry, clearly:

$$Rt_b = \frac{4}{3}\pi r_1^3 \sigma = \text{total mass of the fireball immediately before burnout,} \quad (13.16)$$

where σ is the density of the gas mixture comprising the fireball, with units kg m^{-3}. Combining equations 13.15 and 13.16:

$$r_1 = \left[\frac{3}{4} \times \frac{(W_{prop} + W_{air})^{5/6}\, t_b}{0.68\pi\sigma}\right]^{1/3} \text{ m.} \quad (13.17)$$

Incorporating the stoichiometric numerical relationship between W_{prop} and W_{air} and also an estimated value of σ:

$$r_1 = 2.86\, W_{prop}^{5/18}\, t_b^{1/3} \qquad \text{m,} \quad (13.18)$$

and substituting for t_b from equation 13.10:

$$r_1 = 2.94\, W_{prop}^{1/3} \qquad \text{m,} \quad (13.19)$$

so we have the necessary estimate of r_1 to enable us to predict hazards at various distances d from the fireball via equation 13.12. Combining these equations:

$$Q = 7430\, \frac{W_{prop}^{5/6}}{d^2} \qquad \text{kJ m}^{-2}, \quad (13.20)$$

and from the potential danger viewpoint this is a worst case, since the adiabatic flame temperature has been used as surface temperature and the fourth power of this was used in estimating q_o. If instead the surface temperature is taken as 1725K, equation 13.20 becomes, via the revised value of q_o:

$$Q = 2810\, \frac{W_{prop}^{5/6}}{d^2} \qquad \text{kJ m}^{-2}. \quad (13.21)$$

Now burn injuries classified as second and third degree are caused by amounts of heat of respectively 40 and 125 kJ m^{-2} received over a period of seconds. Consider a fireball resulting from a leak of 10 tonnes (10^4 kg) of LPG. The radius of the fireball immediately before burnout is easily estimated from Equation 13.19 as 63 m. From equation 13.10, the duration of the fireball (to the nearest second) is 5 s. On the basis of the adiabatic flame temperature at the surface, a potential victim would suffer second-degree burns if standing a distance d' from the fireball, where d' is obtainable from equation 13.20 as:

$$d' = \left[\frac{7430\, W_{prop}^{5/6}}{40}\right]^{0.5} = 630 \text{ m} \approx 0.4 \text{ mile.} \quad (13.22)$$

If the surface temperature is 1725K, then from equation 13.21 the value of d' becomes 383 m. For third-degree burns the distances are 350 m if the surface temperature is the adiabatic value and 217 m if the surface temperature is 1725K. Hence serious injuries can be caused to people a considerable distance from the actual fireball. It should perhaps be noted that because of the form of the view factor used to obtain equation 13.7 and carried through the subsequent working, d' is actually a distance from the center of the fireball. The radius is, however, fairly small compared with the calculated d' values. If the leaked quantity is an order of magnitude higher, still quite consistent with records of known cases, the duration of the fireball is 7 s and its radius immediately before burnout is 136 m. Second- and third-degree burns are expected at distances of respectively 1.7 km ($\approx$ 1 mile) and 0.9 km from the fireball for a surface temperature of 2200K.

The above treatment may overestimate the distance from the fireball at which burns will occur, even when the 1725K value for the surface temperature is used, since it is possible that there will be some attenuation of the radiation. The degree of attenuation will depend on conditions, including the humidity. However at distances of a few hundred metres, such as those calculated above, the assumption of transmissivity (fraction of radiation from the fireball transmitted by the air) of unity is a fair approximation. It is possible for attenuation to be of the order of 50 percent at a mile from the fireball, in which case the predictive calculations with uncorrected transmissivity err on the side of safety.

We have outlined this approach for LPG, though in fact it was first applied [7] to liquefied natural gas (LNG) in which case the expression linking heat exposure to distance from the center of the fireball, using the adiabatic temperature approximation, is:

$$Q = \frac{7785\, W_{\text{methane}}^{5/6}}{d^2} \quad \text{kJm}^{-2}, \qquad (13.23)$$

where W_{methane} is the weight of methane leaked. However, it is widely known that the fireball risk with LNG is fairly small because of the manner of its storage: as a cryogenic liquid. In the event of container failure it is released as such, and vapor buildup is slower than that resulting from escape of a hydrocarbon stored at ambient temperature under its own vapor in pressurized vessels.

13.2.3 A case study

The case study chosen involved the derailment of a train pulling 12 tankcars containing an average of 63.6 tonnes of LPG, together with many other types of car. We will examine a carefully documented account [8] in the light of the above principles.

Ten of the 12 LPG-bearing cars were derailed. At derailment the LPG-bearing tankcar closest to the locomotives was ruptured by impact from the tankcar behind it and its contents were very rapidly released and ignited to form a fireball before the other derailed cars had even come to rest. This was the only one of the tankcars to be ruptured by the impact itself. This fireball was therefore coincident with actual derailment and there were no further fires until one hour and eight minutes had elapsed. Several of the derailed tankcars had started to leak LPG through their safety release valves, set to open at a pressure of 2 MPa. Ignition of LPG leaking in this way from one particular tankcar led to a long flame with considerable recirculation and this contacted the outside wall of another derailed tankcar and acted as a torch.

The tankcar experiencing the torch effect from the nearby car released LPG copiously through its release valve and it is estimated that 25 tonnes was discharged. At this point the tank wall structure fractured because of the effects of the flame and there was a BLEVE involving the remaining 35 tonnes of LPG in the tankcar. This took place 5 minutes after initiation of the recirculating flame, 1 hour and 13 minutes after derailment. The BLEVE was estimated as 230 m in height and 150 m in lateral dimension. Approximating spherical geometry, from the weight of LPG the diameter is predicted from equation 13.19 as 200 m, so agreement is semiquantitative.

The pattern of events whereby a flame at a release valve affected a nearby tankcar was repeated and there were further hazards to tankcar walls from gas which had leaked without igniting immediately. There were several more BLEVEs, the final one occurring over four hours after derailment. While there was blast damage, this was due to the pressure effects of the fuel release and not to the combustion. Two of the tankcars released LPG only through relatively small orifices created by localized damage, so the release of LPG was too slow for a BLEVE. Burning in these tankcars was therefore as a steady flame and lasted for about two days. There were no fires in the tankcars which had not been derailed.

13.3 Dust explosions

13.3.1 Introduction

A dust is commonly defined as a material which will pass through a 76 μm space in a sieve and a powder is defined as a material with a maximum dimension less than 1 mm. The term *dust explosion* undoubtedly includes powders, according to this distinction, and there is a dependence of explosion behavior on particle size. Coal-dust explosions predate gas or vapor explosions as industrial hazards. Other susceptible materials include plastics, wood, flour, grain and pharmaceuticals. The author has had recent experience of testing the explosibility of a powdered sugar/lemon essence blend which is sold as such and then added by the consumer to hot water as a drink. Dust explosions tend to start in places such as ducts not normally occupied by people, because airborne dust causes severe discomfort at lower levels than those required for an explosion. Where dust has been allowed to accumulate on surfaces it can be a hazard if there is subsequently a fire nearby. The dust can be dispersed by rapid influx of gas, then ignited by the fire, creating a dust explosion as well as the fire itself. This is a well-known possibility in firefighting.

Minimum ignition energies in dust explosions are, in general, substantially higher than those in gas-phase processes and also depend strongly on particle size. Laboratory examination of dust explosions has made use of small pieces of apparatus such as bombs of various kinds. By contrast, a very recent investigation [9] of dust explosions in cornstarch (derived from maize, average particle size 20 μm) and a bituminous coal (average particle size 30 μm) has used a 5.7 m high, 1.5 m internal diameter, cylindrical arrangement which is a clear advance in size over traditional apparatus. We will review this work in Section 13.3.2 as a means of bringing out some of the important features and principles of dust explosions.

13.3.2 Dust explosion experiments with cornstarch and a bituminous coal [9]

Dust initially held in nine equally spaced cups along the vertical axis of a cylindrical vessel was dispersed by admitting pressurized air to the base of each cup. The air so admitted, and the air initially present, gave a total pressure equivalent to atmospheric pressure. This necessitated the use of different initial pressures according to the duration of the air pulse (dispersion time), typically 500 ms. A reproducible quantity of electrical energy to ignite the dust was applied at times up to 1600 ms after the cessation of the air pulse in each experiment. Transient pressure and temperature measurements were respectively by a pressure transducer and multiple rare-metal thermocouple junctions. Fans were mounted in the vessel and were used in some runs to supplement the turbulence created by the dispersing air.

Obviously for a dust explosion to occur there must be sufficient turbulence to prevent the particles from settling under the influence of gravity. Use of dispersion times longer than those necessary to meet this basic requirement appears not markedly to affect either the peak pressure of the explosion or the explosion constant, defined by:

$$K_{st} = V^{1/3}\left[\frac{dP}{dt}\right]_{max}, \tag{13.24}$$

where K_{st} = explosion constant (bar m s^{-1})
V = system volume
P = transient pressure.

The differential is clearly the maximum rate of pressure rise during the explosion, and dust explosion behavior is often expressed in terms of this differential. With cornstarch dust at a concentration of 291 g m^{-3} in the cylindrical reactor at dispersion times in the range 200 to 800 ms, and with a delay of 800 ms between the cessation of the air pulse and the admittance of ignition energy, values of the peak (maximum transient) pressure were $\approx$ 600 kPa, and values of $K_{st} \approx$ 45 bar ms^{-1}. There was in each case a weak positive correlation with dispersion time (and hence with turbulence) at the higher end of the dispersion time range.

With shorter delays between the air pulse and supply of ignition energy and a fixed dispersion time of 500 ms, much higher values of both the peak pressure and K_{st} were obtainable with cornstarch at 291 g m^{-3}. For example, at 200 ms these quantities were respectively 800 kPa and 400 bar m s^{-1}. The value of K_{st} drops very dramatically in the region of delay time 200 to 400 ms, establishing there an almost invariant value of about 40 bar m s^{-1}. The peak pressure also declines very markedly with ignition delay up to about 500 ms. The reason for the effect of ignition delay on the peak pressure and the explosion constant is clearly that short delay times mean retention of the turbulence created by the air pulse, with consequent enhancement of combustion rates.

Effects of dust concentration and fuel:air ratio on cornstarch explosions were investigated using single values of the air-pulse duration and ignition delay of 500 ms and 800 ms respectively. Under these conditions the minimum concentration for a dust explosion was $\approx$ 100 g m^{-3} and peak pressure rises monotonically to approximately 800 kPa at a concentration of 800 g m^{-3}. The value of K_{st} also rises monotonically with concentration to a value of $\approx$ 70 bar m s^{-1}. From elemental analysis of the material it is known that a stoichiometric mixture of the dust and air corresponds to $\approx$ 250 g m^{-3} and it therefore follows from the observations that a highly fuel-rich initial mixture is required for the strongest explosion effect—in maximum pressure or K_{st} terms—of which the cornstarch at specified initial pressure, dispersion time and ignition delay is capable. The rise in maximum pressure and K_{st} with concentration is steepest below the stoichiometric concentration of 250 g m^{-3} and there is a likelihood of close to complete reaction in this region. At higher concentrations there is less than complete reaction. If this were not so there would be a drop in maximum pressure and K_{st} beyond the stoichiometric concentration, which is not observed. The tendency of dust explosions to display incomplete reaction is well known [10], and probably has to do with the thermal inertia of the ensemble of particles.

The situation for bituminous coal dust is rather different. Using a pulse duration of 700 ms (longer than for the cornstarch particles because of the higher density of the coal particles) and an ignition delay of 500 ms (shorter than for the cornstarch because of the apparently more rapid settlement of the coal particles after the pulse) the minimum concentration for explosion is $\approx$ 200 g m^{-3} while the stoichiometric concentration is 102 g cm^{-3}. Hence very fuel-rich conditions are essential and it is only possible for a proportion of the coal dust to be burnt. The graphs of maximum pressure and K_{st} against concentration each display a maximum at about 380 g m^{-3}. The maximum pressure is $\approx$ 800 kPa—a value obtainable with the cornstarch—but the maximum in K_{st} is $\approx$ 170 bar m s^{-1}, much higher than values observed for cornstarch. Moreover, corresponding graphs for the latter do not display a maximum in the sense of a turning value, as already noted. The decline in explosion severity of the coal dust at higher concentrations is due at least in part to the fact that the very high concentrations promote increasing CO formation at the expense of CO_2 formation in the oxidation of the coal carbon (that is, an even greater degree of incompleteness of reaction) with a resulting drop in quantity of heat released.

Two further aspects of this work on dust explosions merit inclusion in our discussion:

(a) When extra turbulence was brought about by the use of fans, conditions otherwise being as before (500 ms pulse, 800 ms ignition delay), pressure histories for cornstarch at a concentration of 291 g m^{-3} showed an increased maximum pressure and an increased rate of pressure rise. Values of the maximum pressure and of K_{st} across a range of dust concentrations were increased by use of the fans; the enhancement was more marked for K_{st} than for the maximum pressure. However, for coal dust, using a 700 ms pulse and a 400 ms delay, there was barely any effect due to the fans. This is probably a reflection of the sensitivity of the explosion to pulse and delay times, as much as an intrinsic effect; the coal experiments use longer pulses and shorter delay times, so at ignition they have retained considerable turbulence without the need for the fans. This is broadly consistent with our treatment of turbulent combustion in Chapter 4 in which we saw that flame speeds are increased by turbulence only up to a point (Figure 4.3).

(b) Flame speeds were estimated from the time between ignition and traverse by the combustion wave of a fast-response thermocouple a known distance from one of the cups initially holding the powder. Values up to 7 m s^{-1} were measured for the cornstarch, depending on the concentration, and values up to 11 m s^{-1} for the coal dust. This is an order of magnitude higher than the velocities of p.f. flames discussed in Chapter 6. The difference is due to the fact that in dust explosions the wave front is far from planar and also that the turbulence—even that required just to keep the particles airborne immediately before ignition—enhances the propagation rate. When these two factors were semi-quantitatively dealt with in the original coverage [9], velocities of 0.3 m s^{-1} were estimated for the coal dust, in fair agreement with measured values on p.f. (see Section 6.5).

13.3.3 Dust explosion case histories

One of the more recent serious dust explosions to have been the subject of discussion in the scientific literature [10] occurred during the cleaning of a hopper coated with a 1–2 cm deposit of powder of a particular polymer. It was noted in the introductory part of Section 13.3.1 that such hoppers are likely scenes for a dust explosion. An employee wearing protective clothing was standing at the entrance to the hopper cleaning it, a two-stage operation. First the wall of the hopper was knocked sharply to loosen as much of the polymer powder as possible, and powder so removed was collected. Water from a hose was then directed onto the remainder of the powder adhering to the wall and this appears to have generated a cloud which was above the minimum concentration for a dust explosion. A portable light, connected to the mains supply via an extension lead, is believed to have malfunctioned at this point and provided energy for ignition. The employee died from his injuries.

The combination of the dust cloud generated by the action of the water spray and the accidental input of energy by the portable light proved fatal. However, it is far preferable for such procedures always to avoid the buildup of a dust cloud at a concentration capable of explosion. Although minimum ignition energies in dust explosions tend to be higher than those in gas-phase explosions, they are still frequently only a fraction of 1 joule and therefore capable of accidental supply.

Table 13.2 sets out summaries of a number of other industrial dust explosions.

Table 13.2 Case histories of dust explosions

Details of the dust	*Details of the accident*	Ref
Flour (Italy, 1785)	The first industrial dust explosion recognized and recorded as such	12
Powdered resin, 19 tonnes packed in bags and transported by a motor lorry (Scotland, UK)	Collision of the motor lorry with shedding of the load and splitting open of many of the bags, generating a dust cloud. Ignition by an electrical cable also damaged by the collision of the lorry. Twenty-seven people hospitalized	13
Grain dust at a facility for despatching grain for export (Galveston, Texas)	Grain received at the facility by rail and loaded onto ships for export. Creation of dust during removal of grain from railway cars and ignition by a spark. Eighteen people killed and 22 injured; major damage to the facility. At the time of its occurrence in 1977 this accident was the second fatal dust explosion in a USA grain-handling center within a week; the other (in Louisiana) killed 36 people.	11
Polyethylene powder in a silo	Ignition source believed to have been static electricity generated as the powder slid into the silo	14
Flour (W. Germany, 1979)	Fourteen people killed, 17 injured, extensive building damage	15

References

[1] Strehlow R. A., 'Unconfined vapor cloud explosions—an overview', *Fourteenth Symposium (International) on Combustion*, 1189, Pittsburgh: The Combustion Institute (1973)

[2] Wiekema B. J., 'Vapour cloud explosions—an analysis based on accidents, Part I', *Journal of Hazardous Materials* **8** 295 (1984)

[3] Wiekema B. J., 'Vapour cloud explosions—an analysis based on accidents, Part II', *Journal of Hazardous Materials* **8** 313 (1984)
[4] Work cited by Wiekema [3] and discussed independently in: Baker W. E., Cox P. A., Westine P. S., Kulesz J. J., Strehlow R. A., *Explosion Hazards and Evaluation*, New York: Elsevier (1983)
[5] Van den Berg A. C., 'The multi-energy method. A framework for vapour cloud explosion blast prediction', *Journal of Hazardous Materials* **12** 1 (1985)
[6] Williamson B. R., Mann L. R. B., 'Thermal hazards from propane (LPG) fireballs', *Combustion Science and Technology* **25** 141–145 (1981)
[7] Hardee H. C., Lee D. O., Benedik W. B., 'Thermal hazard from LPG fireballs', *Combustion Science and Technology* **17** 189 (1978)
[8] Lewis D., 'Crescent City, Illinois: 21 June 1970', *I.Chem.E. Loss Prevention Bulletin* **101** 22 (1991)
[9] Kumar R. K., Bowles E. M., Mintz K. J., 'Large-scale dust explosion experiments to determine the effect of scaling on explosion parameters', *Combustion and Flame* **89** 320 (1992)
[10] Anon. 'Dust explosion during cleaning operation on polymer hopper—case history', *I.Chem.E. Loss Prevention Bulletin* **95** 1 (1990)
[11] Field P., *Dust Explosions*, Amsterdam: Elsevier (1982)
[12] Palmer K. N., *Dust Explosion and Fires*, London: Chapman and Hall (1973)
[13] Anon., 'Unusual dust explosion', *I.Chem.E. Loss Prevention Bulletin* **51** 9 (1983)
[14] Anon., 'Static electricity and dust explosions', *I.Chem.E. Loss Prevention Bulletin* **36** 1 (1980)
[15] Bartknecht W., (Trans. Bruderer R. E. et al.), *Dust Explosions, Course [sic], Prevention, Protection*, New York: Springer Verlag (1989)

CHAPTER 14

Propellants, Explosives and Pyrotechnics

Abstract

Propellants utilizing ammonium perchlorate are described and the pressure dependence of the rate of burning discussed quantitatively. The mechanism of combustion of composite propellants using ammonium perchlorate as oxidant is treated by means of a suitable example. Liquid propellants and hypergolic propellants are also discussed, giving examples. High explosives are discussed, giving examples, and dealing with detonation velocities and impact initiation. The burning of certain gasless *pyrotechnics is discussed in detail.*

14.1 Introduction

We live in a world bathed in a gas mixture containing 21 percent of oxygen so oxygen is the inevitable choice of oxidant in many combustion applications. However there are combustion processes where it is not possible to use atmospheric oxygen, rocket propulsion being an obvious example. Ways in which combustion can be sustained without atmospheric air are:

(a) use of a fuel which releases heat by decomposition, not combustion. Also in this category are fuels that have sufficient intramolecular oxygen for this, on decomposition, to be the oxidant, in which case fuel and oxidant are actually in the same molecule. Some explosives work this way;
(b) use of an oxidant, other than atmospheric oxygen, held separately from the fuel until required for combustion;
(c) mixing of solid fuel and solid oxidant to form a composite which is stable until ignited, when it displays sustained burning. Certain propellants work this way and the oxidant is often ammonium perchlorate.

We will examine some examples of combustion without atmospheric oxygen, starting with propellants using ammonium perchlorate.

14.2 Ammonium perchlorate

14.2.1 Background

Classical work [1] on the development of propellants using this compound as oxidant with various organic materials as fuel is outlined. It must first be pointed out that the oxidant itself without any fuel decomposes with quite sufficient heat release to sustain a flame, largely by the reaction:

$$2NH_4ClO_4 \rightarrow Cl_2 + \frac{3}{2}O_2 + 4H_2O + N_2O,$$

which is exothermic by 1.3 MJ kg^{-1} of perchlorate. Therefore in discussing propellants using ammonium perchlorate there is the possibility of comparison of the performance of a particular fuel: oxidant blend with that of the oxidant alone. This is shown in Figure 14.1 which gives the rate of burning, expressed as measured surface regression rate, of strands of propellant made from finely milled ammonium perchlorate oxidant with various amounts of polystyrene fuel. Combustion was under 6.9 MPa (almost 70 atm) pressure of nitrogen in a bomb. The flame speed is seen to be ≈ 7 mm s^{-1} with oxidant only, rising to

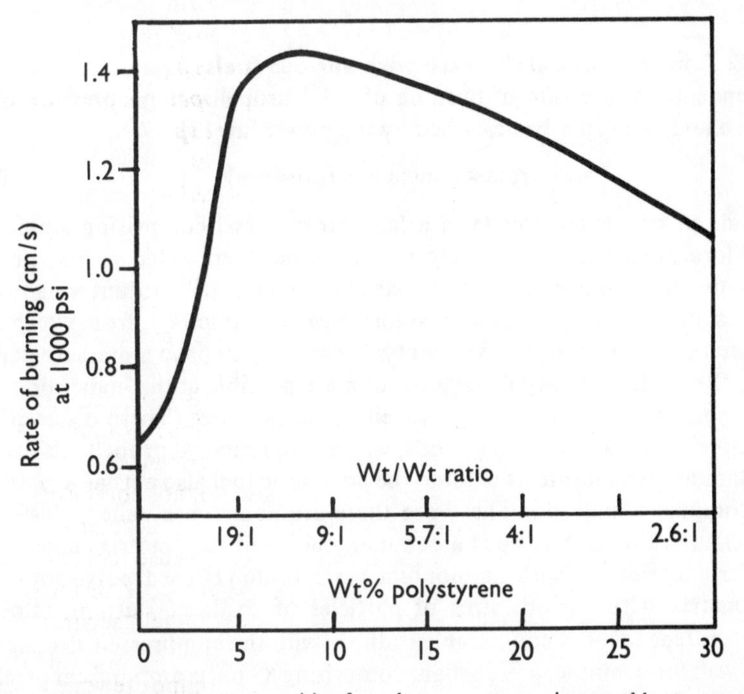

Figure 14.1 Flame speed against composition for polystyrene–ammonium perchlorate propellants (Adams, Newman and Robins [1])
(Reproduced with the permission of the Combustion Institute.)

about twice that at the maximum of the curve. On the basis of oxygen release by the perchlorate according to the above chemical equation and the oxygen requirement of polystyrene combustion it is easily shown that the maximum in Figure 14.1 corresponds approximately to the stoichiometric proportions of the fuel and oxidant.

The pressure of ≈ 70 atm nitrogen was the lowest with which a flame due to ammonium perchlorate decomposition could be sustained and pressures much greater than atmospheric were found to be necessary for combustion of the ammonium perchlorate/polystyrene mixtures in the bomb. The enhancement of burning rates by increased pressure is at least partly attributable to mechanical effects: the creation of cracks in the propellant structure which the flame can penetrate, and the enlargement of existing cracks. Also the precise stoichiometry of decomposition of the perchlorate is sensitive to pressure and so, therefore, is the heat release.

Previous investigators [2] had observed an *upper* pressure limit for the ammonium perchlorate decomposition flame around 30 MPa. One explanation relates to the possibility that intermediate decomposition products diffuse to active sites on the solid where the remainder of the reaction (and heat release) occurs. This diffusion will be restricted by higher pressures. Also, convective heat losses from the propellant surface will be increased by high pressures.

14.2.2 Ammonium perchlorate with various fuels [1]

Dependence of the rate of burning of solid propellants on pressure is well known and can often be described by the power law [3]:

$$\text{rate} = \text{constant} \times (\text{pressure})^n. \tag{14.1}$$

Strands of propellant, made in a laboratory press, comprising ammonium perchlorate oxidant with a variety of organic fuels were tested across a pressure range under various conditions of particle size and fuel:oxidant weight ratio. The results of many such tests are summarized in Table 14.1, from which it can be seen that in general the exponent n is expected to have a pressure dependence, although nil or negative values of n are possible at the higher pressures. The value of n can also be influenced by particle size. These discoveries of course relate to compressed strands, whereas in practical propellants utilizing ammonium perchlorate as oxidant the polymeric fuel also acts as a matrix for the composite material. These are therefore sometimes called plastic-type propellants. Polyisobutene is a common choice of fuel/matrix material and further experiments with ammonium perchlorate [1] used polyisobutene as fuel/matrix with various sizes of particles of oxidant. Burning rates and pressure dependences comparable to those with the compressed strands were observed; for example, a propellant comprising 89 parts ammonium perchlorate (coarse particles), 10 parts polyisobutene and 1 part wetting agent (which assists the fuel/matrix component in binding the composite material), at 6.9 MPa of nitrogen, gave a burning rate of 1.88 cm s^{-1} with $n = 0.67$.

Table 14.1. Summary of strand-burning tests of some solid propellants

Specifications	*Burning rates* (cm s^{-1}) (exponent n)	*Comments*
Ammonium perchlorate only Average particle sizes 90, 70 and 58 μm N_2 at 6.9 to 20.7 MPa	0.6 to 1.3 (≈ 0.5)	Particle size effects insignificant
Ammonium perchlorate and polystyrene		Results shown in Figure 14.1
(a) Fuel < 74 μm Oxidant 90 μm N_2 at 6.9 MPa Up to 28% by weight of polystyrene	Up to 1.4	
(b) Fuel < 74 μm Oxidant 138, 90, 70, 58 and < 26 μm N_2 at 1.4 to 20.7 MPa Stoichiometric proportions of fuel and oxidant	0.6 to 2.9 (0.5 to 0.6 up to 5.9 MPa, lower at higher pressures)	Smaller oxidant particle sizes lead to faster burning at high pressures. Decline in the value of n at high pressures more marked for larger particle sizes
Ammonium perchlorate and paraformaldehyde		Similar particle size effects to ammonium perchlorate/polystyrene
Fuel < 74 μm Oxidant 90 μm N_2 at 1.4 to 20.7 MPa Fuel 12.5, 24.2 and 33.3% by weight	< 0.5 to 1.5 (0.5 up to 9 MPa, zero at higher pressures)	Nil pressure dependence of rate above 9 MPa lost if carbon black partially substituted for paraformaldehyde

(Continued p. 274)

Table 14.1 (Continued)

Ammonium perchlorate and carbon black Oxidant 90 μm N_2 at 1.4 to 7.7 MPa Stoichiometric proportions of fuel and oxidant	2.5 at the highest pressure used and rising steeply (0.5 up to 5 MPa and larger at higher pressures)	Very reactive
Ammonium perchlorate and cellulose acetate Oxidant 138, 90 and < 26 μm Fuel 138, 90 < 26 μm N_2 at 1.4 to 20.7 MPa Fuel 17, 20 and 29%	Up to 1.1 (rate/pressure plot with a plateau at intermediate pressures)	Smaller oxidant particle sizes leading to faster burning, especially for < 26 μm particles
Ammonium perchlorate and sucrose octa-acetate Oxidant 90 μm Fuel < 74 μm N_2 at 1.4 to 20.7 MPa Fuel 18, 20, 25 and 29%	Less than 1, with a rate decreasing with pressure above 7 MPa	A rate lower than that of the oxidant only at higher pressures

14.2.3 Combustion mechanism of composite propellants using ammonium perchlorate [4]

Propellant combustion with ammonium perchlorate as oxidant has been outlined in exploratory work several decades ago but the compound continues to be important in composite propellants. It remains to discuss further the mechanism of composite propellants and factors influencing their performance. Processes occurring in combustion of composite propellants are ammonium perchlorate decomposition, pyrolysis of the fuel/matrix and combustion of the fuel pyrolysis products with the oxidizing gases from the perchlorate decomposition. While these processes are obviously interdependent, in view of the heterogeneity of composite propellants it is possible to discuss the combustion in terms of different oxidant and fuel surface temperatures.

The perchlorate decomposes exothermically at the surface of the propellant in a liquid layer. This decomposition is not complete and the remainder of the ammonium perchlorate sublimes to NH_3 and $HClO_4$. NH_3 is a fuel gas (see

Chapter 5) and $HClO_4$ is a good oxidant for fuel gases, so the sublimation products react together in a flame close to the surface. Because the fuel and oxidant are sublimation products of the same solid, the flame is a premixed one. The surface temperature of the perchlorate bears a direct relationship to the regression rate. Fuel/matrix pyrolysis proceeds via a liquid layer at the surface whose thickness d is given by:

$$d \approx \frac{\alpha}{r_f} \quad \text{m}, \tag{14.2}$$

where α = thermal diffusivity ($m^2\ s^{-1}$)
r_f = regression rate ($m\ s^{-1}$).

The mass-loss rate from the fuel/matrix by pyrolysis can be expressed as:

$$\text{mass lost per square metre per second} = \sigma_f r_f, \tag{14.3}$$

where σ_f is the fuel/matrix density ($kg\ m^{-3}$).

For many polymer materials used in composite propellants with ammonium perchlorate, the regression rate is found to depend on fuel/matrix surface temperature T_{fs} according to the equation:

$$r_f = \left[A \exp\left[-\frac{E_f}{RT_{fs}} \right] \frac{\alpha}{(E_f/RT_{fs})} \right]^{0.5}, \tag{14.4}$$

where A = pre-exponential factor (s^{-1})
E_f = activation energy ($J\ mol^{-1}$).

Heat for the pyrolysis is provided by the decomposition of the perchlorate and the premixed $NH_3/HClO_4$ flame and hydrocarbons formed from the pyrolysis react with gas from the perchlorate decomposition to form a primary flame. Some of the heat from this feeds back to promote further perchlorate decomposition. The height of the primary flame depends on diffusion of the fuel/matrix pyrolysis products and the perchlorate decomposition products towards each other.

The view that propellant combustion can be described with separate, single values of the temperature of the fuel/matrix and oxidant particles can be examined by means of equations such as (14.4) together with supplementary information. Regression rates of fuel/matrix polymer can be determined experimentally as a function of temperature, therefore giving A and E_f, from which T_{fs} can be deduced using the measured regression rate. The same process can be followed for ammonium perchlorate only, giving its surface temperature $T_{s,ap}$ at any particular regression rate. Imagine an extremely fine temperature measurement probe applied to the surface of the propellant which would read either T_{fs} or $T_{s,ap}$ depending on whether it was contacting a fuel/matrix particle or an oxidant particle. If a very large number n of such readings were taken, the average temperature would be:

$$T_{av} = \frac{snT_{s,ap} + (1-s)nT_{fs}}{n}$$
$$= sT_{s,ap} + (1-s)T_{fs}, \qquad (14.5)$$

where s is the fraction of the surface area accounted for by the perchlorate, obtainable from the weight fractions, densities and particle sizes. This is the average temperature to be expected if the idea of separate oxidant and fuel/matrix temperatures is correct.

For a composite propellant using PMMA as the fuel/matrix, the average temperature T_{av}, estimated from equation 14.5, with T_{fs} and $T_{s,ap}$ calculated as indicated was compared with that measured by pyrometry. The pressure does not appear explicitly in equation 14.5 but different values of the surface temperature at different pressures are expected from the pressure dependence of the regression rate. The results of this comparison are summarized in Table 14.2, and agreement is close.

Table 14.2 Measured and calculated T_{av} for ammonium perchlorate/PMMA propellant [4]

Propellant composition	P (atm)	*Regression rate* (cm s^{-1})	$T_{av}(K)$ *measured*	$T_{av}(K)$ Eq.14.5
Ammonium perchlorate 85%	10	0.56	980 ± 20	940
PMMA 15%	40	0.9	1030 ± 20	980

14.2.4 Comparison of ammonium perchlorate with other solid oxidants

Salts such as potassium perchlorate and sodium nitrate can also be used as oxidants in composite propellants. In certain applications the fact that, unlike ammonium perchlorate, they leave a metallic residue on decomposition, may be a disadvantage. Relative performances of NH_4ClO_4, $KClO_4$, $NaNO_3$ and KNO_3 *in use with one particular fuel/matrix material* can be understood from strand-burning experiments [5] similar to those used in the early work on ammonium perchlorate propellants [1]. Oxidants were present in the strands before ignition at 75 percent by weight (with the exception of $NaNO_3$ which was 73.5 percent) and the balance was phenol–formaldehyde fuel/matrix + 0.2% of carbon black. Burning rates at various pressures are given in Figure 14.2 and show one of the main features noted in the earlier work [1] on ammonium perchlorate propellants: a steeper pressure dependence at lower pressures.

Reactants of solid propellants show a considerable spread of theoretical flame temperatures, unlike reactions of hydrocarbon gases in air, which, as we saw in Chapter 5, almost always have a flame temperature under stoichiometric conditions of about 2000° C. For example, a composite with ammonium perchlorate as oxidant and hydroxyl terminated polybutadiene (HTPB) as fuel/binder is capable of flame temperatures of 3000K, the actual value depending of course on the proportions of fuel and oxidant. However, the same fuel with potassium nitrate oxidant is incapable of attaining flame temperatures above 2100K. It follows from the mechanism of propellant combustion outlined above that the flame temperature needs to be distinguished from the surface temperature, examples of which are given in Table 14.2.

14.3 The performance of a liquid propellant in a strand burner [6]

The strand-burner type of experiments described in our discussion of ammonium perchlorate-based propellants can be used for liquid propellants

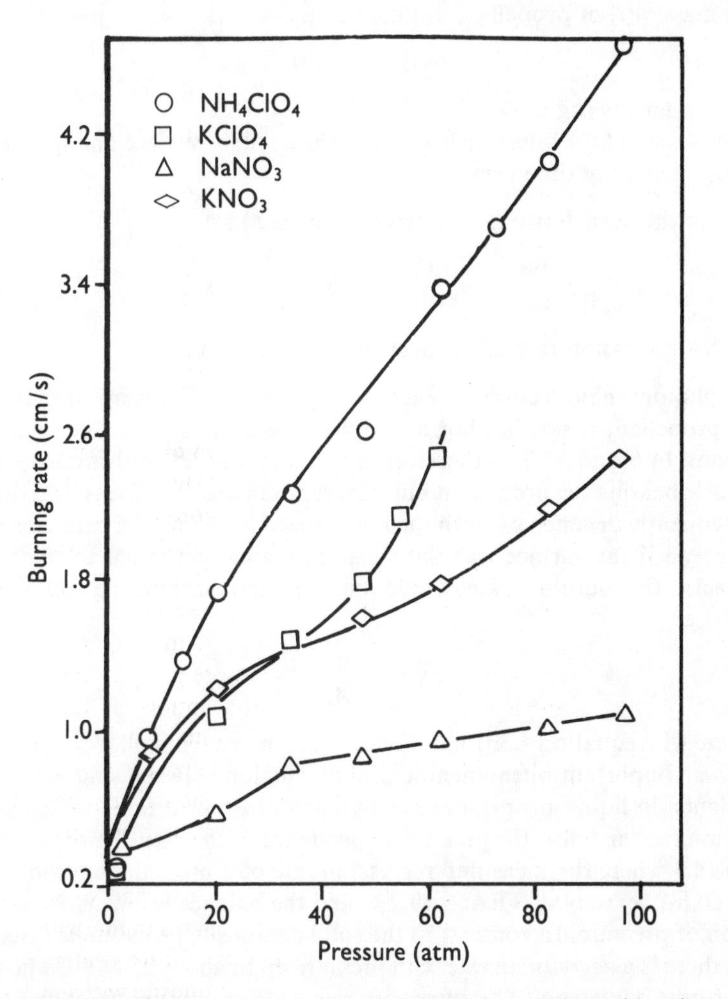

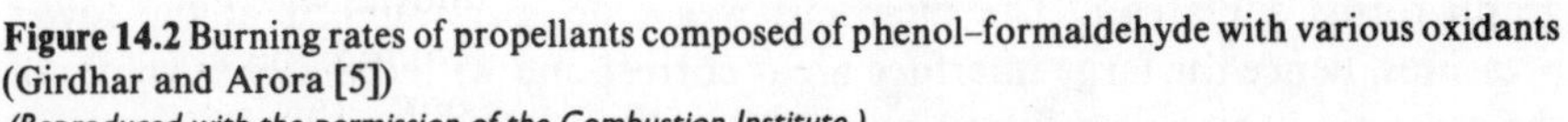
Figure 14.2 Burning rates of propellants composed of phenol–formaldehyde with various oxidants (Girdhar and Arora [5])

(Reproduced with the permission of the Combustion Institute.)

provided that a 'strand' of the propellant is held in a suitable receptacle which does not melt as the propellant combustion takes its course. This technique was applied in recent work [6] on propellants comprising hydroxylammonium nitrate (NH_3OHNO_3, HAN) and triethanolammonium nitrate [$(C_2H_4OH)_3NHNO_3$, TEAN]. In outline, the chemistry of this propellant involves (exothermic) decomposition of HAN to an oxidizing mixture of products which react with the TEAN to form a flame.

Charges of this propellant 4 cm in initial height held in a receptacle of rectangular cross-section 1.8 mm × 1 mm were burnt in a vessel pressurized with argon, ignition being by electrodes. The regression rate was followed by photography. The following analysis applies:

The mass $m(t)$ of propellant at time t is:

$$m(t) = \sigma A_p h(t), \tag{14.6}$$

where σ = density (kg m^{-3})
A_p = area of the interface between the propellant surface and the gas (m^2)
$h(t)$ = height at time t (m).

Therefore the mass-burning rate can be expressed as:

$$\frac{dm}{dt} = \sigma A_p \frac{dh}{dt} = \sigma A_p S \qquad \text{kg s}^{-1}, \tag{14.7}$$

where S = regression rate of the propellant surface (m s^{-1}).

The photographic records reveal that in general the burning surface of the liquid propellant is not flat but has a meniscus and ripples. The area A_p in equations 14.6 and 14.7 is therefore not the cross-sectional area A_r of the receptacle bearing the propellant but is larger than this. The measured value of S consequently depends on both the flame speed itself and the behavior of the surface and if the surface was flat so that its area was equal to that of the receptacle, the burning rate would be the laminar burning velocity S_L. Otherwise:

$$S = S_L \frac{A_p}{A_r}, \tag{14.8}$$

(compare with equation 4.30) and A_p can be estimated from photographs. This leads to an important phenomenological distinction between solid and liquid propellants: in liquid propellants the dynamics of the surface influences the regression rate and also the pressure dependence of the rate. This is shown in Figure 14.3 where the measured regression rate of a propellant composed of HAN, 60.8% by weight; TEAN, 19.2%; and the balance water, is shown as a function of pressure. In contrast to the solid composite propellants discussed above, there is a *decrease* in rate with pressure up to about 25 MPa, where the graph forms a plateau. The meniscus was more pronounced at the lower pressures, hence the large interface areas correspond with the higher values of the regression rate as predicted by equation 14.8.

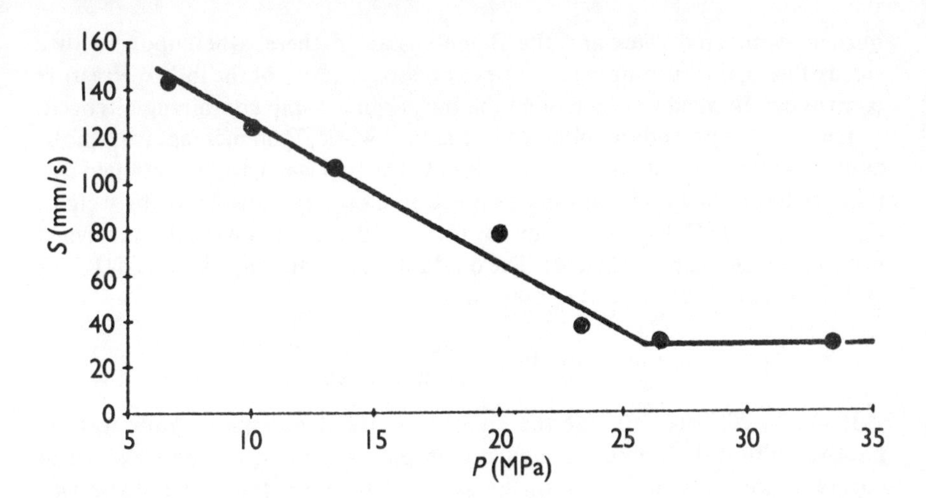

Figure 14.3 Burning rates of a liquid HAN/TEAN propellant as a function of pressure (Vosen [6]) *(Reproduced with the permission of the Combustion Institute.)*

The actual regression of the liquid surface was due to HAN decomposition. The luminous reaction between the decomposition products and TEAN was at the top of the receptacle as a stationary flame. This is because of the narrow diameter of the receptacle; the luminous part of the propellant reaction will propagate along a wider one. These facts provide a basis for discussion of flame structure. The burning rate at any particular pressure of argon is steady as the liquid surface recedes farther from the luminous flame, hence there appears to be negligible thermal feedback between the TEAN oxidation and the HAN decomposition and the two processes under these conditions are decoupled.

With a receptacle of cross-section 5 mm × 5 mm, S values some 70 percent larger than those obtained with the smaller receptacle were obtained and this is thought to be due to there being scope for larger differences between A_p and A_r with a large liquid surface. Indeed, this is supported by direct observation, the larger surface having ripples with higher amplitudes.

14.4 Hypergolic reactions

14.4.1 Definition

Our discussion of propellants so far—liquid and solid—has been of examples where the fuel and oxidant are mixed and ignited by a heat source. However, in some propellants the fuel and oxidant have to be brought into contact only as and when required, since they react on contact without applied heat. This is termed hypergolic behavior.

14.4.2 The hypergolic behavior of thiols with nitric acid

Clearly the strand-burning technique for studying propellant performance in either of the forms mentioned so far will not avail for hypergolic propellants. A suitable adaptation of the method is for the fuel and oxidant to be admitted to a burner at metered rates and the flame sustained there, whereupon, with a steady flame, the burning rate can be expressed in terms of the influx. Pressure control can be achieved by having the burner in a bomb containing nitrogen.

This was the procedure followed in classical work [7] on mercaptans (thiols) as hypergolic fuels with nitric acid oxidant. The fuel was actually obtained as a refinery byproduct and had the composition C_3 mercaptans 27.8% by weight, C_4 mercaptans 65.3%, C_5 mercaptans 6.6%, and C_6 mercaptans 0.3%, termed mixed butyl mercaptan (MBM). The oxidant was red fuming nitric acid (RFN) and the rate of burning was expressed as:

$$\text{specific consumption rate} = \frac{\text{mass flow rate to the burner}}{\text{volume of the burning zone}}. \quad (14.9)$$

It was possible to estimate the quantity in the denominator quite well by photographing the flame through a transparent port in the bomb since at constant flow rate and constant pressure of nitrogen the flame shape was cylindrical. Figure 14.4 shows the specific consumption rate across a wide range of nitrogen pressures in the bomb. The dashed part of the line at lower pressures is subject to greater uncertainty in the measurements than the portions of the graph above 200 psia. A pressure exponent of 0.33 applies between 200 and 430 psia, whereas at higher pressures the exponent is 3.0. Between 200 and 430 psia the flame was described as being blue-white and denser and more turbulent than a Bunsen burner flame. Soot was deposited on the inside surface of the bomb above the flame. There was poorer flame stability in the higher pressure regime, pulsating flames sometimes being observed.

14.4.3 Miscellaneous liquid propellants

Having discussed a small selection of propellants in detail, it is necessary to draw attention briefly to other widely used examples and no coverage of propellant combustion would be at all representative that did not mention hydrazine, N_2H_4. This can operate either by reason of its decomposition flame or as a fuel in a propellant also incorporating a suitable oxidant. In its decomposition reaction, hydrazine is catalyzed by certain metals, for example, chromium, iron and copper. While this is disadvantageous in that the inside wall of a storage tank might be able to act as a catalyst unless treated with a suitable catalyst poison, it is beneficial in that catalysts can be used to promote decomposition at low temperatures. In a bed of catalyst in which iridium is the active metal, hydrazine liquid only just above its freezing point of 2°C will decompose sufficiently rapidly for a flame to occur [8].

Hydrazine decomposition involves first:

$$N_2H_4 \rightarrow NH_3 + \frac{1}{2} N_2 + \frac{1}{2} H_2,$$

followed by ammonia decomposition into its elements. However, this is rarely complete in propellant applications and typical stoichiometry would be:

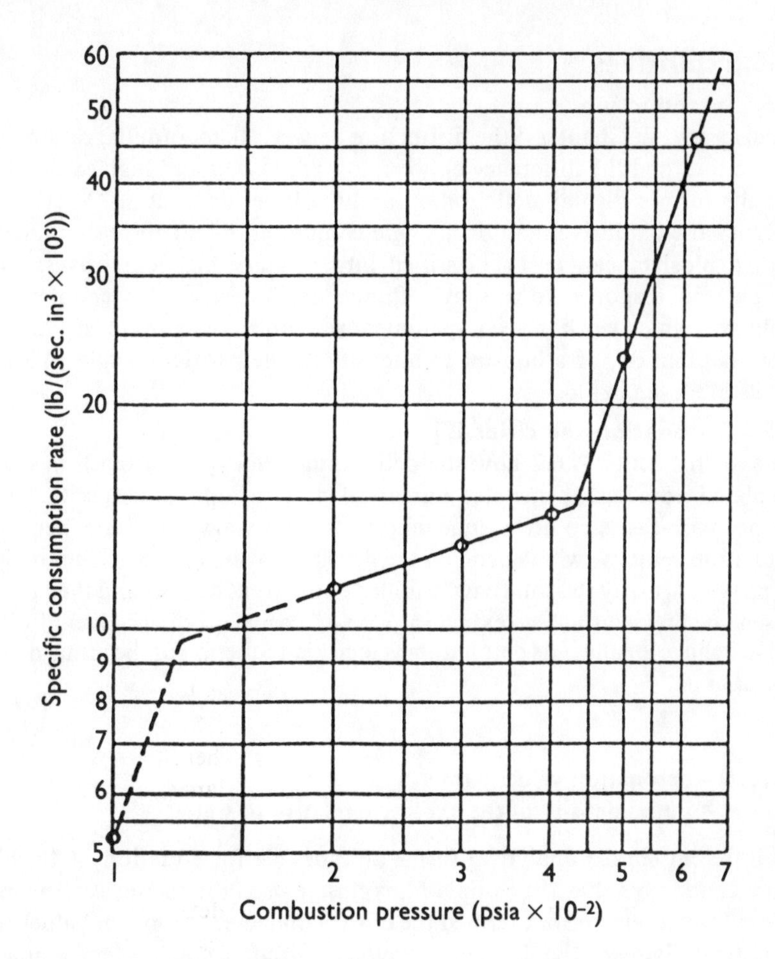

Figure 14.4 Specific consumption rate for MBM/RFN with oxidant:fuel ratio 3.5 at various pressures of nitrogen (McCullough and Jenkins [7])
(Reproduced with the permission of the Combustion Institute.)

$$N_2H_4 \rightarrow \frac{4}{5}NH_3 + \frac{3}{5}N_2 + \frac{4}{5}H_2.$$

The process as written has a calculated adiabatic flame temperature of just below 2000°C. When so-called hydrazine is used as a fuel in the sense that it is oxidized, in fact it often consists of hydrazine itself mixed with methyl derivatives. Propellants comprising hydrazine as fuel and dinitrogen tetroxide N_2O_4 or hydrogen peroxide H_2O_2 as oxidant are hypergolic.

Hydrogen peroxide is used as an oxidant with other propellant fuels including kerosene and alcohols, although it is capable of acting as a propellant simply by catalytic decomposition as follows:

$$2H_2O_2 \rightarrow 2H_2O + O_2.$$

14.5 Explosives

14.5.1 Introduction

We discussed in Chapter 4 the distinction between flames (deflagrations) and detonations and the difference between low explosives and high explosives is that the former display deflagration and the latter detonation. Within each category the explosive may be a single compound or a composite material. High explosives can be subclassified into primary high explosives, which detonate in response to a spark, flame, or impact, and secondary high explosives which work in accompaniment with an auxiliary explosive to bring about detonation, or a booster to intensify the detonation. Table 14.3 gives details of eleven explosives.

14.5.2 Detonation velocities [9]

We saw in Section 4.6.2 how the detonation velocity of a reaction can be calculated from basic principles, numerical input including the specific heat of the product gases. With a solid explosive, the final state and hence the detonation velocity, will depend on the density to which the active ingredients are packed. Clearly the solid can be milled to different extents and then packed loosely, or pressed to some extent into the explosive product. An explosive is said to behave ideally if its detonation velocity is a function of the initial density only, that is:

$$D = D(\sigma_o), \tag{14.10}$$

where D = detonation velocity (m s^{-1})
σ_o = initial density of the explosive relative to water,

and different simple analytical forms are possible for equation 14.10. Ideal behavior requires that the sample of explosive used for the measurement be greater than a certain size, termed the critical diameter. At higher values than the critical diameter the detonation velocity is independent of the diameter.

Many solid explosives, particularly those containing only C, H, N and O as elements, obey a linear form of equation 14.10 at densities above ≈ 0.5, that is:

$$D = D_r + M(\sigma_o - \sigma_{or}), \tag{14.11}$$

where D = velocity with initial density σ_o (m s^{-1})
D_r = velocity with initial density a reference value σ_{or} (m s^{-1}).

The quantity σ_{or} is often taken to be 1.0 and with this choice of σ_{or}, experimentally deduced values of D_r include 5010 m s^{-1} for TNT and 5255 m s^{-1} for picric acid. Respective values of M are 3225 m s^{-1} and 3045 m s^{-1}. For mercury fulminate the situation is slightly different; its detonation velocity at a reference density value of 4 is 5050 m s^{-1} with a value of M of 890 m s^{-1}.

14.5.3 The correlation between detonation velocity and composition for organic explosives

The balance of fuel/oxidant in an explosive has obvious bearing on explosive performance and in single-compound organic explosives relying on intra-

Table 14.3 Compositions and specifications of selected explosives

Explosive	*Composition*	*Comments*
Trinitrotoluene (TNT)	$C_7H_5N_3O_6$ Secondary, high explosive May be combined with ammonium nitrate to form Amatol or with sodium nitrate to form Sodatol. May also be combined with RDX or aluminum (see below)	Partial oxidation of the carbon by intramolecular oxygen (see Section 14.1)
Mercury fulminate	CNO | Hg | CNO Primary, high explosive	Other metal fulminates (e.g. those of Cd, Na) explosive
Tetryl	$C_7H_5O_8N_5$ Secondary high explosive	
Picric acid	$C_6H_3N_3O_7$ Secondary, high explosive	
Nitroglycerine (NG)	$C_3H_5N_3O_9$ Secondary, high explosive or a low explosive, depending on conditions. A liquid, and difficult to use as such	Can be gelatinized with nitrocellulose. An ingredient of dynamites and gelignite (see below)
Cyclotrimethylene trinitramine (RDX)	$C_3H_6N_6O_6$ Secondary, high explosive	RDX/TNT both present in some explosives, e.g. Cyclotol, composition B

(Continued p. 284)

Table 14.3 (Continued)

Dynamites	Contain NG + combustible absorption material, e.g. wood pulp; KNO_3 or $NaNO_3$; $CaCO_3$ or $MgCO_3$ as nonexplosive ingredients	Grade depends on percentage NG, e.g. 40%
Gelignite	Nitroglycerine, chemically treated cotton, ground wood waste, KNO_3	
Gunpowder	KNO_3, S, charcoal low explosive	
Aluminized explosives	For example: RDX 45% TNT 30% Al 25% Al also used with NH_4NO_3	Al oxidized and is an additional fuel
Hexanitrostilbene (HNS)	$C_{14}H_6N_6O_{12}$ Secondary, high explosive	See Section 14.5.4

molecular oxygen there is a straightforward way of estimating the detonation velocity from knowledge of the oxidizing and reducing valencies within the structure [10].

With carbon +4, oxygen −2 and hydrogen +1, for a particular explosive the parameter G_x can be calculated as follows:

$$G_x = \frac{Q - P}{QP} \times \text{m.w.}, \qquad (14.12)$$

where $Q = \Sigma$ reducing valencies
$P = \Sigma$ oxidizing valencies
m.w. = molecular weight of the explosive (g mol^{-1}).

The corrections for functional groups are: OH, −1; OCH_3, −1; COOH, −2; NH_2, +1.5; N-N, +2. With G_x so calculated, the experimental detonation velocity results for aliphatic explosives lie on the line:

$$D(\text{mm } \mu s^{-1}) = 8.6610 - 0.1917 G_x, \qquad (14.13)$$

and those for aromatic explosives on the line:

$$D(\text{mm } \mu\text{s}^{-1}) = 9.1539 - 0.1976G_x, \qquad (14.14)$$

where the values of D are for the maximum theoretical density of the explosive sample which is the actual crystal density of the compound. These results are based on experimental data for 14 aliphatic and 22 aromatic explosives. There was no significant scatter of the points about either line. For a family of explosives with the same oxygen-containing part, for example, nitro, a linear correlation between D and G_x would be expected without corrections to G_x.

14.5.4 Impact initiation of solid explosives

The traditional way of examining and classifying explosives for response to mechanical impact is by the H_{50} index. Samples of the explosive of 20–50 mg are subjected to a falling weight and the height H(m) from which the weight will ignite 50 percent of sample of the explosive in successive 'go/no-go tests' is the H_{50} value. TNT alone on a sandpaper surface with a 2.5 kg weight has an H_{50} value of 1.48 m [11]. In understanding mechanisms of impact initiation the hot-spot concept discussed in Chapter 3 is relevant, such hot spots in the case of impact initiation having a mechanical origin by viscous heating, friction or shear.

In investigating plastic yield and resulting flow of explosives on impact, high-speed photography techniques (one frame per 7 μs) were used to follow the response to impact of 25 mg samples of hexanitrostilbene (HNS) [12]. A 5 kg weight was used from a height of 1 m; the H_{50} of this explosive is ≈ 0.5 m, therefore the tests are expected to lead to 'go' results. Further experiments made use of a strain gauge, and oscilloscope traces of the resulting signal enabled the pressure to be followed.

The photographic records revealed that after approximately 400 μs the layer of explosive had become plastic and was flowing. The diameter of the layer then increased dramatically, indicating very rapid flow once the plastic phase had begun. The features of the photographs about 10 μs later can be interpreted in terms of propagation of gaseous reaction products into the sample residue. The pressure–time trace is shown in Figure 14.5 and initially there is a sharp rise due to the impact, followed by a drop as the explosive becomes plastic. (Experimental conditions were not identical in the photographic work and the strain-gauge work so the times of onset of plasticity are not expected to be exactly the same.) After this drop there is a sharp rise due to reaction, the peak being of the order of 1.4 GPa (≈ 14 000 atm).

Both the photographic and the pressure records therefore indicate a close connection between the loss of solid form of the explosive layer due to the impact, that is, the plastic yield of the explosive, and the initiation of a reaction whose gaseous products cause a sharp pressure rise. When a polymer was added to the HNS there were two drops in the initial part of the pressure curve, one signifying plastic yield of the HNS and the other the same effect for the polymer. When other explosives were tested, many, including RDX, had

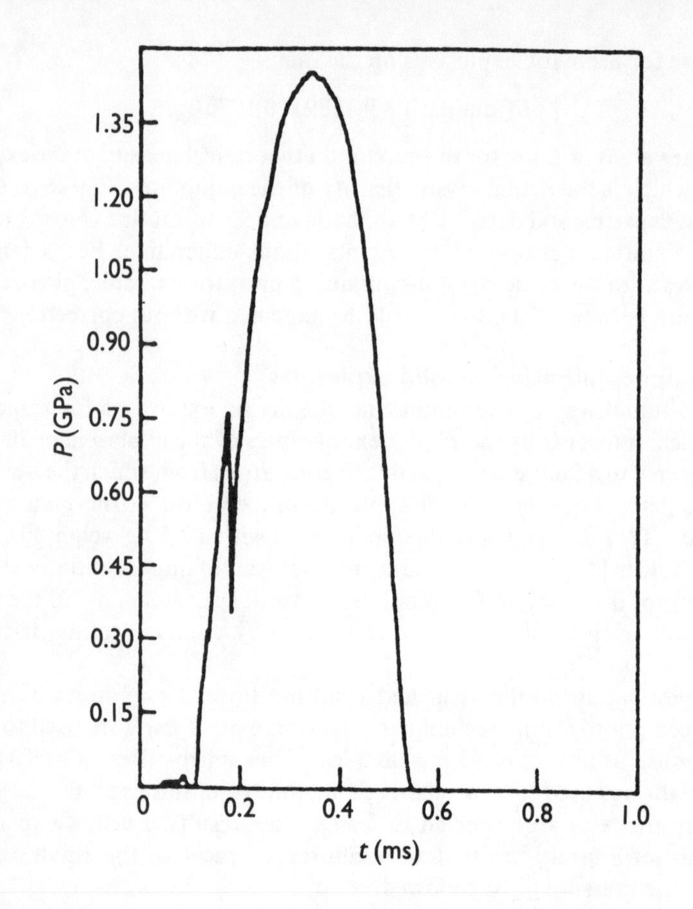

Figure 14.5 Pressure–time trace for HNS under mechanical initiation (Krishna and Field [12]) *(Reproduced with the permission of the Combustion Institute.)*

behavior similar to that of HNS: plastic yield and rapid flow associated with onset of reaction.

TNT by contrast did not show rapid flow. The melting points of HNS and TNT are respectively 588° K and 354° K, and from the principles of elastic behavior of solids, a positive correlation is expected between the melting point and Young's modulus E_Y, defined as:

$$E_Y = \frac{\text{(force/area)}}{\text{(deformation/original dimension)}}. \quad (14.15)$$

Hence for given impact, a low E_Y causes high solid deformation preceding the plastic phase. HNS, with high melting temperature and high E_Y, enters its plastic phase at high stress and with rapid velocity. The reverse is true for TNT with its relatively low melting temperature and E_Y value; it enters its plastic phase at low stress and correspondingly low flow velocity.

14.6 Pyrotechnics

14.6.1 Introduction

Pyrotechnics use reactions of the type discussed for propellants and explosives and their application is often recreational, for example fireworks, but can be functional as in military signalling. Compounds of particular metals can be used to produce particular colors, for example green with copper, yellow with sodium. We will describe the principles of pyrotechnic burning against a background of some recent research work.

14.6.2 Detailed treatment of selected pyrotechnics [13]

Considering the overall reaction in a pyrotechnic comprising tungsten and potassium dichromate:

$$W + K_2Cr_2O_7 \rightarrow K_2WO_4 + Cr_2O_3.$$

Reactants and products are solid and it is therefore plausible that the behavior can be described by a heat-balance equation only, without considering mass transfer. This ideal is termed a gasless pyrotechnic and the following equation for heat balance will apply to it:

$$\sigma c \frac{dT}{dt} = k \frac{d^2T}{dz^2} - U(T - T_o) + w, \tag{14.16}$$

where T = temperature at time t and coordinate z along the axis of propagation (K)

σ = density (kg m^{-3})
k = thermal conductivity (W m^{-1} K^{-1})
U = effective heat loss coefficient (W m^{-3} K^{-1})
T_o = ambient temperature (K)
w = rate of heat release by the reaction (W m^{-3}).

Equation 14.16 can be made dimensionless, requiring the definition of dimensionless time. This is:

$$\tau = \frac{t}{t_o}, \tag{14.17}$$

where t_o is rather ingeniously defined by:

$$(t_o)^{-1} = A_o \exp[-E_o/ RT_{ref}], \tag{14.18}$$

where T_{ref} is a reference temperature (K) which can be assigned a value of the order of adiabatic temperatures in pyrotechnic reactions, $\approx$2000K. A_o is a pre-exponential factor (s^{-1}) and E_o an activation energy (J mol^{-1}) which can be chosen so as to be typical of pyrotechnic reactions, giving a suitably scaled τ.

The dimensionless distance coordinate Z is:

$$Z = \frac{z}{z_o}, \tag{14.19}$$

where

$$z_o = (\alpha_o t_o)^{0.5}, \tag{14.20}$$

and α_o is a thermal diffusivity ($m^2\ s^{-1}$), again chosen to be typical of pyrotechnics. The dimensionless temperature θ is:

$$\theta = \frac{T - T_o}{T_{ref} - T_o}. \tag{14.21}$$

The rate of heat release w is by definition:

$$w = \sigma Q \frac{d\Phi}{dt} = \frac{\sigma Q}{t_o} \frac{d\Phi}{d\tau}, \tag{14.22}$$

where Q = exothermicity ($J\ kg^{-1}$)
Φ = fractional extent of reaction.

By means of these substitutions, equation 14.16 becomes:

$$\frac{d\theta}{d\tau} = \frac{\alpha}{\alpha_o} \frac{d^2\theta}{dZ^2} - \beta\theta + \frac{\theta_{ad}}{T_{ref} - T_o}\left[(1 - \Phi) \frac{A \exp(-E/RT)}{A_o \exp(-E_o/RT_o)}\right], \tag{14.23}$$

where α, A and E are respectively the *actual* thermal diffusivity, pre-exponential factor and activation energy. Also:

$$\beta = \frac{Ut_o}{\sigma c}, \tag{14.24}$$

and

$$\theta_{ad} = \frac{Q}{c} = \text{adiabatic temperature rise.} \tag{14.25}$$

The burning velocity of a pyrotechnic can be obtained from this model if propagation is identified with the rate of advancement in the z-direction of the point of highest heat release rate, w.

In numerical solution of equation 14.23 the values in Table 14.4 are used for the arbitrary constants A_o, T_{ref}, D_o and E_o, and for specific pyrotechnics, actual values of A, E and θ_{ad} are available from analysis of experimental temperature measurements during burning and D also from experiment. Table 14.4 lists the input to numerical solution of equation 14.23:

Table 14.4 Quantities used in solution of equation 14.23

*Pyrotechnic composition**	A (s^{-1})	E ($kJ\ mol^{-1}$)	α ($m^2\ s^{-1}$)
50% W – 50% $K_2Cr_2O_7$	570 ± 300	12.3 ± 2.6	(1.46 ± 0.08) × 10^{-7}
3.75% W – 37.5% $K_2Cr_2O_7$ – 25% $BaCrO_4$	190 ± 150	23.1 ± 7	(2.1 ± 1.5) × 10^{-7}
11% B – 89% MoO_3	220 ± 140	12.9 ± 3.2	(6.3 ± 2) × 10^{-7}

*Percentages on a weight basis

For all pyrotechnics considered:

$A_o = 580\ s^{-1}$
$\alpha_o = 12.0\ kJ\ mol^{-1}$
$D_o = 1 \times 10^{-7}\ m^2\ s^{-1}$
$T_{ref} = 2000\ K$,

from which $t_o = 3.55$ ms (from Equation 14.18)
$z_o = 1.88 \times 10^{-5}$ m (from Equation 14.20)
$\beta = 0.001$*
$T_o = 333K$

Ignition at $z = 0$.

Numerical solutions to equation 14.23 are in the form $\theta(\tau, Z)$, and in Figure 14.6 the Z-component at 7 ms ($\approx 2t_o$) intervals is shown for the 50% W—50% $K_2Cr_2O_7$ pyrotechnic. The adiabatic temperature is that corresponding to θ_{ad}

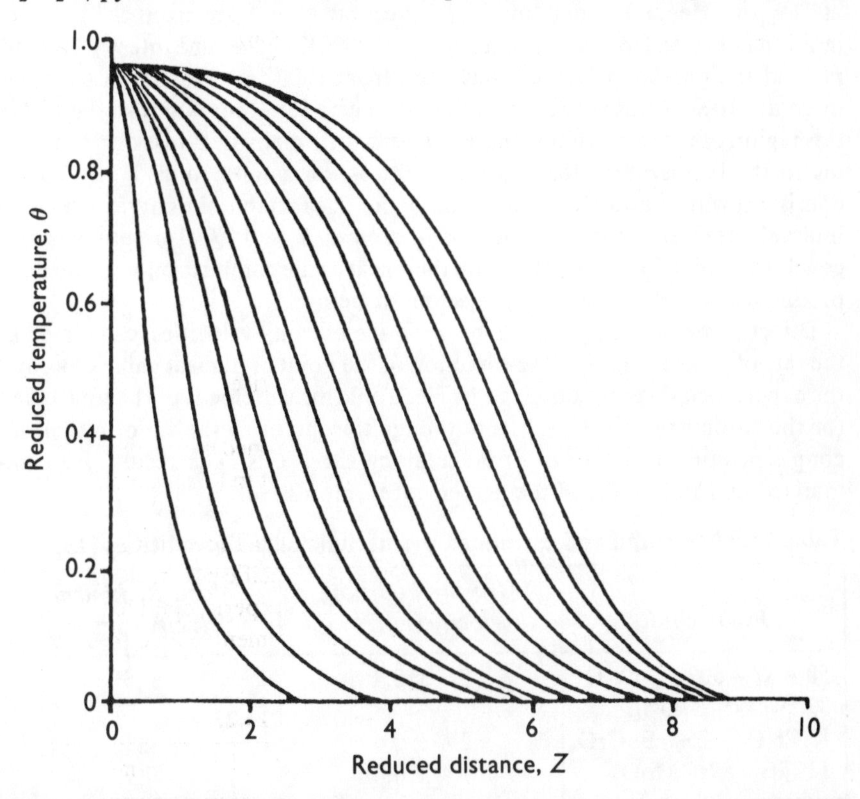

Figure 14.6 Temperature profiles at 7 ms intervals in a pyrotechnic composed of 50% W - 50% $K_2Cr_2O_7$ (Boddington et al. [13])
(Reproduced with the permission of the Combustion Institute.)

*From the above value of t_o and incorporating a value of U appropriate to the value of T_o and the steel channels, used to hold the pyrotechnic mixture, used in the measurements with which the calculated results are compared. These channels provide a means of having $T_o >$ room temperature.

with T_{ref} chosen as 2000K as indicated. The graphs can be understood as follows: Time t (or τ) zero corresponds to ignition of the pyrotechnic. At all Z, the value of θ is zero before ignition. The first of the family of curves in Figure 14.6 is for 7 ms after ignition. At $Z = 0$ on this curve, $\theta = 0.97$, indicating a temperature of 1950K.

Considering now the fourth curve from the left, pertaining to 4×7 ms = 28 ms after ignition, we can read off this curve that $\theta = 0.4$ when $Z = 2.4$. Substituting in equations 14.19 and 14.21 with the given z_o and T_{ref} values reveals that the actual temperature will be 1000K at a distance of 45 μm from $z = 0$. The curve strikes the horizontal axis ($\theta = 0$, $T = T_o$) at $Z = 5.4$ which corresponds to $z = 100$ μm.

Consider now the curve farthest right. This is the temperature profile 91 ms after ignition. Reasoning as in the above two paragraphs, the temperature 91 ms after ignition is 1830K at 47 μm, 1000K at 117 μm and drops to ambient at 169 μm. Hence the part of the pyrotechnic $\approx$ 100 μm from the point of ignition has risen from about ambient to $\approx$ 1000K in the time interval 28 to 91 ms and the part about 45 μm has risen from 1000 to $>$ 1800K in the same interval. However, at very low Z-values the curve for 91 ms is almost flat and in this region of the pyrotechnic there has been little if any heat loss during the 91 ms so the temperature there is $\approx$ 1950K, as it was after only 7 ms. This is consistent with conditions close to adiabatic. The fact that the curves at all time intervals are very 'bunched' at low Z around $\theta \approx 0.97$ also supports this conclusion; reaction products retain their heat as the combustion reaction itself progresses farther along the pyrotechnic structure.

Direct comparison is possible between the burning velocities, calculated as the rate of progression in the z-direction of the point of highest value of w, and the experimental values obtained by electronic measurement of the time taken for the combustion to travel from the ignition point to the tip of a thermocouple positioned a known distance along the z-axis. The results are summarized in Table 14.5 and the agreement is impressive.

Table 14.5 Measured and calculated pyrotechnic burning velocities [13]

Pyrotechnic	*Measured burning velocity* (mm s^{-1})	*Calaculated burning velocity* (mm s^{-1})
50% W – 50% $K_2Cr_2O_7$	5.7 ± 0.4	6.5
37.5% W – 37.5% $K_2Cr_2O_7$ – 25% $BaCrO_4$	2.4 ± 0.4	2.1
11% B – 89% MoO_3	7.4 ± 0.8	7.9

Other trends shown experimentally and predicted by the model outlined include:

(a) A dependence of burning velocity of the W – $K_2Cr_2O_7$ pyrotechnic at a particular ambient temperature on proportion of the two ingredients, with a maximum at $\approx$ 70% W.

(b) A positive correlation of burning velocity of the 50% W – 50% $K_2Cr_2O_7$ and 11% B – 89% MoO_3 pyrotechnics with T_0.

(c) Reduction of the burning velocity by addition of chromic oxide, a reaction product, to an initial reactant composition of 50% W – 50% $K_2Cr_2O_7$. Addition of 25–30% of chromium oxide is sufficient to prevent combustion altogether. The reason is not simply dilution and reduction of the effective exothermicity, but has a mechanistic basis. When temperature profile curves are used to estimate Arrhenius parameters such as those in Table 14.4, values of A (the pre-exponential factor) determined for the pyrotechnic plus various amounts of chromic oxide are not the same as the value for simple 50% W – 50% Cr_2O_7.

All the experiments were performed under conditions such that a value of β of 0.001 was appropriate, this being quite close to adiabatic conditions. Burning velocities calculated (as the rate of advance along the z-axis of the point of maximum heat-release rate, as described) for the 50% W – 50% $K_2Cr_2O_7$ at other β-values show that up to $\beta = 0.25$, there is steady burning, though at higher values of β the velocities are lower. At even higher values of β, while there may be commencement of combustion, there will be eventual failure and not steady propagation.

References

[1] Adams G. K., Newman, B. H., Robins A. B., 'The combustion of propellants based on ammonium perchlorate', *Eighth Symposium (International) on Combustion*, 693, Baltimore: Williams and Wilkins (1961)

[2] Friedman R., 'Deflagration of ammonium perchlorate', *Sixth Symposium (International) on Combustion*, 612, New York: Reinhold (1959)

[3] Langton N. H. (Ed.), *Rocket Propulsion*, University of London Press (1970)

[4] Lengelle G., Brulard J., Moutet H., 'Combustion mechanisms of solid propellants', *Sixteenth Symposium (International) on Combustion*, 1257, Pittsburgh: The Combustion Institute (1977)

[5] Girdhar H. L., Arora A. J., 'Effect of pressure on burning rate of phenol–formaldehyde composite propellants', *Combustion and Flame* **34** 303 (1979)

[6] Vosen S. R., 'Hydroxylammonium nitrate-based liquid propellant combustion—interpretation of strand-burner data and the laminar burning velocity', *Combustion and Flame* **82** 376 (1990)

[7] McCullough F., Jenkins H. P., 'Studies of the rates of combustion of hypergolic fluids', *Fifth Symposium (International) on Combustion*, 181, New York: Reinhold (1955)

[8] Dipprey D. F., 'Liquid Propellant Rockets', *Chemistry in Space Research* (R. F. Landel and A. Rembaum, Eds), 465, New York: Elsevier (1972)

[9] Cook M. A., *The Science of High Explosives*, Reprint of American Chemical Society (1958) Edition by Krieger Publishing Company, New York.

[10] Sundararajan R., Jain S. R., 'A simple method of estimating the detonation velocity from chemical composition of organic explosives', *Combustion and Flame* **45** 47 (1982)

[11] *Lawrence Livermore National Laboratory Explosives Handbook*, cited by Krishna and Field

[12] Krishna V., Field J. E., 'Impact initiation of hexanitrostilbene', *Combustion and Flame* **56** 269 (1984)

[13] Boddington T., Cottrell, A., Laye P. G., 'A numerical model of combustion in gasless pyrotechnic systems', *Combustion and Flame* **76** 63 (1986)

Appendix—Thermocouple Temperature Measurement

Abstract

The emf distribution in a thermocouple is discussed, and heat transfer effects on thermocouple readings are treated in some detail. A quantitative discussion is given of the suction pyrometer, which promotes accuracy by radiation shielding and by enhanced convection heat transfer from the gas surrounding a thermocouple to the thermocouple tip. Calibration drift is also discussed.

1 Introduction

Clearly, temperature measurement is very important in combustion. Thermocouple temperature measurement has been mentioned several times in this book and we will outline some important points about thermoelectric thermometry as they relate to combustion measurements. We do not intend to deal with the widely known and well-documented fundamental principles of thermocouple operation or the details of circuitry. It suffices to note that there are eight *letter-designated* thermocouple types: B,S,R,N,K,T,J and E. Types B, S and R are noble-metal, sometimes denoted rare-metal, thermocouples containing platinum and rhodium and the remainder are base-metal thermocouples. Of the base-metal thermocouples, Type K with one thermoelement made of chromel (containing Ni, Cr and Si) and the other of alumel (containing Ni, Si, Mn and Al) is probably the best known. It was developed around the turn of the century and continues to be widely used.

2 The distribution of thermocouple emf [1]

A thermocouple measures the temperature at its tip, that is, where the two different thermoelements are in electrical contact. However, it cannot be concluded that the emf occurs at the thermocouple tip: it does not. Figure A.1 shows that the emf is in fact distributed along the thermocouple length. The total emf in a circuit comprising the measuring thermocouple and a reference junction in ice (or equivalent instrumentational compensation) and therefore application of standard tables of thermocouple emf's are independent of the distribution, depending only on the temperature of the tip.

3 Heat transfer effects on thermocouples

Consider a thermocouple in a gaseous environment. To what errors will it be subject? If the temperature of the tip of the thermocouple is to be a good indication of that of the gas, there is a need for good convective heat transfer from gas to thermocouple tip. The convection coefficient will, if there is laminar or turbulent flow causing forced convection, depend on the Reynolds and Prandtl numbers:

$$\mathrm{Nu} = f(\mathrm{Re}, \mathrm{Pr}), \qquad \text{(A.1)}$$

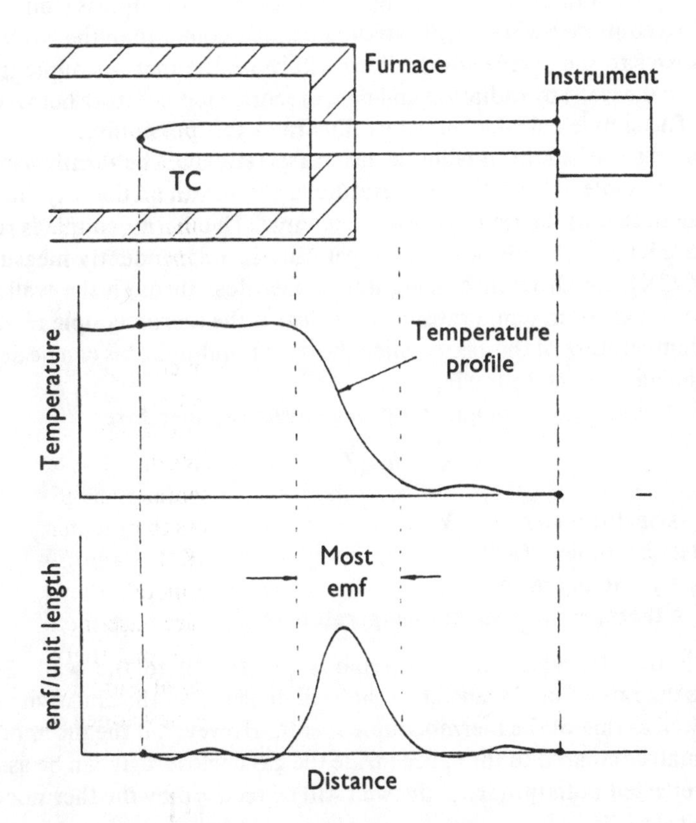

Figure A.1 Schematic of thermocouple emf distribution (Bentley [1])
(Reproduced with the permission of the Institute of Instrumentation and Control Australia.)

where Nu = Nusselt number = $\frac{hd}{k}$ (A.2)

h = convection coefficient (W m^{-2} K^{-1})
d = dimension of the body at the surface of which convection is occurring (m)
k = thermal conductivity of the gas (W m^{-1} K^{-1}).

Equation A.1 is a generalized form of a large number of engineering correlations for forced convection under various conditions and the particular form of it that is appropriate to a thermocouple in a flowing gas depends in principle on the orientation of the thermocouple with respect to the direction of gas flow. Therefore the convection heat transfer to the thermocouple tip, and hence the thermocouple reading, is subject to some possible margin of uncertainty on these grounds. Thermocouple materials are of course themselves conductors of heat so there is also the possibility of heat leakage down the wires by conduction.

Radiation heat transfer has the potential to introduce serious errors in thermocouple readings. This can happen when a thermocouple is standing in a hot gas in a container whose walls are significantly cooler than the gas because of heat losses to the surroundings. This will cause the thermocouple itself to lose heat to the walls by radiation and hence record a temperature below that of the gas. The simple calculation below illustrates this possibility.

Imagine a sheathed thermocouple inside a gas stream. The sheath is made of a conducting material and the temperature of the sheath all the way along it is very close to that of the tip of the thermocouple. The thermocouple is reading 600°C (873K) and the inside-wall temperature is independently measured as 400°C (673K), the difference being due to heat loss through the wall to the surroundings at room temperature. How close is the thermocouple reading to the true temperature of the gas in which the tip is standing? This can be deduced by heat balance in the following way:

Heat is transferred from gas to tip by convection, therefore:

$$q_1 = hA(T_g - T_t), \tag{A.3}$$

where q_1 = heat transfer rate (W)
A = sheath area (m^2)
T_g = gas temperature (K)
T_t = thermocouple sheath temperature (K).

Heat is transferred from the thermocouple sheath to the wall. Strictly speaking the rate of heat transfer to the wall depends on the emissivity of the wall, as well as that of the thermocouple sheath. However, if the thermocouple is very small compared to the space inside the gas enclosure, it can be assumed that no reflected radiation from the wall will be received by the thermocouple, in which case [2]:

$$q_2 = \epsilon\Omega A(T_t^4 - T_w^4), \tag{A.4}$$

where Ω = Stefan's constant
ϵ = emissivity of the thermocouple
T_w = wall inside temperature (K).

Since the sheath attains a steady temperature it follows that:

$$q_1 = q_2. \tag{A.5}$$

Therefore

$$T_g - T_t = \frac{\epsilon\Omega}{h}(T_t^4 - T_w^4), \tag{A.6}$$

so the amount by which the reading T_g is in error can be estimated. Putting $\epsilon = 0.5$ as a value for an oxidized metal surface and $h = 100$ W m^{-2} K^{-1} gives, on substitution into equation A.6:

$$T_g - T_t = 107\text{K}. \tag{A.7}$$

The quantity T_t is of course valid as a measure of the temperature of the thermocouple; the question is how different that is from the temperature of the surrounding gas. Through radiation effects, the thermocouple has created its own slightly cooler environment and the error of 107K if the thermocouple temperature is taken to be that of the gas is obviously unacceptable.

4 Radiation shields and the suction pyrometer

The suction pyrometer can be used in situations where radiation errors occur of the type described above. Its action can be understood by reference to Figure A.2 which is a schematic diagram of a cylindrical thermocouple sheath (diameter d_1, surface temperature T_1) surrounded by a concentric cylindrical

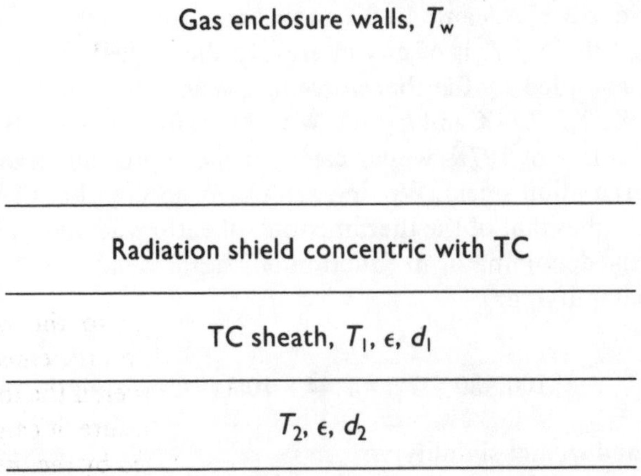

Gas enclosure walls, T_w

Radiation shield concentric with TC

TC sheath, T_1, ϵ, d_1

T_2, ϵ, d_2

Figure A.2 Schematic of heat transfer from a thermocouple with a radiation shield

radiation shield (surface temperature T_2, diameter d_2). We make two simplifying assumptions:

(a) That the sheath and the radiation shield have the same emissivity ϵ and that ϵ has a value characteristic of an oxidized metal surface, say ≈ 0.5.

(b) That the metal radiation shield has a very thin wall so that conductive thermal resistance through it is negligible.

Let us perform a heat balance on this arrangement under steady conditions. If the gas temperature is T_g, the convective heat transfer rate q_1 at the thermocouple is:

$$q_1 = h \times \pi d_1 L(T_g - T_1) \qquad \text{W}, \tag{A.8}$$

where L = length of the thermocouple in the gas.

Radiative transfer between the surfaces of concentric cylinders of the same axial length is a well-known standard case [3] from which the heat transfer rate q_2 from the sheath of the thermocouple to the shield is given by:

$$q_2 = \frac{\Omega \pi d_1 L(T_1^4 - T_2^4)}{1/\epsilon + (d_1/d_2)[(1/\epsilon) - 1)]} \qquad \text{W}, \tag{A.9}$$

and for the radiation heat transfer rate q_3 from the shield to the wall of the enclosure:

$$q_3 = \Omega \epsilon \pi d_2 L(T_2^4 - T_w^4) \qquad \text{W}. \tag{A.10}$$

Returning to our previous example of $T_g = 980\text{K}$ and $T_w = 673\text{K}$ we can use equations A.8 to A.10 to assess the improvement in the thermocouple as a measure of the gas temperature reading brought about by the shield. We first note that since conditions are steady:

$$q_1 = q_2 = q_3 = q. \tag{A.11}$$

Hence equations A.8, A.9 and A.10 are three equations in three unknowns: T_1, T_2 and q, though only T_1 is of any interest to the discussion since that is the temperature recorded by the thermocouple. Also returning to our example with $T_g = 980\text{K}$, $T_w = 673\text{K}$ and $h = 100 \text{ W m}^{-2} \text{ K}^{-1}$, for which the thermocouple reading was in error by 107K, we can calculate the improvement to be expected by using the radiation shield. We first arbitrarily specify that the diameter of the shield is 3 times that of the thermocouple sheath, whereupon, with $\epsilon = 0.5$, the value of the denominator in equation A.9 has a value of 7/3. Equating q_1 and q_2, and simplifying:

$$100(980 - T_1) = 2.44 \times 10^{-8} (T_1^4 - T_2^4). \tag{A.12}$$

Equating q_2 and q_3 and simplifying:

$$T_2^4 = \frac{1}{4.5}(T_1^4 + 3.5T_w^4). \tag{A.13}$$

Combining these to eliminate T_2^4:

$$4.18 \times 10^{12} = 0.78T_1^4 + (4.1 \times 10^9)T_1. \qquad (A.14)$$

This equation in T_1 can be solved by trial and error and the interested reader can easily verify that the solution is $T_1 = 897K$, which is the temperature that the thermocouple would record. Since $T_g = 980K$ the error is now 83K, a clear improvement on the value without the shield, but still significantly low.

However, the suction pyrometer does not rely solely on radiation shielding but also uses suction to enhance the convection heat transfer from gas to thermocouple. Suppose that the effect of this is to double the value of h from 100 to 200 W m^{-2} K^{-1}. The equivalent of equation A.14 with this value of h is then easily derived as:

$$8.19 \times 10^{12} = 0.78T_1^4 + (8.2 \times 10^9)T_1, \qquad (A.15)$$

and the trial and error solution is $T_1 = 930K$ so the reading is 50K low. The efficiency of a suction pyrometer is defined as:

$$\text{efficiency} = \frac{\text{error with TC only} - \text{error with pyrometer}}{\text{error with TC only}}. \qquad (A.16)$$

Therefore in the above example the efficiency is 53 percent. The efficiency of the shield only, without suction, is 22 percent.

The configuration discussed above of a sheathed thermocouple and a single radiation shield, each with fairly high emissivity, and suction, is in fact capable of $> 90\%$ efficiency at gas temperatures of up to about 600° C. The performance obviously depends, through h, on the speed of gas flow past the thermocouple and pyrometer characteristics are commonly expressed as efficiency against velocity, values up to 700 ft s^{-1} being used [4]. Improved performance may be gained by using multiple concentric radiation shields and this will be necessary to obtain good efficiencies at temperatures of $\approx$ 1000° C or higher. Alternatively, refractory shields, with lower emissivities than oxidized metal ones, can be used at the higher temperatures.

5 Calibration loss

5.1 Bare-wire thermocouples

We have analyzed thermocouple errors in terms of heat transfer, but errors due to drift of the thermocouple calibration itself are possible, especially in high-temperature work. One contributor to drift, even under quite mild chemical conditions, is oxidation of the thermoelement materials. Clearly the base-metal types are more susceptible than the noble-metal types. The extent of drift depends on:

(a) the peak temperature;
(b) the length of time spent in the high-temperature environment;
(c) whether there is steady exposure to the peak temperature or a temperature cycle;
(d) the diameter of the thermoelement wire.

In general, for base-metal thermocouples in bare wire configuration, emf drifts of the order of tens or hundreds of μV can occur as a result of exposure to air only, at elevated temperatures, for times of the order 1000 to 2000 hours, and the drift may be up or down. It is questionable whether a drift of this magnitude is significant in practical terms. For example, at 850° C a Type K thermocouple (reference junction at 0° C) has an emf of 35.34 mV. A drift upward of say 50 μV would cause the emf as measured to be 35.39 mV, corresponding to a temperature reading of 851° C, or a drift downward by 50 μV would cause the emf to read 35.29 mV, making the temperature reading 849° C. This difference is likely to be less than the intrinsic uncertainty in the thermocouple specification at manufacture. While there is always the need to pay attention to long-term drift in the calibration of thermocouples, it is when the environment is 'aggressive' that there is most likelihood of calibration loss, and mineral-insulated metal-sheathed thermocouples can be used in these situations as discussed in Section 5.2.

5.2 Mineral-insulated metal-sheathed thermocouples

The mineral-insulated metal-sheathed (MIMS) design of thermocouple provides some protection of the thermoelements from their environment. In this configuration, the thermoelements are contained in a sheath with mineral packing (usually magnesium oxide) with good thermal conductivity and good electrical resistance. The sheath material is commonly stainless steel or inconel, although other sheath materials are available and are used in certain applications. When new, a type K MIMS thermocouple is expected to have an accuracy at higher temperatures of ± 0.75% of the value in Celsius of the temperature being measured, that is, an uncertainty of ± 7.5° C in an actual temperature of 1000° C. This of course is an uncertainty in the temperature of the tip of the thermocouple; to what extent it is a true measure of the temperature of interest is a separate matter and depends on heat transfer, as discussed in Appendix Section 3.

MIMS thermocouples, although protected by the metal sheath from the effects of whatever environment the thermocouple is standing in, nevertheless do often show calibration drift because their own immediate environment, the sheath, may also have chemical effects on the thermoelements. For example, manganese present in a sheath can, during usage, migrate to the thermoelements, causing changes in their composition and hence calibration drift. In discussing the performance of MIMS thermocouples in very vigorous environments, sheath deterioration or total failure also has to be considered. For example, a MIMS thermocouple standing in a bed of solid fuel in a pot furnace such as was discussed in Section 6.6.2 will withstand only a limited number of firings of the pot furnace before the sheath breaks. Hence the development of new sheath materials for particular applications is an important area of current research into thermocouple performance.

References

[1] Bentley R. E., 'The distributed nature of emf in thermocouples and its consequences', *Australian Journal of Instrumentation and Control* **38** 128–132 (1982)
[2] Kern D. Q., *Process Heat Transfer*, New York: McGraw-Hill (1950)
[3] Holman J. P., *Heat Transfer*, SI Metric Edition, New York: McGraw-Hill (1989)
[4] Land T., Barber R., 'Suction Pyrometers, Theory and Practice', *Journal of the Iron and Steel Institute* **184** 269 (1956)

Index